# Liquid Chromatography in Clinical Analysis

# Biological Methods

***Liquid Chromatography in Clinical Analysis,*** edited by **Pokar M. Kabra** and **Laurence J. Marton,** *1981*

***Metal Carcinogenesis Testing:*** *Principles and In Vitro Methods,* by **Max Costa,** *1980*

# Liquid Chromatography In Clinical Analysis

Edited by

Pokar M. Kabra

and

Laurence J. Marton

*University of California School of Medicine*
*San Francisco, California*

The Humana Press Inc. • Clifton, New Jersey

## Dedication

This volume is dedicated to George Brecher, M.D. for a lifetime of contributions and devotion to Laboratory Medicine and for having the wisdom to encourage us to establish our LC laboratory.

Library of Congress Cataloging in Publication Data
Main entry under title:

Liquid chromatography in clinical analyses.
(Biological methods)
Includes bibliographical references and index.
1. Liquid chromatography. 2. Chemistry, Clinical-Technique. I. Marton, Laurence J. II. Kabra, Pokar M. III. Series. [DNLM: 1. Chromatography, Liquid. QD 79.
C454 L765]
QP519.9.L55L54 616.07′5′028 80-29377
ISBN 0-89603-026-1
Crescent Manor
P.O. Box 2148
Clifton, NJ 07015


Printed in the United States of America.

# Preface

Liquid chromatography is widely used in clinical laboratories for monitoring a variety of therapeutic agents. In addition to its usefulness in the areas of pharmacokinetics and toxicology, liquid chromatography is currently being developed for the routine analysis of a number of endogenous constituents. The present book is designed to serve as a reference for, and stimulus to, scientists involved in patient care monitoring.

In most instances, the authors review the fundamental concepts underlying their respective approaches to the use of liquid chromatography, and continue with detailed presentations of the specifics of a particular method. This is done so that readers may gain insight into the potential problems facing them in any application area, based upon the cumulative experience of individuals who have been pioneers in the field. The general concepts and approaches described here change only slowly, and their proper understanding will serve the biomedical scientist well even as specific methodology changes rapidly.

*Liquid Chromatography in Clinical Analysis* is an outgrowth of a course sponsored by the Department of Laboratory Medicine of the University of California in conjunction with the University's Extended Programs in Medical Education.

We sincerely thank the contributors to this volume for their dedication to quality, Mr. William Kerr, an outstanding hospital administrator, for his willingness to explore new techniques, and our wives and children for their support and understanding.

San Francisco
February, 1981

Pokar M. Kabra
Laurence J. Marton

# Contents

*Chapter 1*

Introduction to Liquid Chromatography

STEPHEN R. BAKALYAR

***Chapter** 2*

## Instrumentation for Liquid Chromatography

RICHARD A. HENRY AND GENRIKH SIVORINOVSKY



***Chapter** 3*

## Liquid Chromatography Column Technology

RONALD E. MAJORS

# Part II

## Therapeutic Drug Monitoring and Toxicology

***Chapter** 4*

### Why Measure Drug Levels?

LEWIS B. SHEINER



***Chapter** 5*

### Anticonvulsants

POKAR M. KABRA, BRIAN E. STAFFORD, DONNA M. MCDONALD, AND LAURENCE J. MARTON

## *Chapter 6*

## Theophylline and Antiarrhythmics

F. L. VANDEMARK



## *Chapter 7*

## Antibiotics

JOHN P. ANHALT

## *Chapter 8*

## Tricyclic Antidepressants

GARY J. SCHMIDT



## *Chapter 9*

## Antineoplastic Drugs

WOLFGANG SADÉE AND YOUSRY MAHMOUD EL SAYED

***Chapter** 13*

Steroids

FELIX J. FREY, BRIGITTE M. FREY, AND LESLIE Z. BENET

***Chapter*** *14*

Proteins

FRED E. REGNIER AND KAREN M. GOODING



***Chapter*** *15*

Bilirubin and Its Carbohydrate Conjugates

NORBERT J. C. BLANCKAERT

## *Chapter 16*

## Porphyrins

George R. Gotelli, Jeffrey H. Wall, Pokar M. Kabra, and Laurence J. Marton



## *Chapter 17*

## Organic Acids by Ion Chromatography

William E. Rich, Edward Johnson, Louis Lois, Brian E. Stafford, Pokar M. Kabra, and Laurence J. Marton

## *Chapter 18*

## Major and Modified Nucleosides, RNA, and DNA

CHARLES W. GEHRKE AND KENNETH C. KUO

## *Chapter 19*

## Polyamines

LAURENCE J. MARTON

# Part 1

## Introduction To Liquid Chromatography

# Chapter I

# Principles of Liquid Chromatography

Stephen R. Bakalyar

*Rheodyne, Inc., Berkeley, California*

## I. Introduction

This article reviews the basic principles of high performance liquid chromatography (HPLC). The introductory section provides an overview of the HPLC technique, placing it in historical context and discussing the elementary facts of the separation mechanism. The next section discusses the nature of resolution, describing the two principal aspects, zone center separation and zone spreading. The third section takes a detailed look at how HPLC is used in practice to achieve a separation. It discusses the three key variables that need to be adjusted: retention, efficiency, and selectivity. A fourth section is concerned with various relationships of practical importance: flow rate, temperature, and pressure. A final section discusses future trends in HPLC.

There are many synonyms for HPLC terms. These will be indicated in parentheses when a term is first introduced.

### A. Overview of HPLC

High performance liquid chromatography is a technique that was developed during the 1960s, was steadily improved during the 1970s, and promises considerable further improvement and extension in the 1980s. Like all chromatographic techniques, it operates by separating the various chemical species in a mixture from each other. What sets HPLC apart from other methods is its ability to accomplish this with great speed, sensitivity, and precision, and its applicability to an enormous variety of compounds.

HPLC is, first of all, liquid chromatography (LC); the mobile phase is a liquid. It can be used to separate far more compounds than the complementary technique of gas chromatography (GC) because only a minority of chemical compounds has the good volatility required by GC. Unless a compound is volatile, it cannot exist in the gas mobile phase of GC.

Before the development of HPLC, gas chromatography had far more speed, sensitivity, and precision than classical LC. This was true because GC was from the outset a highly instrumented technique: mobile-phase flow rate was controlled by a pressure regulator; sophisticated detectors could quantify less than a nanogram of solute; and various electronic controls assured very reproducible peak retention times and peak areas. In contrast, classical LC in all its forms—column, thin layer, and paper chromatography—used gravity or capillary action to control mobile phase flow and was restricted by the availability of detectors of only limited capability. The important exception to this was the amino acid analyzer. Demand for this specific application was so high that hardware was developed that was specifically optimized for this separation.

Thus, the first high performance liquid chromatographs were in fact the amino acid analyzers of the 1950s. In a way, the gas chromatograph was the real progenitor of modern, general purpose HPLC. This is so because it is from the experience with gas chromatography that two vital factors developed. First, GC provided a useful and general theory of chromatography that in turn was the intellectual stimulus for a deeper and more fundamental look at liquid chromatography to see how it could be improved. Second, GC was also an outstanding commercial success for a number of companies, and thereby provided an economic stimulus to them to attempt to accomplish for LC what had been done for GC.

We have said that HPLC is, first of all, liquid chromatography, and that this distinguishes it from gas chromatography. Secondly, HPLC is column chromatography. This distinguishes it from the companion LC techniques of thin layer chromatography (TLC) and paper chromatography (PC). The difference is all important, for with the two-phase system confined to a tube, the mobile-phase flow rate can now be controlled, and pressure can be generated if it is required to cause flow. As will become evident, mobile-phase flow rate is one of the most important HPLC variables. Furthermore, the effluent from the column is easily directed to a flow-thru detector operating on-line. In contrast, the thin layer and paper detection techniques require a separate step subsequent to the separation process.

HPLC, then, is an analytical method that combines the latest instrumentation, proven theory, and the wealth of chemical interaction knowledge reaching back through the entire history of liquid chromatography in all its forms.

## B. Basic Facts of the HPLC System

The column in HPLC contains a two-phase system. The mobile phase (carrier or solvent) flows past the stationary phase (packing, sorbant, particles). The packing occupies roughly 60% of the volume in the column, and the mobile phase flows throughout the remaining 40%.

The sample is a solution of solutes in a solvent, often the same solvent used for the mobile phase, but in any case one that is miscible with the mobile phase. The sample is injected into the mobile phase and proceeds into the column. The solutes distribute (partition) between the two phases. If a solute does not interact with the packing at all, it travels down the column at the same linear velocity as the mobile phase. If a solute interacts with the packing, its velocity is decreased. The stronger the solute-packing interaction, the slower the zone velocity. It is the task of the analytical chemist to choose a combination

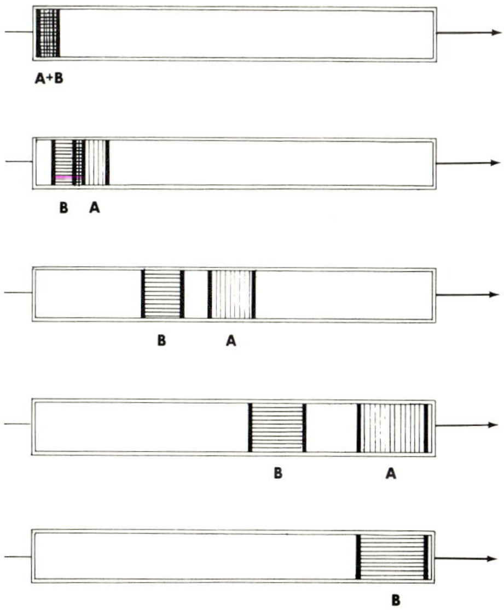


FIG. 1. Characteristics of zone migration.

of stationary and mobile phases so that different compounds in the sample have differing interactions with the packing, and are thereby separated as they travel down the column.

Figure 1 is a schematic series of snapshots taken at different times during the chromatography of two sample components, A and B. Because B interacts more strongly with the stationary phase, it travels more slowly. The longer the distance (time) over which this differential migration is allowed to continue, the greater the distance between the center of the two solute zones. If the mobile phase continues to flow (elution continues), the zones emerge from the column into the detector at different times. The detector signal then produces a chromatogram when sent to a recorder. The retention time in the column is characteristic of the compound, i.e., is the data for qualitative analysis. The peak height or area is characteristic of the amount of compound, i.e., is the data for quantitative analysis.

## II. Nature of Resolution

### A. Zone Separation vs Zone Spreading

Figure 1 shows that two phenomena take place simultaneously in the column. Firstly the zones move apart from each other, the more so the longer the distance of travel. More precisely, the *centers* of the zones become more separated. Secondly the zones broaden (spread) as they travel down the column. Shortly after the zones start to travel, the zone centers grow apart, but the full widths of the zones still overlap. It is the nature of chromatography that zones separate faster than they spread; thus, given sufficient column length, zones can be completely separated, i.e., resolved.

Note that for any given location in a column both zones have the same width when they arrive there. At the column outlet all zones have the same width. They occupy the same volumetric space in the column. However, we find that the solute zones as they appear on the chromatogram are not of equal width. It is important to understand why.

Figure 2 shows the chromatogram that would result from the detector signal illustrated in Fig. 1 (we assume that the two solutes were originally injected at equal concentrations). Peak B has a different concentration profile from peak A. This occurs because as B elutes from the column it is traveling more slowly than A. The time duration from peak onset to peak end is consequently longer. The maximum concentration is more dilute than in peak A because at any one moment more of the B solute molecules are actually residing in the

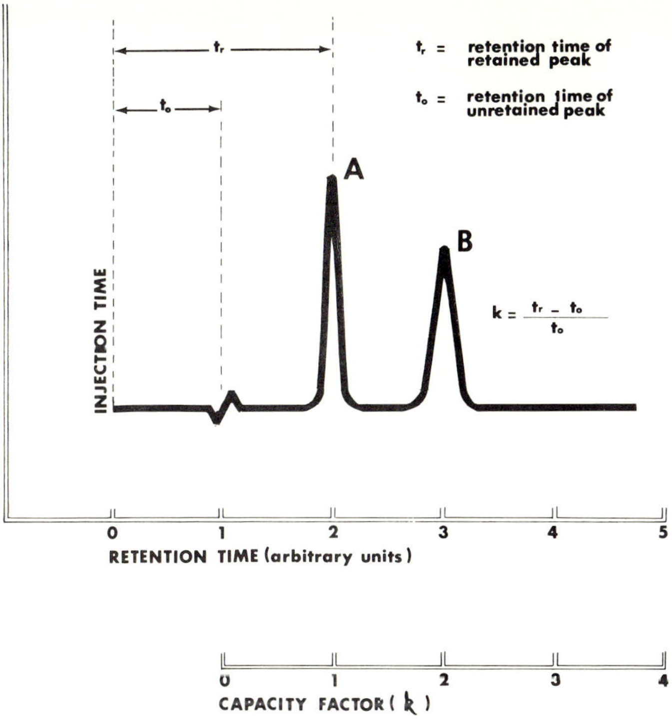


FIG. 2. Appearance of the chromatogram.

stationary phase than in the mobile phase, and it is only the mobile phase that the detector monitors.

So we see that a characteristic of chromatograms is that zones with longer retention times (retention volumes) are wider and shorter, i.e., more dilute. Recall, however, that we have assumed that the zones started at the same concentration and that a true mass concentration detector was used. In practice, one peak may be much taller than another simply because of the detector's greater sensitivity to it. This relationship between zone width and elution time also assumes that the mobile-phase eluting strength remains constant during the analysis (isocratic conditions). Later we shall examine what happens when the eluting strength is programmed, termed gradient elution.

Returning now to the subject of resolution, we see that it has a dual nature. It is directly proportional to zone center separation, $\Delta X$, and inversely proportional to zone width, $W$. The formal expression for resolution explicitly states this fact:

$$Rs = \Delta x / w \qquad (1)$$

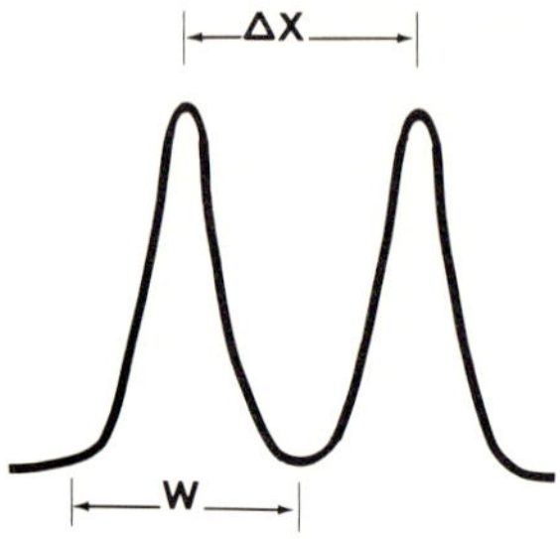


FIG. 3. Definition of resolution.

Figure 3 illustrates this. As a consequence of this dual nature, there are two fundamental ways of improving resolution. Figure 4 shows two incompletely resolved peaks. Resolution can be achieved either by increasing zone center separation while holding width constant, or by decreasing zone width while holding zone center separation constant. The former method improves selectivity, the latter efficiency.

The selectivity depends on the chemical nature of the mobile and stationary phases. In gas chromatography the mobile phase is a relatively inert gas, and only the stationary phase can be changed to effect improvement in selectivity. In liquid chromatography either or both of the phases can be changed. The various types of packings will first be discussed, followed by comments on mobile phases.

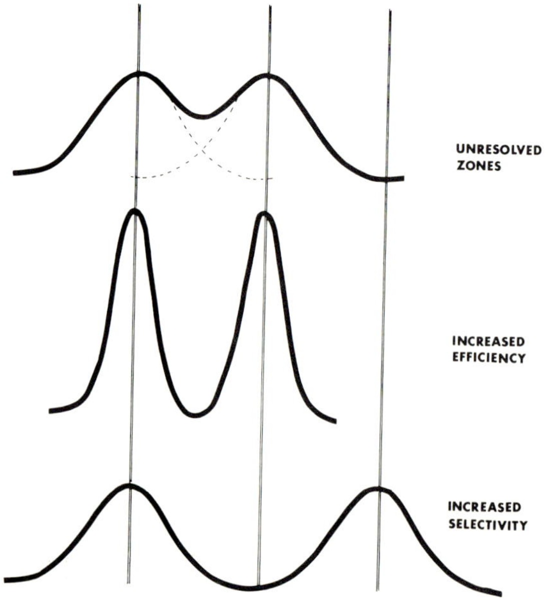


FIG. 4. Improving resolution with efficiency and selectivity.

## B. Stationary Phase Selectivity

*1. Affinity vs Exclusion Packings.* If a porous packing particle has no chemical attraction for the solutes, but is wetted by the mobile phase, it is an exclusion packing (gel permeation, gel filtration). Separation occurs because of differences in the molecular size (molecular weight) of solutes. Large molecules are excluded from the pores, spend less time in the stagnant mobile phase trapped in the pores, and migrate faster through the column. Small molecules are included in the pores, spend more time there, and are retained relative to the larger molecules.

However, even the smallest of solutes cannot be retained longer than the total volume of the column (we are assuming ideal behavior of the exclusion packing). No chemical affinity is involved, just the experiencing, via diffusion, of the volumes associated with the moving mobile phase (roughly 40%) and the stagnant mobile phase in the pores (roughly another 40–50%). So the maximum difference in elution volumes between earliest and latest eluting peaks is small.

In contrast, packings that exhibit a chemical affinity for solutes can provide enormous differences in peak elution times. For this reason, complex sample mixtures are usually separated on affinity packings.

*2. Adsorption vs Partition Packings.* When the stationary phase is a surface we speak of adsorption chromatography. Silica is an example. When the stationary phase is a bulk liquid, it is partition chromatography. Almost all modern HPLC uses adsorbents or surface-modified adsorbents (bonded phases), because the bulk liquid partition phase tends to strip off and slowly dissolve in the mobile phase. (GC stationary liquid phases do not have this problem since the solubility in the gas mobile phase is low.) The bonded phases have organic groups covalently bonded to the adsorbent surface. These bonded groups can be of high, medium, or low polarity, and can even include ionic groups.

*3. Normal vs Reverse Phases.* Historically the earliest stationary phases were more polar than the mobile phase. For example, a mobile phase containing a few percent of methylene chloride and isopropanol in heptane might be used with a silica packing. The silanol groups on the silica surface are more polar than the mobile phase solvents, and polar solutes would interact with the silica and be retained.

More recently systems have been developed where the relative polarity of the two phases is reversed; this is termed reversed phase LC. For example, a mobile phase containing a few percent methanol in water might be used with a nonpolar packing—a silica with hydrocarbonaceous layer bonded to it. The aqueous mobile phase is more polar than the packing. The greater the solute polarity, the less it is retained on the stationary phase. This solute behavior is opposite to that in normal phase chromatography, and yields advantages in the separation of certain classes of compounds, so much so that it is currently used for the majority of applications.

Among the many reasons for the popularity of reversed phase LC are: (1) An unequalled range of solute polarities can be chromatographed, from low molecular weight polar ionic species, such as amino acids, to medium molecular weight polycyclic aromatic hydrocarbons. (2) The bonded nonpolar stationary phases reach equilibrium rapidly and their chromatographic properties are relatively stable. (3) Most samples of biochemical and clinical interest are already aqueous solutions. They can be often injected directly into a reversed phase column without extraction procedures.

The most popular reversed phase packings in the United States and Europe are 5 or 10 $\mu$m diameter silica particles to which hydrocarbon chains have been covalently bonded, most commonly 2, 8, or 18 carbon atoms long. Each of these in turn is bonded by different techniques by the various manufacturers. And the native silica starting material differs. The result is that different reversed phase packings have subtle physicochemical differences that provide different chromatographic selectivities.

Other bonded phases have markedly different selectivities from the hydrophobic reverse phases. Among the chemical functionalities available are cyano, diol, amino, quaternary amine, and sulfonic acid. Columns of the same functionality from different manufacturers exhibit different selectivities, again because the starting materials and synthetic methods differ. Clearly there are many different HPLC columns to choose from when trying to improve selectivity. For a good review of columns see reference *1,* and chapter 3 of this volume.

## C. Mobile Phase Selectivity

Just as subtle differences in packings provide different selectivities, so also do small changes in mobile phase composition. A detailed discussion is beyond the scope of our treatment here, but we will briefly describe the situation for reversed phase HPLC.

With reversed phase the eluting strength of the mobile phase increases as its polarity is decreased. The weakest mobile phase is water. Adding organic solvents—typically methanol, acetonitrile, or tetrahydrofuran (THF)—decreases the polarity and makes nonpolar solutes more soluble in the mobile phase so that they elute sooner.

(Remember that with reversed phases, in contrast to the results with such normal phases as the silica adsorbents, the least polar solutes elute last.)

The retention of compounds is therefore controlled by the percent of organic modifier added to the water. THF being less polar than methanol, it requires less THF than methanol to produce a given solvent strength. Nevertheless, it is often found that two such mobile phases of approximately the same eluting strength have different selectivities. For example, we might find that a mobile phase of 40% THF elutes the peaks in about the same time as one of 50% methanol. However, the relative retention of the peaks may be different. The order of elution may even change for a few peaks. Mobile phase selectivity, therefore, can be changed simply by changing the type of solvent modifier used. For more information on this powerful technique see references *2–4*.

The manipulation of modifier concentration controls what are termed the primary equilibria in the column. Another method of controlling selectivity is to manipulate what are termed the secondary equilibria. These equilibria effect the solutes directly, changing their polarity. Acidic compounds are made more hydrophobic by lowering the pH, basic compounds by raising the pH. Compounds with a formal charge, such as sulfonic acids, can be rendered hydrophobic by complexing them with ion pairing agents of the opposite charge.

The ability to control retention and selectivity by adjusting both primary and secondary equilibria is another reason why reversed phase HPLC has become the dominant technique.

## D. Column Efficiency

Changing the distance between zone centers, i.e., the selectivity, is one way to improve resolution. Changing the zone widths is a second way. The degree to which a column keeps zones narrow is termed the efficiency. Zone spreading is caused by three concurrent phenomena: longitudinal diffusion, multiple flow paths, and resistance to mass transfer.

Longitudinal diffusion along the column axis is an obvious source of zone spreading. This is an insignificant contribution in practice because analyses are completed in a time period that is short compared to solute diffusion rates. (In contrast, diffusion rates of solutes in the mobile phase of gas chromatography are about $10^5$ times faster, and molecular diffusion is a major contributor to zone spreading.)

Multiple flow paths (flow velocity inequalities or eddy diffusion) throughout the packed bed cause some molecules to travel faster than others. A uniform packing structure minimizes this effect. When

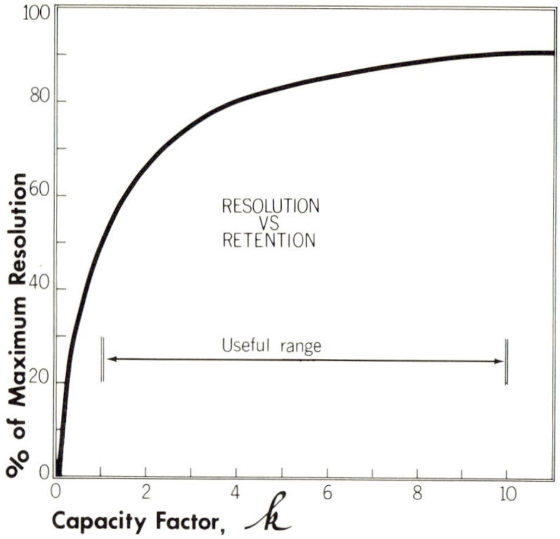


FIG. 5. Resolution vs retention.

It is of little value to remember this expression, but only to appreciate its significance. Figure 5 plots the relationship. Resolution is seen to increase rapidly as the zone becomes retained. At $k = 1$, 50% of the maximum resolution is wasted, so mobile phase polarity should normally be adjusted to operate above this value. But above $k = 10$, resolution increases only slowly, so there is little gained at higher retention, and a significant loss of separation speed.

Once a column has been chosen, the first task is to adjust the mobile phase eluting strength so that retention times for the peaks of interest are in the range of $k$ values between about 1 and 10. Operating outside of this range will needlessly squander either resolution or analysis speed. What if the range of polarities of the solutes is so broad that all peaks do not elute within the useful retention range? This has been termed the general elution problem. If the eluting strength is adjusted so that early eluting peaks are adequately retained (adequately resolved), late eluting peaks require an unacceptably long time to elute, and when they do, the peaks are sometimes so dilute as to be undetectable.

The solution to this situation is gradient elution. The eluting strength of the mobile phase is programmed, increasing in strength throughout the analysis. This is analogous to temperature programming in gas chromatography. All solutes elute as relatively narrow, tall peaks in a reasonable time.

## C. Efficiency

In a previous section, efficiency was described as the degree to which zones are kept narrow as they move down the column. It is clear from

Fig. 4 that this is an important factor in resolution. Here we will expand on the concept of efficiency.

Figure 1 shows that zones become increasingly broader as they travel through the column. The width increases in proportion to the square of the distance traveled, $w \propto L^{1/2}$. The value of $w$, whether in millimeters, milliliters, or seconds for a particular column, is a function of many variables. These variables can all be lumped together into one constant of proportionality, $w = (HL)^{1/2}$. The plate height $H$ is a "goodness factor" that indicates how efficient the column is. It is also called the height equivalent to a theoretical plate. The smaller the value of $H$, the smaller the zone width. High resolution columns thus have smaller plate heights than low resolution columns. Stating the expression explicitly for $H$ we have:

$$H = w^2 / L \tag{5}$$

The plate height is the rate of zone spreading per unit length of column. It thus allows comparison of packings even though the columns are of different length.

The three factors that cause zones to spread were previously described as multiple flow paths, longitudinal diffusion, and resistance to mass transfer. The last two are time-related phenomena, so it is not surprising that their contributions to efficiency are flow rate dependent. Figure 6 shows a typical plot of plate height vs flow rate. Remember that smaller $H$ values mean narrower peaks and thus better resolution. It is the resultant of the sum of all three factors. As flow rate is reduced, more time is allowed for the diffusion-controlled transfer of

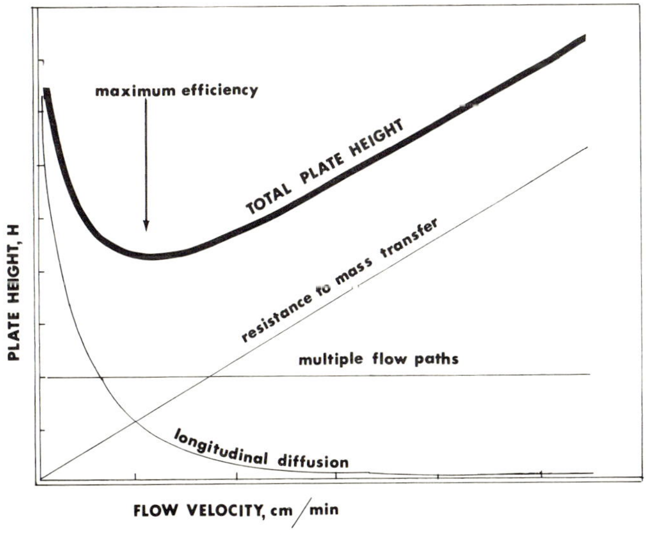


FIG. 6. Efficiency vs flow rate.

solutes between the two phases, thus the contribution of the resistance to mass transfer term decreases. However, more time is also allowed for longitudinal diffusion, so its contribution to the total plate height increases. The multiple flow path term is independent of flow rate. Most HPLC practiced today operates at flow rates on the ascending part of the $H$ vs flow rate curve, i.e., at flow velocities above the minimum on the curve.

The important practical significance of the plate height vs flow rate curve is that resolution and speed are opposed to each other, at least for a given column with a fixed length. One can always be improved at the expense of the other, simply by changing the flow rate.

We said that $H$ indicated the degree of spreading per unit length of column. It is useful to be able to describe a column's separating power by taking into account both the plate height and the column length. Such a measure is called the number of theoretical plates, $N$. It is proportional to column length and inversely proportional to the plate height H:

$$N = L/H \tag{6}$$

High resolution columns thus have a larger number of plates than low resolution columns. This relationship should seem right, because we previously stated that zone center separation is proportional to column length, and zone width is directly related to the plate height. Another aspect of this expression that should make sense is that the height equivalent to a theoretical plate has dimensions of length. The total number of plates in a column is therefore the column length divided by the height of a plate.

Modern LC columns have theoretical plate heights in the range of 0.01–0.1 mm. A 25 cm column with a plate height of 0.02 mm therefore has about 12,500 plates:

$$N = L/H = 250\ \text{mm}/0.02\ \text{mm/plate} = 12{,}500\ \text{plates} \tag{7}$$

The number of plates in a column is easily determined from measurements made on the chromatogram, using the following expression:

$$N = 16(t/w)^2 \tag{8}$$

The $w$ is the width of the peak at its base, expressed in time units. The $t$ is the retention time of the peak, in the same time units. This expression will not be derived, but is related to the previous equations. Just measuring the peak width is not an adequate indicator or the column's separating power. Recall that peaks become wider on the chromatogram (not in the column) the later they elute. So the elution time must be factored out. The $t/w$ ratio in Eq. (8) does this.

### D. Selectivity

The discussion of the selectivity in a previous section does not require further development, other than to introduce a quantitative measure. The selectivity or separation factor between two peaks is simply the ratio of the two capacity factors, the later eluting peak appearing in the numerator:

$$\alpha = k_2/k_1 \tag{9}$$

When two peaks elute at the same time, the system exhibits zero selectivity. Remember that either the column or the mobile phase can be changed to achieve better selectivity.

## IV. Control and Monitoring Parameters

### A. Pressure and Flow Rate

The small channels between packing particles resist the flow of liquid. It takes energy to overcome this resistance, i.e., a source of pressure at the column inlet. The larger the pressure, the larger the resulting flow rate. Most liquid chromatographs use metering pumps that can deliver a specified flow rate regardless of pressure (up to the pressure limit of the pump). This is appropriate, since the important chromatographic variable that should be under control is the flow rate, not the pressure. We observe a pressure at the column inlet as a consequence of flowing through the column. This pressure is termed the column inlet pressure or the column pressure drop, $\Delta P$, the difference between inlet and outlet pressure.

In addition to flow rate, $F$, pressure drop depends on several factors, as stated in the following expression:

$$\Delta P \propto FL\eta / d_p^2 \tag{10}$$

The pressure is directly proportional to column length, $L$, and mobile phase viscosity, $\eta$. It is inversely proportional to the square of the diameter of the packing particles, $d_p$. For example, a 1 mL/min flow rate through a 4.6 mm ID × 25 cm long column of 5 $\mu$m particles produces an inlet pressure of roughly 1500 psi with methanol and 6800 psi with the more viscous isopropyl alcohol.

Most binary solvent mixtures have viscosities that vary with the composition. Mixtures of water and methanol are the most extreme example. When programming from pure water to pure methanol, the pressure first rises and then falls, although the flow rate from the metering pump remains constant.

### B. Temperature

Solvent viscosity decreases as temperature increases. So one benefit of elevated temperature operation is that it reduces the pressure required to achieve the desired flow rate, providing more reliable operation of pump, injector, and column seals. However, a more significant benefit of elevated temperature is that it improves resolution by increasing efficiency. This follows from the fact that diffusion rates increase with increasing temperature, and it has been pointed out that resistance to mass transfer is the dominant cause of zone spreading.

Adjusting temperature is also a way of controlling selectivity, although the effects are usually not as great as those achieved by adjusting mobile phase composition.

Finally, the control of temperature at a constant value improves the reproducibility of retention times because retention is temperature dependent.

## V. Future Trends

For some applications there is a desire to improve the speed of analysis further. Up to now such improvements have been made by reducing the particle size. The smallest commercial packings at the present time are about 5 μm diameter. It may be that 2 or 3 μm particles will become available. However, we are approaching at least two limits. The pressure drops generated by such small particles become excessive, placing great demands on the hardware. Secondly, these large pressures are really an indication of the energy spent in pushing the mobile phase through the column. This energy is converted to heat. Since the heat can be lost from the column wall, the fluid closer to the wall is cooler than that in the center of the column. This temperature gradient results in a viscosity gradient that in turn causes nonequal flow velocity in the column. This causes poorer efficiency, the very thing we are trying to improve by using smaller particles in the first place.

Another way of achieving faster analyses may be to use shorter columns and lower flow rates. There is some debate on just how to optimize column performance, but the chances are good that column dimensions have not yet reached their theoretical optimum. Certainly the trend has been to shorter columns. A few years ago 1 m and 50 cm columns were common. Today it is rare to use a column longer than 30 cm; 10, 15, and 25 cm columns are common.

## References

1. Majors, R. E., *J. Chromatogr. Sci.* **15,** 334 (1977).
2. Bakalyar, S. R., *Amer. Lab.* **10,** 43 (1978).
3. Karger, B. L., Gant, J. R., Hartkopf, A., and Weiner, P. H., *J. Chromatogr.* **128,** 65 (1976).
4. Bakalyar, S. R., McIlwrick, R., and Roggendorf, E., *J. Chromatogr.* **142,** 353, (1977).
5. Giddings, J. C., *Dynamics of Chromatography,* Part I, Dekker, New York, 1965.

# Chapter 2

# Instrumentation for Liquid Chromatography

Richard A. Henry*

*Scientific Systems, State College, Pennsylvania*

*and*

Genrikh Sivorinovsky

*Altex Scientific, Berkeley, California*

## I. Introduction

High performance liquid chromatography (HPLC) is one of the most rapidly growing and potentially largest branches of analytical chemistry. Although further advances in HPLC are to be expected, current methodology is already far enough advanced to insure its use in the clinical laboratory.

The basic components of a high performance liquid chromatograph are shown in Fig. 1. The primary function of the solvent reservoir is to hold the composition of the mobile phase constant during the operation of the instrument. The flow stream from the solvent reservoir usually travels through a filter or series of filters that remove particles that could damage the pump or column.

A central component of a modern LC instrument is the pump. Three principal types of pumps—pneumatic, syringe-type, and

*Currently, with Applied Science Laboratory, State College, Pennsylvania.

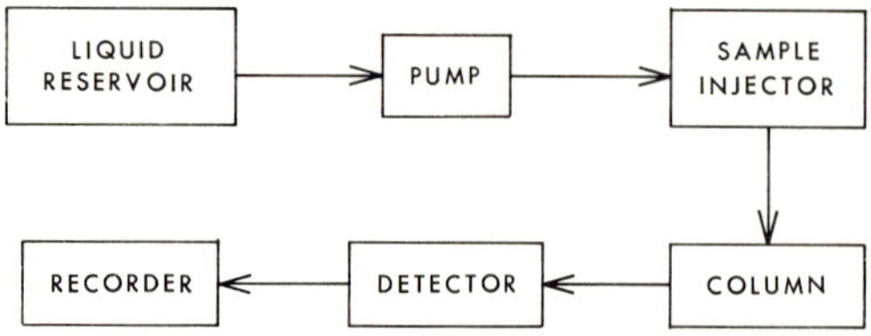


FIG. 1. Block diagram of a liquid chromatograph.

reciprocating piston—have been used in HPLC. These pumps are designed to maintain a constant, pulse-free flow rate at very high pressures. An injection device is also a very important component in an LC system. It is used to introduce the sample at the head of the column with minimum disturbance of the column packing. Recent improvements in the reliability and performance of pumps and injectors are largely responsible for the current wide acceptance of LC in clinical and other routine analytical laboratories.

The actual separation occurs on a narrow column tightly packed with small particles of packing material. The column has been called the heart of the liquid chromatograph because the success or failure of a chemical analysis by HPLC depends critically on the proper choice of column and operating conditions.

Detectors are used for distinguishing the presence and measuring the amount of solute eluting from the column. The most frequently used device is the ultraviolet photometric detector. Infrared, refractive index, flame ionization, fluorescence, electrochemical, atomic absorption, mass spectrometry, and many other detectors have been used in the analysis of the LC column effluent.

Results of the chromatographic separation are usually displayed in the form of a chromatogram on a strip-chart recorder. If proper column and mobile phase selection has been made, each Gaussian shaped peak represents a zone of pure solute that is free of interfering substances and can be easily analyzed or collected.

## II. Pumps and Reservoirs

Solvent reservoirs are used for holding the composition of the mobile phase or solvent used in HPLC constant during the time of analysis and can also be used for degassing. Degassing the mobile phase is often advisable because bubbles of air can influence flow precision, and bubble formation in the detector flow cell can cause high noise level. In addition, oxygen dissolved in the mobile phase can cause chemical changes in oxygen-sensitive samples and reduce the sensitivity of fluorescence detection (*1*). One of the simplest and most effective ways of degassing solvents is by aspirator vacuum. A better approach is the continuous sparging of the mobile phase with a flow of inert gas such as helium. Gases such as nitrogen and oxygen are replaced by smaller concentrations of less soluble helium.

Columns for modern HPLC are packed with 5–10 μm particle size packing material that offers a high resistance to flow. Column pressure drop is described by the equation,

$$P = \eta L v / \theta d_p^2$$

where $\eta$ = fluid viscosity, $L$ = length of column, $v$ = liquid velocity, $d_p$ = particle diameter, and $\theta$ = dimensionless structural constant of about 600 for packed beds in HPLC. HPLC separations require pressures in the range of 200 to over 6000 psi. The most common range is 750 to 3500 psi.

Applications that require very high pressures are not common in clinical chemistry; however, pumps with high pressure ratings tend to have fewer problems operating at lower pressures than those operated close to their design limits. Also, high-pressure pumps allow the chemist to explore higher flow rates in order to decrease analysis time, which can be very important in the clinical laboratory.

Several distinctly different pump designs have been offered during the last decade; however, there now appears to be a definite trend toward motor-driven pumps with small reciprocating pistons. The reasons for this overwhelming acceptance of the small displacement volume reciprocating pump are summarized in Table 1. Considerations such as these probably have been responsible for the similarity of the pumps recently introduced by competitive manufacturers. Perhaps

Table 1
Properties of Reciprocating Pumps for HPLC

| Property desired | Easy | Hard |
|---|---|---|
| Ease of operation | x | |
| Economy | x | |
| Compatibility with gradient elution | x | |
| High pressure operation | x | |
| Rapid solvent change | x | |
| Continuous operation | x | |
| Reproducible flow (long-term) | x | |
| Uniform flow (short-term) | | x |

the only inherent limitation of motor-driven reciprocating pumps is that it is difficult to obtain uniform "pulseless" flow rates over the short term. Pulseless flow is desirable because most detectors are flow-sensitive, hence chromatographs with pulseless pumps show lower baseline noise and better detection limits. Also, pumps with uniform, pulse-free flow give more uniform solvent composition in gradient elution.

Figure 2 shows a timing diagram for a single-piston pump. The diagram shows three segments that occur in each pump cycle. Two of the cycles are solvent delivery and refill. The third, or compression, segment varies in size according to solvent compressibility, pressure, and the amount of solvent within the pump chamber.

Historically, the single-piston reciprocating pump design was less than satisfactory for the majority of HPLC applications because there were long periods when no liquid was being delivered to the column. A typical flow profile for such a pump is shown in Fig. 3A. Note that at least half the time for any particular pump cycle is consumed by filling the cylinder and compressing the solvent. Clearly, flow uniformity could be increased if the time required for these operations could be decreased.

Figure 3B shows a flow profile from a single-piston pump with rapid refill and compression. Of course, there are limits imposed by other considerations. For instance, rapid refilling of the pump can cause cavitation, which is the formation of gas or solvent vapor

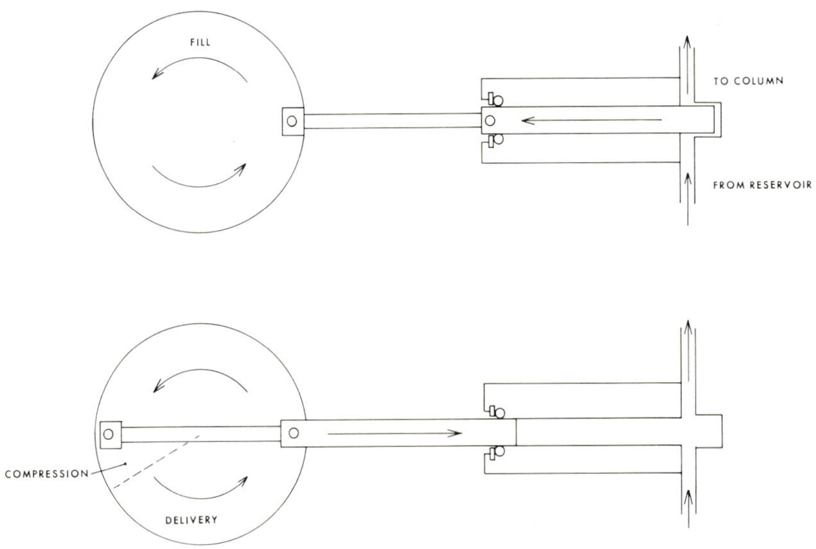


FIG. 2. Timing diagram for a single piston pump.

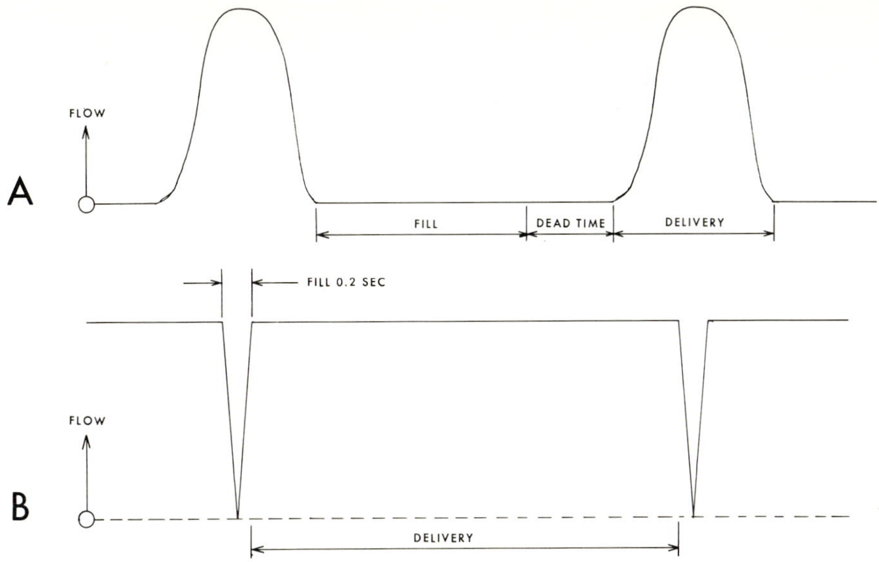


FIG. 3. Flow profile for a single piston pump.

bubbles in the pump chamber. Currently, both Altex Division of Beckman Instruments and Varian Instrument Company offer very popular single-piston reciprocating pumps with rapid refill.

A side view of the Altex 110 pump is shown in Fig. 4. The cam is driven by a small motor that is usually found in computer-tape drives. These motors are well known for long life and ability to change speed rapidly. The cam drives the piston roller that is also lubricated for long life. Another feature of the pump is a 5 $\mu$m filter placed to protect the ball and seat of the outlet check valve from particles that can cause the valve to leak. These particles probably originate from seal wear as well as from other sources.

A schematic of the Varian pump is shown in Fig. 5. Important features include a stepping motor, fast-refill piston, and cam-driven inlet valve. The Varian pump also employs a filter to protect the outlet check valve.

Pulse dampeners of different forms have been added to these and other single-piston pumps for further reduction of pulsation, but pulse dampeners add internal volume, which can be undesirable when changing solvents or performing gradient elution.

Several models of dual-piston reciprocating pumps have been designed to minimize flow pulsations. Uniform flow rate is achieved in a dual-piston pump by means of two cam-driven reciprocating pistons as shown in Fig. 6 for a pump designed by Perkin-Elmer. Each piston

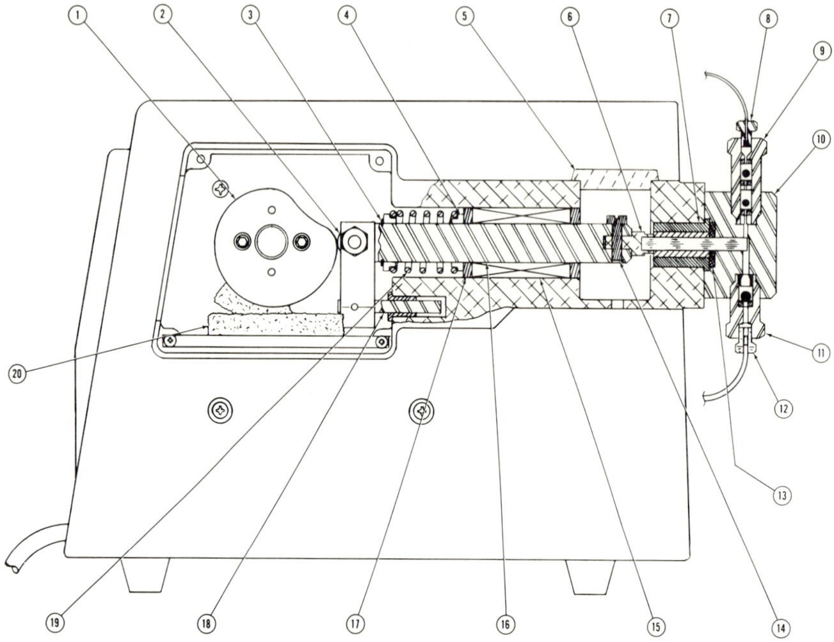

1 CAM
2 CAM FOLLOWER
3 O-RING
4 WASHER (2)
5 INSPECTION COVER
6 PISTON ASSEMBLY
7 PISTON GUIDE ASSEMBLY
8 HIGH PRESSURE TUBING ASSEMBLY
9 OUTLET CHECK VALVE
10 LIQUID HEAD
11 INLET CHECK VALVE
12 LOW PRESSURE TUBING ASSEMBLY
13 PISTON SEAL
14 PISTON RETAINING SCREW
15 BEARING
16 CAM FOLLOWER SHAFT
17 OIL SEAL (2)
18 GUIDE ROD ASSEMBLY
19 SPRING
20 CAM LUBRICATION ASSEMBLY

FIG. 4. Diagram of Altex 110 pump.

of the pump has inlet and outlet check valves, and the two pistons are driven from opposite sides of the same cam so that their motion is approximately 180 degrees out of phase. The shape of the cam is such that the sum of the forward velocity of both pistons is at all times a constant value when the cam is rotated at a constant angular velocity. This produces a constant solvent flow rate. Altex Division of Beckman Instruments, Waters Associates, Spectra-Physics, Laboratory Data Control, Tracor, Micromeritics, and others manufacture similar dual-piston reciprocating piston pumps. Hewlett-Packard offers a reciprocating diaphragm pump that achieves pulse-free flow by operating at very high frequency.

Most clinical assays by means of HPLC have involved the determination of a single compound or groups of similar compounds.

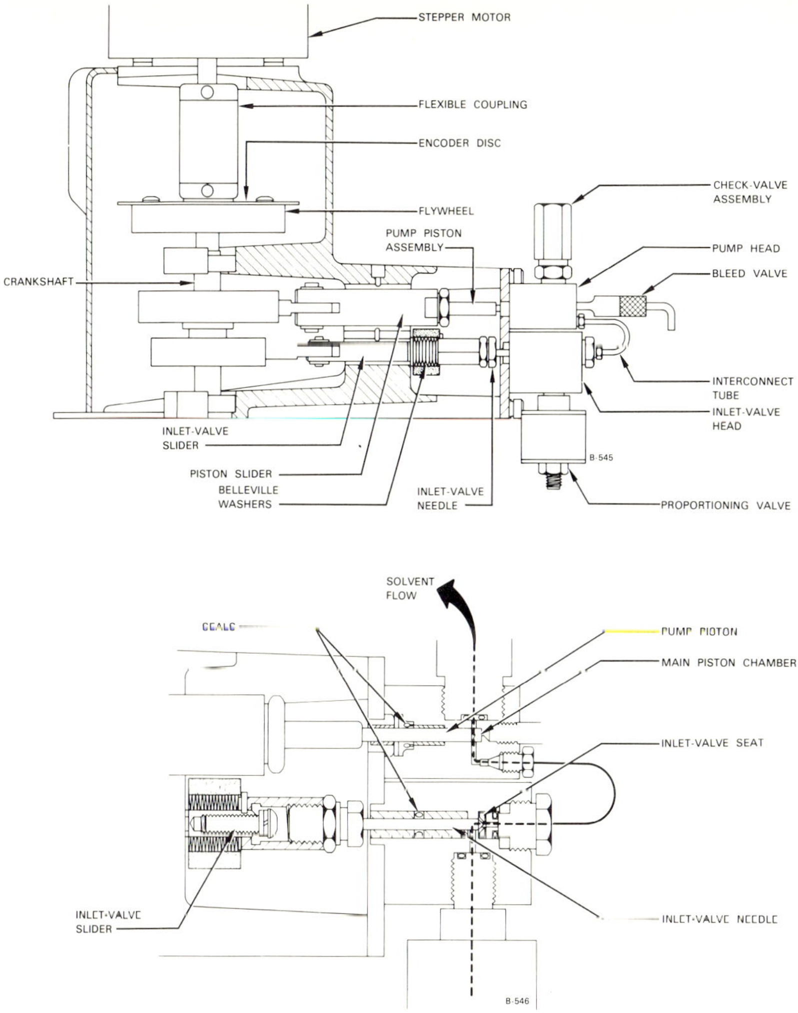


FIG. 5. Diagram of Varian pump.

Such assays are best carried out by isocratic elution, where the same mobile phase conditions are maintained throughout the analysis.

In other cases, the clinical chemist or pharmacologist may have to determine several physiological metabolites or drugs of widely different elution properties. In this situation the isocratic technique may not be adequate for reasons that include excessive separation time and peak broadening. Because of these problems, gradient elution or solvent-programming chromatography has been used in an increasing

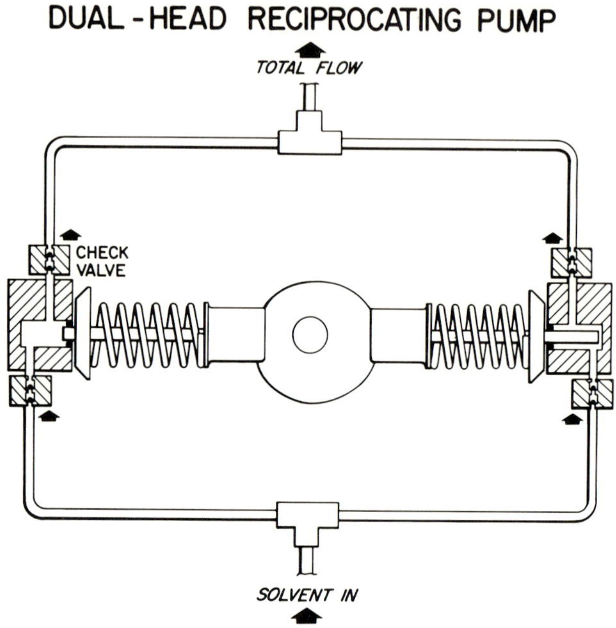


FIG. 6. Diagram of Perkin-Elmer dual-piston pump.

number of applications. In an analysis using the gradient mode, the composition of the mobile phase is gradually changed in such a way that the solvent strength increases throughout the separation. Using a properly programmed and designed mobile phase gradient, compounds with both widely different and simlar chemical structures can be separated in a relatively short period of time. This feature makes the technique especially useful when dealing with mixtures of unknown composition (*2*).

Sophisticated gradient liquid chromatographs are commercially available in two principal designs, shown in Fig. 7. In dual-pump systems, the gradient is formed on the high pressure side of the pumps, with each pump delivering a different solvent. Gradient composition is controlled by simultaneously varying the flow rates of the two pumps to maintain a constant flow. In single-pump systems, solvents are mixed and the gradient is formed on the low pressure side of the pump. Since solvents are proportioned and mixed prior to entering the pump, this design requires only a single high pressure pump that operates at a constant flow rate. Both types of gradient system require a high pressure mixing chamber, as shown in Fig. 8.

A variant of a one-pump gradient system is a step-gradient, which has been successfully used in the separation of several pharmacological

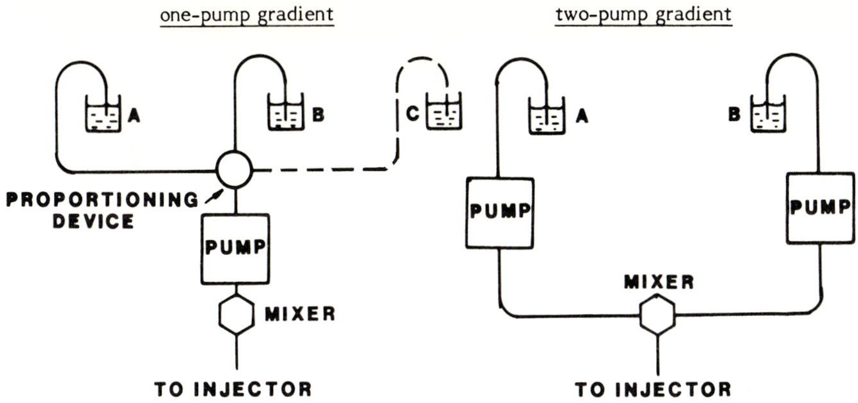


FIG. 7. Single-pump and dual-pump gradient systems.

preparations. The chromatogram in Fig. 9, the active ingredients in cough syrup, was obtained using a simple-to-operate, low-cost, step-gradient liquid chromatograph. A three-way slider valve was inserted into the pump inlet line just before the check valve. The system was initially equilibrated with mobile phase A, and the slider valve was switched concurrent with injection so that mobile phase B was delivered to the pump. At the end of the analysis, the slider valve was

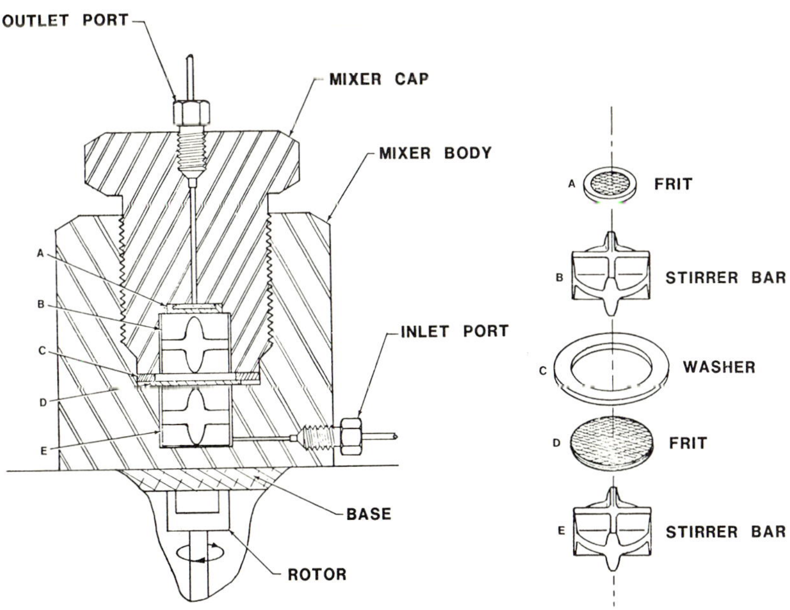


FIG. 8. Diagram of a typical high pressure gradient mixing chamber.

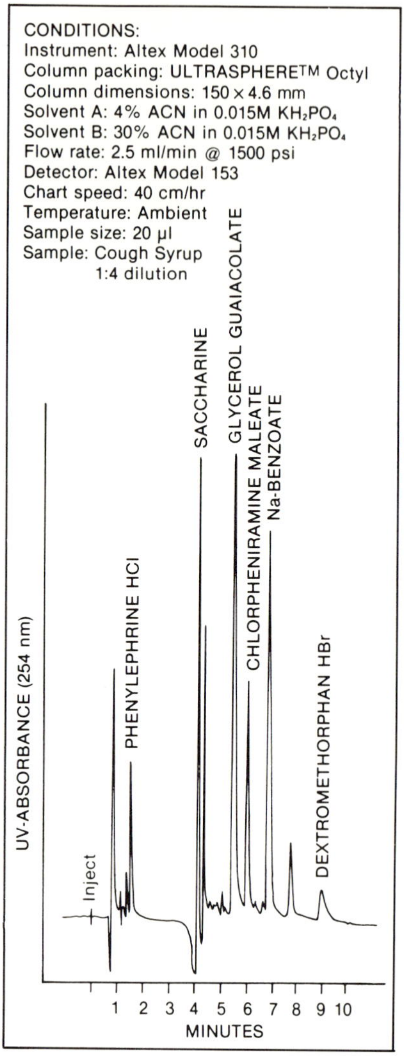


FIG. 9. Step gradient analysis of cough syrup ingredients.

returned to the original position and the system re-equilibrated for 5 min with mobile phase A prior to the next injection. In order to use this technique of a step-gradient at the pump inlet, the pump internal volume should be very small. Otherwise, a step-gradient can be generated after the pump by means of a high-pressure switching valve with a loop containing the strong solvent.

## III. Sample Injectors

There are two general modes of introducing the sample at the head of the column: (a) stopped-flow, and (b) with solvent flowing. In stopped-flow injection, flow is stopped, sample is introduced at atmospheric pressure, the system is closed and flow reinitiated. This technique is rarely used in modern liquid chromatography and has found little application in biochemical or clinical HPLC. Septum sample injectors commonly used in gas chromatography are also very rarely used in modern LC because they do not hold very high pressure, septa are not compatible with all LC solvents, and small particles tear from the septa and tend to cause plugging of the inlet frit of the column.

In modern LC, the preferred sample injection device has become a loop injection valve with four, six, or seven ports. One popular injection valve design is shown in Fig. 10. Several methods can be used to inject sample with loop valves. Most commonly, the valve is equipped with a removable needle port for sample injection with a standard blunt-needle syringe. Alternatively, a luer adapter can be connected to the sample port to allow use of a luer-tip syringe to fill the loop, or to draw sample into the loop from a vial. The sample port can also be connected to an automated sample delivery system for unattended operation. In this case, the valve must be equipped with an automatic actuator.

Loop sample injection valves manufactured by Altex, Rheodyne, Valco, Waters, and others allow the biochemist or clinical chemist to make precisely controlled injections and are therefore favored for quantitative analysis. The most precise form of sample injection is full loop loading. Sample loops are available from 5 to 2000 μL or greater to permit maximum flexibility in full loop injection. Although 5 and 10 μL loops are most common, larger loop sizes can often provide higher precision and sensitivity without serious loss of resolution. Partial loop loading is also possible; however, care must be taken to obtain reproducible results. The speed with which the needle plunger is inserted can have a significant effect on the form the sample assumes within the loop, as shown in Fig. 11. Slow insertion of the needle plunger can displace air, which may be trapped behind the needle seal, into the sample loop. This trapped air keeps the sample in a neat slug. Fast insertion of the needle may break the sample into segments which are interspersed with air. If the needle port has been flushed with solvent between injections, as is necessary when different samples are being loaded, solvent, rather than air, may be trapped behind the needle seal. In this case, sample moves through the loop in a laminar

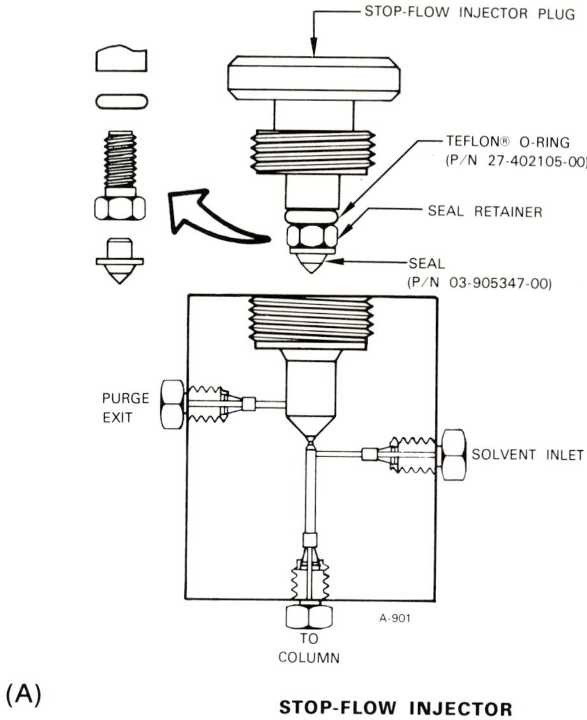


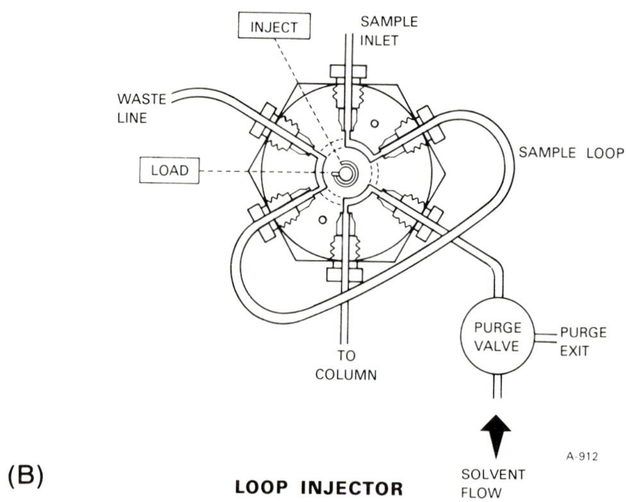


FIG. 10. A, Diagram of Varian stop-flow injector; B, diagram of Varian loop injector; C, flow pattern of Altex loop injector.

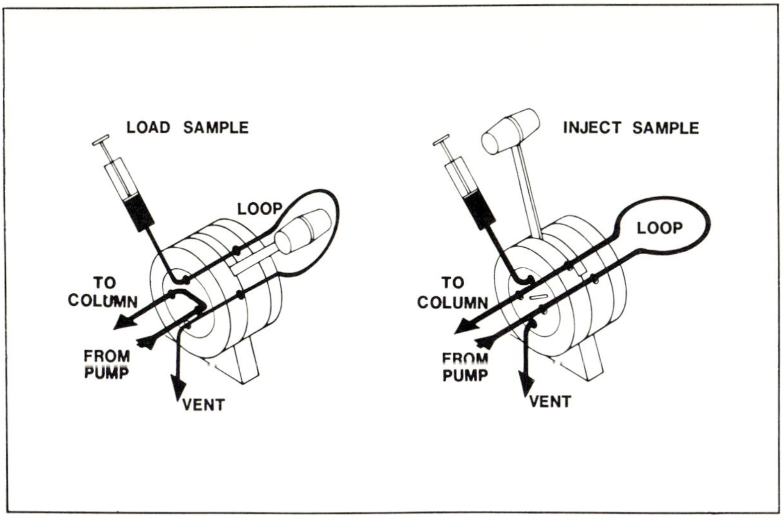


210 Valve Flow Pattern

FIG. 10. Continued

flow pattern rather than as a square front. When the desired sample volume is a high percentage of the total loop volume, this laminar profile can allow part of the sample to flow through the loop and out the vent, rather than be injected. To obtain maximum precision in partial loop loading, the injected sample volume should be less than

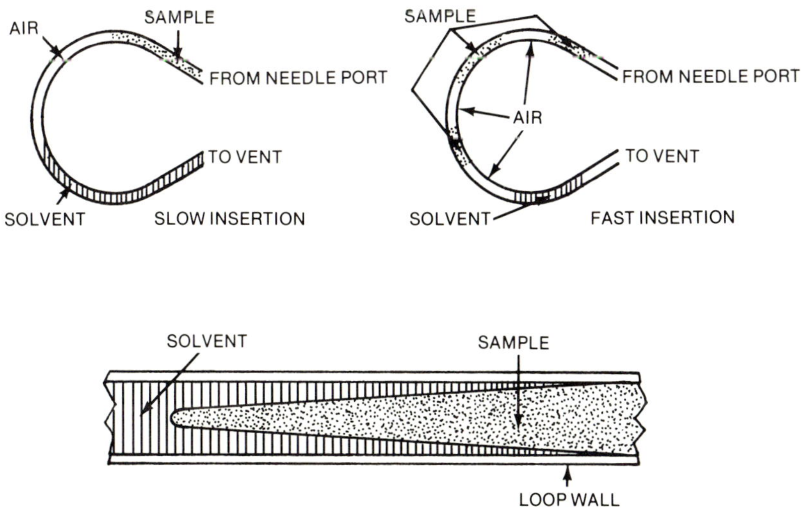


FIG. 11. Illustration of sample behavior in a sample loop.

75% of the total loop volume or the loop should be flushed with air prior to filling (*3, 4*).

Many manufacturers of modern liquid chromatography equipment have realized the need for the automation of sample injection for both method development and routine laboratory tests. Technicon, DuPont, Waters Associates, Altex Scientific, Perkin-Elmer, Varian, Micromeritics, and others have introduced autosamplers. Usually, the sample is contained in a small vial with a septum cap. The injection sequence starts with a needle penetrating the top of the vial, followed by a means of forcing or drawing the liquid from the vial into the sample loop of an injection valve. The amount of sample is controlled in such a way that the valve loop is overfilled and flushed with excess sample. An electronic signal actuates the valve, placing the sample in the eluent stream to the column. A more complete discussion of autosamplers and automation is beyond the scope of this article. The reader is referred to the manufacturers for further description, specifications, and performance data. Remember that the most important requirement for automated, unattended operation is reliability.

## IV. Columns

The part of the system where separation takes place is a narrow column, tightly packed with small particles called packing material. Columns for HPLC are almost invariably made from stainless steel tubing with a smooth inner surface. The tube should be cleaned to remove grease and other material left from the manufacturing process. This is normally done by washing the tube with organic solvents. The tube should next be passivated with a concentrated solution of nitric acid to remove loose metal particles and build up an inert oxide coating on the inner surface of the tube. Cleaning and passivating is usually done by the supplier of chromatography-grade tubing.

A lot of attention is currently being focused on packing materials, both to enhance the range of molecular types that can be separated and to maximize the stability of the particles and columns. We will here limit our description of the columns by introducing only the hardware, because a detailed discussion of column theory and modes of separation will be presented elsewhere.

Columns for HPLC may be divided into analytical and preparative. Analytical columns have a typical inside diameter of 2–6 mm, with the length being dependent upon the type of packing material. For pellicular packing, the usual length is 50–100 cm. For

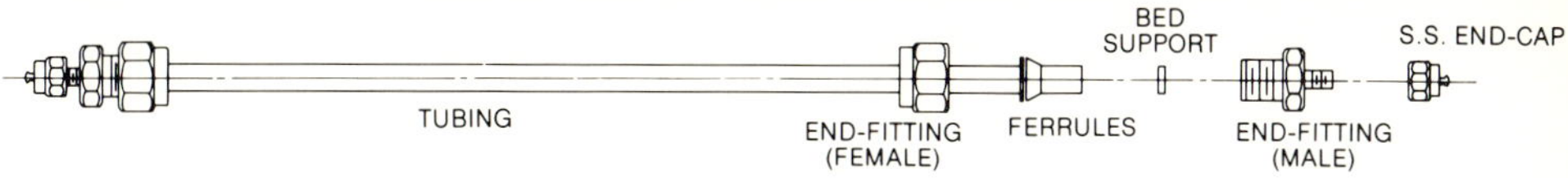


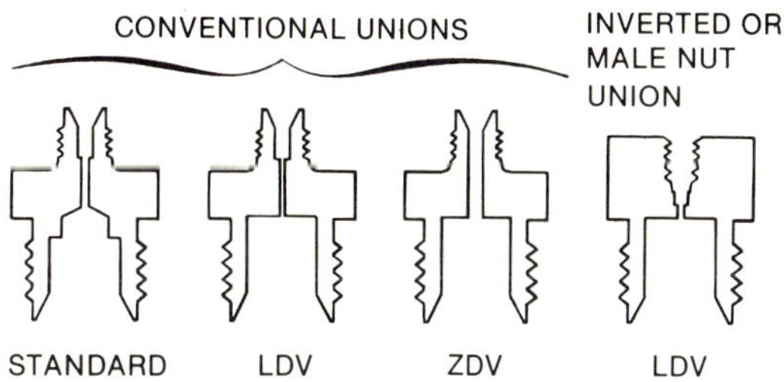


FIG. 13. Summary of various column fitting designs.

microparticulate porous packing, a length of 10–30 cm is more common. Preparative columns generally have diameters of 7 mm or larger and lengths from 25–100 cm. A typical stainless-steel column with end fitting, ferrules, bed supports and end-caps for protection against drying is shown in Fig. 12. Columns made by different manufacturers usually do not have the same end fittings. A summary of various end fitting designs is given in Fig. 13. In many cases, the experimenter or laboratory technologist must design adapters for interfacing LC system and columns. Standardization of column fittings and liquid chromatograph connections would be very well accepted by chromatographers, but it is not likely to occur in the near future.

## V. Detectors

Table 2 lists the most popular detectors in modern liquid chromatography with an indication of detection limits for favorable substances. The most popular LC detectors are concentration dependent. According to this, the minimum detectable injected quantity is a combination of minimum detectable concentration (g/mL) of compound in a flow cell multiplied by average peak volume, which can be estimated by width at half-height. For example, if the detector has a minimum detectable concentration for a compound of 5

Table 2
Detectors Used in HPLC

| Detector features | Nondestructive/concentration sensitive | | | | Destructive/mass sensitive | | | |
|---|---|---|---|---|---|---|---|---|
| | Ultra-violet | Fluores-cence | Refractive index | Infrared | Electro-chemical | Flame ioniza-tion | Mass spectro-metry | Atomic absorption |
| Selective | Yes | Yes | No | Yes | Yes | No | Yes | Yes |
| Gradient capability | Yes | Yes | No | Yes | No | Yes | Yes | Yes |
| Minimum detectable concentration, g/mL | $10^{-10}$ | $10^{-11}$ | $10^{-7}$ | $10^{-7}$ | $10^{-11}$ | $10^{-7}$ | $10^{-10}$ | $10^{-10}$ |
| Minimum detectable quantity level | ng | ng–pg | $\mu$g | $\mu$g | ng–pg | $\mu$g | ng | ng |

parts per million (5 μg/mL) and the average chromatographic peak volume is 0.2 mL, the minimum detectable quantity is then 1 μg. The peak volume is a measure of the degree to which the injected sample is diluted by the mobile phase prior to entering the detector cell. This dilution or band spreading is a function of column geometry and column efficiency. Thus, minimum detectable concentration characterizes the detector, while minimum detectable quantity characterizes the total chromatograph–detector system. Very short and efficient columns lower the minimum detectable quantity.

There are two general types of detectors: *selective and universal*. A detector is classified as *selective* if its response differs widely with molecular structure, and as *universal* if its response is similar for most compounds. Selective detectors are sensitive and are especially useful for trace analysis, while universal detectors are more valuable for scouting unknown samples for major components. Universal detectors must also be employed for certain classes of compounds that do not respond to selective detectors. Absorbance and fluorescence detectors are commonly used selective detectors. Refractive index is the most common universal detector.

In modern HPLC, by far the most frequently used detector is the ultraviolet photometer. There are two major types available: fixed wavelength, with low or medium pressure mercury source, and spectrophotometer. In all cases, low volume detector cells (~ 8–20 μL) are employed in order to limit extra-column band broadening.

In fixed wavelength photometric detectors, 254 nm and other strong lines from the mercury source are selected in most commercial instruments. This detector gives the lowest noise level, but flexibility and selectivity are lost by not being able to work at any wavelength. Nevertheless, the fixed wavelength detector is extremely useful because its source of energy corresponds to strong absorption bands of most aromatic compounds.

A block diagram of one fixed wavelength detector optical geometry and control circuitry is illustrated in Fig. 14. The sample and reference flow cell passages lie on two optical axes radiating from the low pressure mercury light source. Both passages are very close to each other. The light projected through the reference passage to the reference sensor is essentially the same view of the source as the light projected through the sample flow passage onto the sample sensor. The sensors usually are silicon diodes or special cadmium sulfide photoresistors with a phosphor screen in front of them to make them UV sensitive. A removable interference filter is positioned directly in front of the sensors to insure that they respond only to monochromatic light and obey Beer's law. The stronger lines from the low pressure

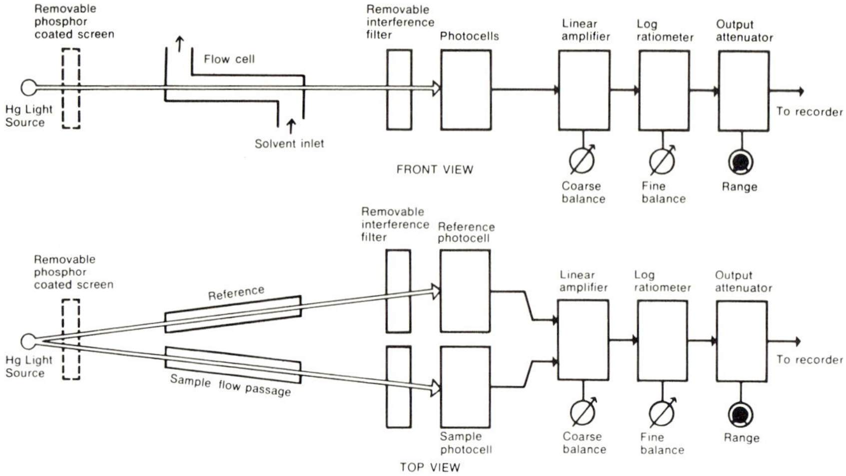


FIG. 14. Block diagram of Altex fixed wavelength detector.

mercury source (254, 313, 365, 405, 436, 546, and 578 nm) can be directly monitored with the appropriate interference filter installed. These lines have very narrow bandwidths (typically 0.2 nm) making Beer's law hold true even if the compound being measured does not have an absorbance maximum at the wavelength being used.

For other wavelengths between 280 and 660 nm, a phosphor screen is inserted between the lamp and flow cell. The screen fluoresces at a longer wavelength when excited by the strong 254 nm radiation from the lamp and acts as a wavelength converter.

A block diagram representation of a unique dual wave-length detector is illustrated in Fig. 15. A mercury light source excites a phosphor-coated block that emits radiation at a longer wavelength. Both light sources shine through sample and reference cell. The small apertures of the flow cell passages act as a pinhole camera and produce a physical separation between the two light sources at the photocells. One pair of photocells, in conjunction with a high-quality interference filter, is used to monitor at 254 nm, while a second pair of photocells with another filter is used to monitor at the second wavelength, usually 280 nm. Both sets of photocells are continuously monitoring both the sample and the reference side of the flow cell. This allows "real-time" absorbance ratioing, which can be used to identify compounds.

There are also some new developments in light sources that can allow fixed wavelength detectors to operate between 200–254 nm; however, these products currently do not have clear cost and performance advantages over variable wavelength detectors that employ a continuum source and filter or monochromator.

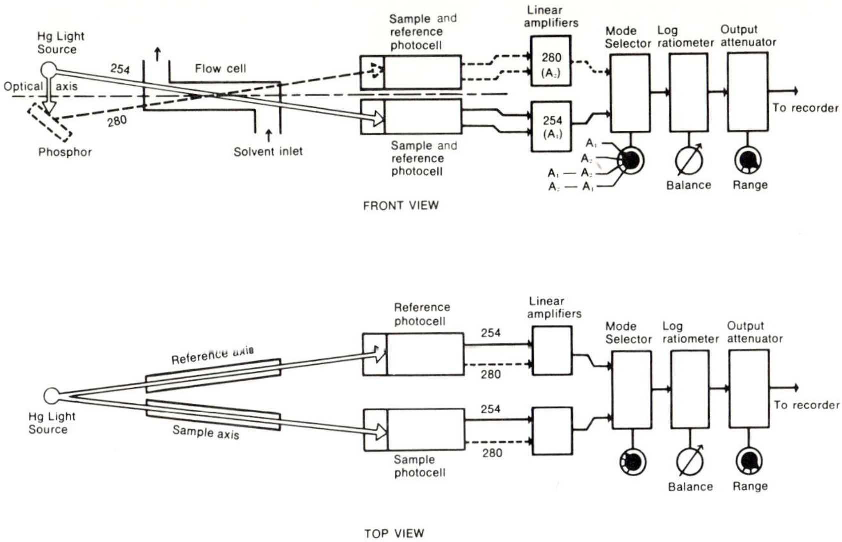


FIG. 15. Block diagram of dual wavelength detector (Altex).

In situations where compounds of interest absorb far below 254 nm, such as some anticonvulsant drugs, a variable wavelength detector is the right choice. Also, the components of a mixture may have very different absorption spectra, so that by varying the wavelength of detection, the peak sizes can be adjusted relative to one another in order to minimize interference effects. When operating under gradient elution conditions, the wavelength of a variable wavelength detector can be set to a value where both solvents have low and equal absorption, hence the baseline remains relatively flat throughout the run. Variable wavelength photometric detectors employ a continuum source, such as a deuterium lamp (190–400 nm output) and a monochromator to isolate narrow wavelength bands. One product has recently been introduced that uses filters to isolate the desired wavelengths. A diagram of a typical variable wavelength detector is shown in Fig. 16. The visible region is also available in most variable wavelength detectors. The visible light source is usually a tungsten lamp. Several variable wavelength detectors have the additional benefit of being able to scan the full absorption spectrum of the sample, which can be isolated in the flow cell by stopping the flow of the chromatograph. New developments should soon allow the full compound spectrum to be obtained under flowing conditions.

The ability to select wavelength greatly enhances the applicability and selectivity of the photometric detector. Fortunately, most of the solvents used in HPLC have wide windows in the UV–VIS region,

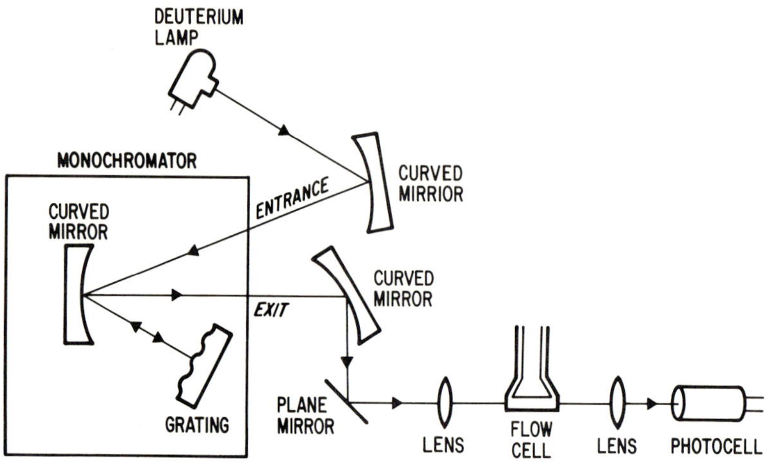


FIG. 16. Diagram of variable wavelength detector (Perkin-Elmer).

making them compatible with UV detectors even at very low wavelengths. Many solvents allow the clinical chemist to operate at the level of 210 nm. Water and acetonitrile are important solvents that can be used down to 195 nm.

Another type of optical detector widely used in HPLC is the differential refractometer. Being a bulk property detector, the differential refractometer responds to all substances and is a good example of a universal detector. Refractive index (RI) detection limits are several orders of magnitude higher than obtained with the UV detectors, and RI cannot be used with solvent gradients. The biochemist or clinical chemist would only turn to RI detectors when non-UV active compounds, such as lipids, prostaglandins, sugars, etc., are being investigated. RI detectors also find frequent use in preparative scale operation, where sensitivity is less important.

Fluorescence detection is becoming increasingly popular and is especially applicable to substances of clinical interest. Many molecules, especially those with rigid structures, have the property of absorbing light and emitting it, essentially instantaneously, at a longer wavelength. In favorable cases, detection levels are two orders of magnitude lower than the UV detector can achieve. Two main advantages of fluorescence detectors for HPLC are better detection limits for many compounds, and extreme selectivity. The selectivity makes the fluorescence detector virtually immune to baseline drift, even during gradient elution. Compounds that do not have natural fluorescence can be modified with a fluorescing "tag" compound. Most of the nonhalogenated solvents that are used with photometric detectors can be used with the fluorescence detector. Halogenated solvents, such as methylene chloride or chloroform, should be used with care because they tend to diminish fluorescence, or "quench" it.

Another type of photometric detector is the infrared (IR) detector. Practically all molecules absorb infrared radiation at a wavelength corresponding to a particular functional group. However, detection is possible only if the HPLC solvent system is transparent in the region of interest. There are many solvents that have windows that permit detection of certain functional groups, but the infrared technique is more difficult to use than conventional UV detection methods. The most useful solvents in modern HPLC contain hydroxyl groups (water, alcohols) that absorb strongly in the 3000–3700 and 1000–1200 $cm^{-1}$ regions and limit the number of functional groups that can be detected in the reversed-phase LC mode. Also, extinction coefficients for infrared absorption are much smaller than UV–visible absorption, which significantly decreases the sensitivity of the system. Infrared detectors have not yet found much application in the biomedical field.

The commonly used optical detectors described above are nondestructive and allow sample collection for further qualitative characterization. They can also be used in series to supplement one another and provide qualitative information on the identity of various components of the chromatogram. The chromatographer should be aware that there is always the possibility that light-sensitive compounds can undergo photolytic reactions in the flow cell of the optical detector while illuminated by relatively strong UV light. An example of this interference could be partial photolysis of retinol in vitamin assay. However, the time of illumination in HPLC microvolume cells is very short and photolysis even when it occurs is extremely low and does not affect detector function.

Electrochemical, flame ionization, mass spectrometric, and atomic absorption are examples of destructive LC detectors, since at least part of the sample is altered in the process of detection.

Electrochemical (EC) detectors are most frequently used in the amperometric mode where the sample reacts by means of oxidation or reduction at an electrode surface to produce current flow. They are among the most selective detectors and in favorable cases compare to fluorescence in sensitivity. In the biomedical field, EC detectors have been applied to the analysis of catecholamines and their metabolites, and the determination of trace amounts of tyrosine metabolites (5). Their primary disadvantage lies in the fact that polar solvents with electrolytes must be used. Also, relatively few compounds are electroactive and therefore detectable at available potentials, and electrode surfaces can be altered during operation, which can markedly affect reproducibility. Nevertheless, much interest exists in the amperometric detector for biochemical analysis. Coulometry and conductivity detectors are related and have been utilized in modern liquid chromatography, but they have not yet found major application in the biomedical field.

Flame ionization (FI) detectors were originally invented for gas chromatography, and there have been several attempts to adapt flame ionization detection to modern HPLC, with only limited success. Flame detectors operate by burning the organic samples and measuring ionic fragments as they pass through an electrode system. FI detectors are not very specific and respond to virtually any organic compound. The main problem in adapting to LC is getting rid of the mobile phase without losing too much of the sample. Currently, no FI detectors are commercially available, owing primarily to their mechanical complexity, high cost, and lack of reproducibility.

Mass spectrometry (MS) can be used to identify almost any chemical compound of molecular weight less than 1500, including many biochemicals of importance and essentially all therapeutic agents and their metabolites. The main advantages of MS as an analytical technique are high sensitivity and specificity in identifying or confirming compounds. The enhanced sensitivity results primarily from the electron multipliers and the action of the analyzer as a mass filter to reduce background interference. The excellent specificity results from characteristic fragmentation patterns, which can give information about molecular weight and molecular structure. When coupled with chromatography, MS can become a quantitative detector that can be operated in specific or nonspecific (total ion) modes and also yield qualitative information.

As in the flame ionization detector, a major challenge in mass spectrometry is the interface to handle or remove the large excess of mobile phase solvent in the column effluent. In one approach, the column effluent is coated on a thermally stable belt that passes through a vacuum chamber where solvent is removed. The sample on the belt is then passed through a heater where it is volatilized into the source. Major problems with this approach include difficulty with aqueous buffered mobile phase, loss of sample, and high cost. The use of very small ID columns with low flow rates can make the LC/MS interface much easier.

Atomic absorption (AA) spectroscopy is highly specific and sensitive for metals and thus, when combined with a liquid chromatograph, should provide an excellent method for selective detection of organometallic compounds such as enzymes, vitamins, and certain drugs. Components that elute simultaneously with the analyte species, but do not contain the particular metal monitored, will not interfere (*6*). Recently Vickrey has described an LC/AA system that could be applied to monitoring metalloenzymes and some other organometal compounds in clinical chemistry (*7*).

Since the stability and separation power of even the best LC column is limited, there is always the possibility of mistaking a peak identified only by its relative retention. Especially in complex samples, it is possible for another compound to have the same or a very similar retention time. One simple way to solve this problem is to determine absorbance ratio (AR) at two or more wavelengths to characterize the compound (*8*). If wavelengths are well chosen, AR provides a very specific and reliable technique for peak identification. Absorbance ratios are equal to extinction coefficient ratios and are independent of concentration of compound in the detector flow cell. Identity or purity of individual peaks can be confirmed when the AR of an unknown is compared with the AR of individually injected pure standards. Although AR values determined for pure standards by static spectrophotometry should be similar to those obtained using LC flow-through photometric detectors, the best results will be obtained by comparing unknown and standard values calculated from the LC detector. AR can be measured in several ways:

1. Ratio peak maximum absorbance in two consecutive runs at different wavelengths.
2. Ratio peak maximum absorbance in one run with two detectors at different wavelengths connected in series or parallel.
3. Ratio absorbance at front and back of a peak to test identity and purity.

Detectors will soon be available that can give AR values and other qualitative information such as multiwavelength detection with scanning (*9*) automatically on-the-fly. Keep in mind that response ratios can also be calculated for detectors, such as fluorescence and electrochemical, to give qualitative data, and that detectors can be connected in parallel to minimize bandspreading.

## VI. Data Processing

The output of the LC detector is usually presented on a strip-chart recorder. Modern recorders use integrated circuits for reliability and feature null-balance, servo-type systems for high precision and accuracy. It is a common fault to couple an expensive LC system to an inexpensive recorder and expect ultimate performance.

A typical chromatogram obtained on a strip-chart recorder is shown in Fig. 17. From this chromatogram the clinical chemist may easily calculate retention time or volume, and peak height or area of each compound. Data handling and quantitation is a very important step in HPLC analysis, especially in biochemical research and in the clinical laboratory, where results determine steps to be taken in the treatment of illness or the development of new drugs.

To help correct for less than 100% recovery of a compound, an internal standard similar to the compound of interest can also be added to the test tube before isolation is begun. A good internal standard will correct the assay for losses of compound in the isolation steps and improve the precision of the results by compensating random errors associated with aliquot taking, derivatization, and instrumental performance. An internal standard ideally should accompany the compound of interest in a constant ratio throughout isolation or preparation steps, and then be separated on the HPLC system. Homologs or analogs of the compound differing by the addition of a methyl group or halogen will often be excellent internal standards as long as they are not naturally present in the sample.

After obtaining a chromatogram of the compounds of interest and an internal standard, the investigator quantitates the results by measuring peak areas or peak heights and preparing a standard or calibration curve. Both have their uses, and selection depends upon the type of analysis being performed and experimental conditions (*10*).

Peak height is measured as the distance from baseline to peak maximum, and should not be used when peaks are visibly distorted or when the column is overloaded. Peak heights are relatively independent of flow rate variation when nondestructive detectors are employed and can yield precisions of 1–2% provided that temperature and mobile phase composition are carefully controlled.

Peak area measurements are less dependent upon operator and other variations when flow rate is controlled, but they must be measured by an electronic integrator for best results. Many modern data systems can record the chromatogram and print a full alphanumeric quantitation report. One example is shown in Fig. 18 for anticonvulsant analysis. Data systems for automatic quantitation range from simple, inexpensive units with printer output that operate in conjunction with strip-chart recorders, to multichannel products with printer/plotters that eliminate the need for a separate recorder. If electronic area integration is employed, it is good practice to compare results with peak heights as the method is being developed. Although

CONDITIONS
Instrument: Model 330
Column: ULTRASPHERE™ ODS with
Vydac precolumn
Mobile phase: 15 mM $KH_2PO_4$ pH 6,
with 6 x $10^{-3}$ Triethylamine:
50% Acetonitrile
Flow rate: 2.5 m L/min.
Detector: Model 153 at 254 nm
Chart speed: 40 cm/hr
Temperature: 50° C
Sample size: 20$\mu$L
Sample: Phenothiazines, concentration
of 0.005 mg/mL

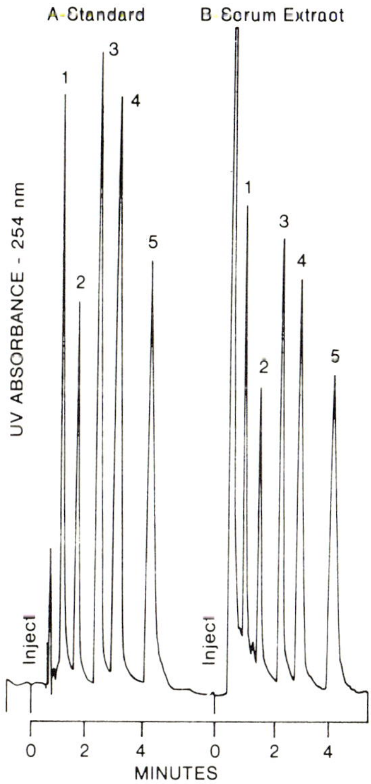


PEAK IDENTIFICATION
1. Mesoridazine Benzenesulfonate
2. Promazine Hydrochloride (Sparine)
3. Chlorpromazine Hydrochloride (Thorazine)
4. Thioridazine Hydrochloride (Mellaril)
5. Trifluoperazine Hydrochloride (Stellazine)

FIG. 17. Separation of phenothiazines.

areas are potentially more precise and accurate than heights, and are easier to automate, large errors can be introduced if integrator parameters are not set properly.

RUN 1 REF STD
SENSITIVITIES 700 20

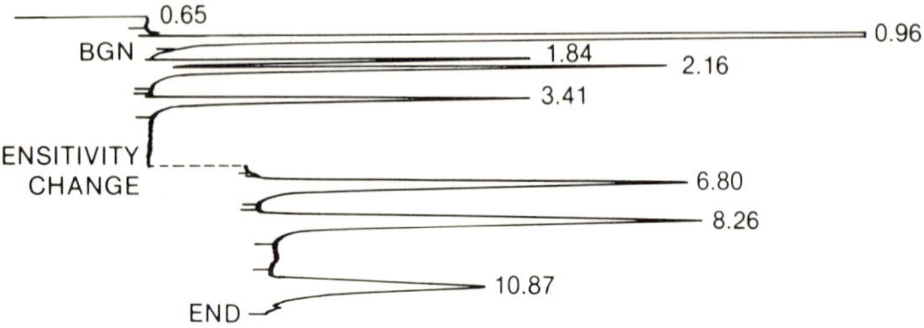


INST 2 METH 17 FILE 174
RUN 1 REF STD 11 : 26.3 11 / 3 / 78
SENSITIVITIES 700 20

| TIME | AREA | BC | RRT | RF | CONC | NAME |
|---|---|---|---|---|---|---|
| 1.84 | 2.8988 | V | 0.270 | 0.898 | 24.8139 | ETHOSUXIMIDE: |
| 2.16 | 4.8876 | | 0.317 | 0.515 | 23.9937 | PRIMIDONE: |
| 3.41 | 5.6521 | | 0.501 | 0.464 | 24.9989 | PHENOBARBITAL: |
| 6.80 | 5.2454 | | 1.000 | 1.000 | 50.0000 | HEXOBARBITAL: |
| 8.26 | 6.6275 | | 1.214 | 0.407 | 25.7119 | DILANTIN: |
| 10.87 | 4.2184 | | 1.598 | 0.668 | 26.8604 | TEGRETOL: |

RUN 8931 PATIENT
SENSITIVITIES 700 20

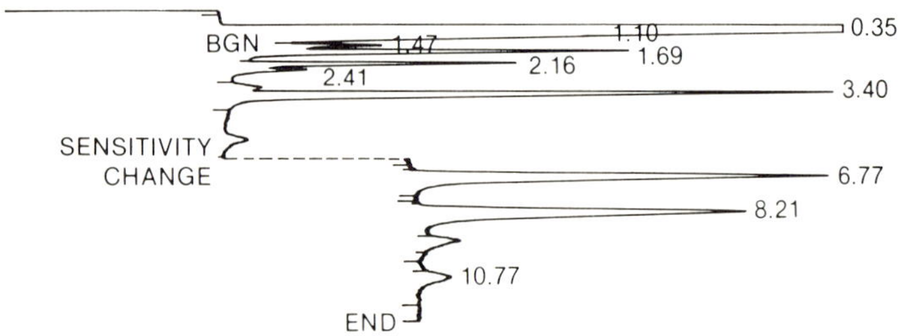


INST 2 METH 17 FILE 182
RUN 8931 PATIENT 13 : 31.0 11 / 3 / 78
SENSITIVITIES 700 20

| TIME | AREA | BC | RRT | RF | CONC | NAME |
|---|---|---|---|---|---|---|
| | | | 0.267 | 0.898 | | ETHOSUXIMIDE: |
| 2.16 | 2.3288 | V | 0.319 | 0.515 | 12.3486 | PRIMIDONE: |
| 3.40 | 7.8012 | | 0.502 | 0.464 | 37.2701 | PHENOBARBITAL: |
| 6.77 | 4.8561 | | 1.000 | 1.000 | 50.0000 | HEXOBARBITAL: |
| 8.21 | 4.5276 | | 1.212 | 0.407 | 18.9735 | DILANTIN: |
| 10.77 | 3.5623 | | 1.590 | 0.668 | 3.8676 | TEGRETOL: |

FIG. 18. Separation of anticonvulsants [P. Kabra, University of California, San Francisco, *Altex Chromatogram* **2** (3) (May, 1979)].

## VII. The Liquid Chromatograph as a System

Modern liquid chromatographs are systems built with the different components described above. As with stereo components, there are two main approaches to combining LC components into a system—modular and integrated. Modular LC systems consist of easily recognized components that have individual controls for stand-alone operation. The primary advantages of modularity include easy service, flexibility (components can be shared by different systems), and easy upgrade when new components are available. Many modular systems are designed for optional control and automation by a central microprocessor module. Integrated LC systems employ the same components housed in a space-saving single cabinet. Primary advantages include low bench space requirement, optimum electronic and flow interfacing, and attractive appearance. Modular designs are often preferred during the development of a technique when components from various manufacturers can be combined for optimum results, and by experienced chromatographers who require versatility in method development and research applications. Intetgrated systems are usually more popular for dedicated application and with chromatographers who are less oriented toward hardware. Many LC systems attempt to combine the best features of modular and integrated design.

Several companies such as Bioanalytical Systems, Dionex, and Technicon offer LC systems that are optimized for a single analysis. This trend should continue as more reliable LC methods are developed and the technique finds its way into the hands of less experienced users.

## References


1. Bakalyar, S. R., Bradley, M. P. T., and Honganen, R., *J. Chrom.* **158,** 277 (1978).
2. Savage, M., *Amer. Lab.,* May 1979.
3. Bakalyar, S. R., Rheodyne Technical Note 1, Injection Valves, Sept. 1979.
4. Hewett, G., and Shackelford, C., *Altex Chromatogram* **2** (2), 6 (1979).
5. Kissinger, P. T., et al., *Clin. Chem.* **23,** 8 1449 (1977).
6. Ettre, L. S., *J. Chrom. Sci.* **16,** 396 (1978).
7. Vickrey, T. M., Buren, M. S., and Howell, H. E., *Anal. Lett.* **A11,** 12 1075 (1978).
8. Yost, R., Stoveken, J., and MacLean, W., *J. Chrom.* **134,** 73 (1977).
9. Saitoh, K., and Suzuki, H., *Anal. Chem.* **51,** 11 1683 (1979).
10. Bakalyar, S. R., and Henry, R. A., *J. Chrom.* **126,** 327 (1976).

## Suggested Additional Literature

J. Giddings, *Dynamics of Chromatography,* Dekker, New York, 1965.

L. Snyder, *Principles of Adsorption Chromatography,* Dekker, New York, 1968.

E. Clarke, *Isolation and Identification of Drugs,* Pharmacology Press, London, 1969.

J. Huber, in *Advances in Chromatography,* A. Zlatkis, ed., Preston, Evanston, Ill., 1969.

J. Kirkland, ed., *Modern Practice of Liquid Chromatography,* Wiley, New York, 1971.

P. Brown, *High-Pressure Liquid Chromatography: Biochemical and Biomedical Applications,* Academic Press, New York, 1973.

D. Davis, and B. Prichard, *Biological Effects of Drugs in Relation to Their Plasma Concentration,* University Park Press, Baltimore, 1973.

J. Dove, J. Knox, and J. Loheac, *Applications of High Speed Liquid Chromatography,* Wiley, New York, 1974.

R. Majors, *Bonded Stationary Phases in Chromatography,* Ann Arbor Science Press, 1974.

C. R. Jones, "Assay of drugs and other trace compounds in biological fluids," in *Methodological Development in Biochemistry,* E. Reid, ed., Longman, London, 1975.

"HPLC Packing and Prepacked Columns," Machery-Nagel & Co., 1975.

P. Dixon et al., eds., *High Pressure Liquid Chromatography in Clinical Chemistry,* Academic Press, London, 1976.

R. Frei, and J. Lawrence, *Chemical Derivatization in Liquid Chromatography,* Elsevier, New York, 1976.

"A Users Guide to Chromatography," Regis, 1976.

B. Karger, ed., *Modern Liquid Chromatography in Clinical Chemistry,* ACS Symp. Series, No. 36, 1976.

N. A. Parris, *Instrumental Liquid Chromatography,* Elsevier, Amsterdam, 1976.

L. R. Snyder, and J. J. Kirkland, *Introduction to Liquid Chromatography,* Wiley, New York, 1974.

C. F. Simpson, *Practical High Performance Liquid Chromatography,* London, 1976.

C. Gardner-Thorpe, et al., eds., *Antiepileptic Drug Concentrations in Children on Multiple Therapy,* Pitman Medical, Kent, England, 1977.

R. Hamilton, and P. Sewell, *Introduction to High Performance Liquid Chromatography,* Halsted Press, New York, 1977.

J. A. Nelson, "Some clinical and pharmacological applications of high speed liquid chromatography," in *Advances in Chromatography,* J. C. Gidding, ed., Dekker, New York, 1977.

E. Johnson, and R. Stevenson, *Basic Liquid Chromatography,* Varian, 1978.

L. Snyder, B. Karger, and R. Giese, "Clinical liquid chromatography" in *Contemporary Topics, in Analytical and Clinical Chemistry,* Vol. 2, D. M. Hercules et al., eds., Plenum, New York, 1978.

K. Tsuji and W. Morowich, eds., *GLC and HPLC Determination of Therapeutic Agents, Part I,* Chrom. Sci. Series, Vol. 9, Dekker, New York, 1978.

C. Pippenger, J. Penry, and H. Kutt, *Antiepileptic Drugs: Quantitation and Interpretation,* Raven, New York, 1978.

# Chapter 3

## Liquid Chromatography Column Technology

Ronald E. Majors
*Varian Instrument Group*
*Walnut Creek, California*

### I. Introduction

The last decade has seen tremendous advances made in LC columns and column technology. From the development of the pellicular packings in the late 1960s to the 5- to 10-$\mu$m microparticles of the 1970s, the column has not only increased the speed, resolution, and sensitivity of the technique, but has also influenced the design of instrumentation. The development of LC columns has reached a state where we are beginning at last to understand their advantages and disadvantages, their properties, their limitations, their optimum use, as well as their occasional misuse. The purpose of this chapter is to review the development of HPLC column technology, to summarize the current state of affairs, and to briefly extrapolate into the future.

There are hundreds of HPLC packings and packed columns now available on the market. This chapter will make no attempt to categorize or tabulate the large number of commercially available products. For those interested in details of commercial packings, reference *1* gives more than adequate coverage. Here we will cover the generic names with only occasional reference to specific products.

## II. Types and Differences in Packings

Basically, as depicted in Fig. 1, there have been three types of column packings used in HPLC: large porous particles (1a), pellicular particles (1b), and microparticles (1c). The larger porous particles ($d_p > 40\ \mu m$) were used in the early days of HPLC for analytical columns, but are now mainly used as an inexpensive packing for preparative columns or as precolumns for mobile phase presaturation or cleanup. The pellicular packings with average $d_p$ in the 40 $\mu$m range consist of a solid glass bead with a thin porous outer shell that may be silica, alumina, or ion exchange resin, or a silica layer to which a "liquid" phase has been chemically bonded. In the late 1960s and early 1970s, the pellicular were the standard packings. However, with the production of commercial quantities of the microparticles and the development of techniques to pack them, the microparticulate packings have now displaced the pelliculars in popularity. Compared to pelliculars, microparticles of 5 and 10 $\mu$m sizes offer the advantages of at least an order of magnitude in column efficiency, sample capacity,

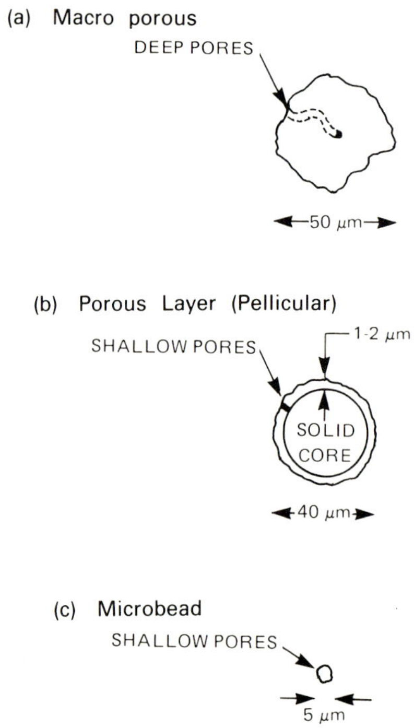


FIG. 1. Types of packing particles used in HPLC (reprinted by permission of Varian Associates).

Table 1
Typical Properties of HPLC Column Packings

| Property | Pellicular | Microparticles |
|---|---|---|
| Average particle size, μm | 30–40 | 5–10 |
| Best HETP[a] values, mm | 0.2–0.4 | 0.01–0.03 |
| Typical column lengths, cm | 50–100 | 10–30 |
| Typical column diameters, mm | 2 | 3–5 |
| Pressure drop, psi/cm[b] | 0.5 | 5 |
| Sample capacity, mg/g | 0.05–0.1 | 1–5 |
| Surface area (LSC), $m^2$/g | 10–15 | 400–600 |
| Bonded-phase coverage, total wt. | 0.05–1.5 | 5–20 |
| Ion exchange capacity, μeq/g | 10–40 | 2000–5000 |
| Ease of packing | Easy, dry pack | Difficult, slurry pack |
| Best use | Guard columns | Analytical and semipreparative columns |
| Cost | | |
| Bulk packing | \$4–5/g (LSC) | \$3–5/g (LSC) |
| | \$7–9/g (BPC) | \$10–16/g (BPC) |
| Prepacked columns | \$120–140 (LSC) | \$225–250 (LSC) |
| | \$170–190 (BPC) | \$250–300 (BPC) |

[a]HETP = height equivalent to a theoretical plate.
[b]Columns of equal dimensions (4.0 mm id) operated at flow of 1 mL/min and mobile phase viscosity of 0.3 cP.

and speed of analysis. However, they generally require more sophisticated column-packing techniques. Table 1 summarizes comparative data for both types of packings.

Owing to the greater efficiency of the microparticles, shorter column lengths will achieve the same separation as on a pellicular column or, conversely, a more difficult separation can be carried out on a microparticulate column of equivalent dimensions and at a lower pressure. It is no small wonder that the use of pellicular packings is rapidly declining relative to the microparticles. They still find use as a packing material for guard columns, since they can be conveniently dry packed.

Both irregularly shaped and spherical microparticulates are available. All ion exchange resins and most exclusion chromatographic packings are spherical in shape. Many of the irregularly shaped silicas are an offshoot from the manufacture of thin layer silicas, but are of smaller particle size and of narrower distribution. The spherical silicas are specially synthesized and carefully sized for

HPLC use. Column efficiencies of spherical and irregular particles of similar diameter (the diameter of irregularly shaped silicas are harder to define) are roughly equivalent, but column pressure drops per unit length for spherical packings are 15–20% lower.

## III. Techniques for Packing LC Columns

In order to achieve the high performance separations of modern HPLC, it is imperative that the column be packed in the proper manner. Porous particles over 30 $\mu$m in diameter and pellicular beads can be packed by dry packing techniques (*2, 3*) similar to those employed in gas chromatography. The microparticles require slurry techniques (*4*). For ion exchange resins of the polystyrene–divinylbenzene type, an aqueous buffer slurry is used and the column is packed by pumping the slurry from a large internal diameter column or reservoir placed ahead of the analytical column into the empty analytical column. The spherical beads usually pack well, provided the packing pressure is not so great as to deform the semirigid particles. Packing pressure should usually be kept under 4000 psi for 4–8% crosslinked resins.

The microparticulate silica adsorbents and chemically bonded silicas require high-pressure slurry techniques. The slurry is usually made up at the 5–15% wt/vol concentration. Many slurry solvents have been reported in the literature, as can be seen in Table 2. With the improvements over the years in particle sizing techniques, with resultant narrower particle size distributions, the matching of solvent density and packing particle density (to prevent size segregation) is not so critical. However, for best column performance and stability, it is important to optimize the type of slurry solvent to the specific type of packing. For example, a slurry solvent that works well for the polar adsorbent silica is not necessarily the best solvent to use for the hydrophobic octadecylsilane packings. Although manufacturers of columns do not publish their proprietary packing procedures, it is generally recognized that different packings require different slurry solvents. Packing pressures vary from 8000 to 12,000 psi, depending on slurry solvent, particle size, and column length and internal diameter.

Should the chromatographer wish personally to pack microparticulate columns, several companies provide bulk packings in 10 g bottles. Silica gel, bonded phases, and ion exchange packings can be obtained. High performance gels for exclusion chromatography are usually sold only in prepacked columns.

Table 2
Slurry Packing Solvents

| Type | Typical solvents | Reference |
|---|---|---|
| Balanced density | Tetrabromoethane, tetrachloroethylene, diiodomethane | *5–7* |
| Ammonia stabilized | 0.001 *M* aqueous ammonia | *8* |
| Balanced viscosity | Cyclohexanol, polyethylene glycol 200 | *9* |
| Other | Carbon tetrachloride | *10–11* |
| | Methanol | *11–12* |
| | Methanol + acetate salt | *13* |
| | Acetone | *14* |
| | Dioxane–methanol | *15* |
| | Tetrahydrofuran–water | *16* |
| | Isopropyl alcohol | *17* |
| | Chloroform–methanol | *18* |

## IV. Prepacked Columns for HPLC

The consistently successful packing of microparticulate columns is still considered somewhat of an "art." Also, considering the expense of the high pressure packing apparatus recommended for producing optimum columns (most liquid chromatographs can not meet the flow rate requirements), for the laboratory that only uses a few columns a year it is advisable to consider the purchase of prepacked, pretested columns. Manufacturers have improved in production technique, and columns are more reproducible than several years ago. Most of the commercial columns are of guaranteed performance and will arrive with a chromatogram actually run on the purchased column with standard test components. The column can be tested upon receipt and from time to time to monitor the state of the column. Most microparticulate columns, if properly used, will last several months without deterioration in performance, and much longer if a guard column is used and periodically replaced.

The most common size of prepacked analytical microparticulate columns is 4–4.6 mm in internal diameter and 25 or 30 cm in length. Recently, there has been a tendency toward shorter columns (10–15 cm) packed with 5 $\mu$m particles. If only several thousand plates are required, such columns provide faster analyses with less solvent usage than conventional columns.

Recently, short (10 cm) polyethylene analytical columns of large internal diameter (8 mm) containing silica or $C_{18}$ reverse phase have become available. These cartridges, called "radially compressed columns" (*19*), have no end fittings and utilize a special holder. The holder is filled with a hydraulic fluid that, when compressed around the outside of the column, exerts a pressure on the polyethylene walls, presumably compressing the plastic material into the packing, thereby eliminating "wall effects" (*20*). The columns are said to be more reproducible than conventional stainless-steel columns.

## V. Preparative Columns

Conventional analytical columns are somewhat limited in their sample capacity. For example, for a moderately difficult separation ($R_s \approx 1.2$), a 30 cm × 4 mm column packed with silica gel could handle 2–5 mg of sample under isocratic conditions before overloading occurs. For an easy separation ($R_s \sim 5$), several tens of milligrams could be injected. Overloading results in skewed peaks and loss of resolution, hence sample purity, a prime concern in preparative separations. Sample capacity increases with the square of the column internal diameter owing to the increase in volume of stationary phase. Likewise, sample capacity increases with the surface area of adsorbent (liquid–solid chromatography), bonded phase coverage (bonded-phase chromatography), or exchange capacity (ion exchange chromatography).

An important consideration in choosing a column for use in preparative separations is the amount of sample required for further use. The sample requirements dictate the internal diameter and length of column needed. In turn, the column dimensions dictate the flow rate needed to perform the separation in a reasonable time. In preparative LC, to keep separation times equivalent to the separation performed on an analytical column, the linear velocity must remain constant. Constant linear velocity means that the flow rate must increase with the square of the column radius ratio. Thus, if the separation was performed on an analytical column (4 mm id) at the normal flow rate of 2 mL/min, then when the column internal diameter is increased to 10 mm, the flow rate should be increased to 12.5 mL/min. Going one step further, if the column diameter is increased to 25.4 mm (1″ id), the flow rate needed would be 81 mL/min. Such a flow rate is well beyond the capability of most analytical chromatographs. Obviously a compromise must be made for those who are required to carry out both analytical *and* preparative work. For very large id columns (1–2 in.) and high sample requirements (10+ g), dedicated preparative chromatographs are available.

Semipreparative microparticulate (10 $\mu$m) columns are available in 8–9 mm internal diameters and up to 50 cm lengths. Their dimensions appear to offer the best compromise between the flow rate capability of analytical chromatographs and the sample capacity of the column. For easy separations, up to 500 mg have been injected on such columns while maintaining adequate sample purity. Since, in preparative chromatography, resolution and resultant sample purity is of prime concern, speed is less critical and linear velocities (and therefore flow rates) are slightly lower for these semipreparative columns. Usually, flow rates in the 4–8 mL/min range are used for 8 mm id columns.

If sample quantities in the range of grams are required, modern liquid chromatographs are capable of controlling autoinjectors and fraction collectors. Thus, by running repetitive injections on semipreparative columns, an analytical chromatogram can give larger quantities of pure sample without the need to purchase a second preparative chromatograph.

## VI. Protecting Microparticulate Columns

Microparticulate columns packed with 5 and 10 $\mu$m particles act as superb filters for impurities and particulate matter introduced from samples, mobile phases, wear particles from pump seals and injector valve cores, as well as other moving parts of the chromatograph. Since columns are usually expensive, some care must be exercised in their use. Mobile phase filtration through a membrane (not paper) filter is often recommended. In-line filters (0.5–2 $\mu$m) in the solvent reservoir or inlet line of the pump can cut down on particulates fouling up pump seals or passing through the pump and lodging elsewhere in the hydraulic system and/or column.

Precolumns are devices installed prior to the injector and after the pump. They can serve to reduce mobile phase particulates or pump seal fragments from getting into the injector or analytical column. Precolumns are usually of 5–10 cm in length and 4–8 mm id, and are dry packed with inexpensive large particle packing (30–70 $\mu$m), usually silica gel (for normal phase work), reverse phase packing, or ion exchange resin. Here the large particle size is not harmful since sample does not pass through the column. For isocratic work, the additional volume added to the hydraulic system when a precolumn is used is unimportant, but in a gradient system it can increase the time for the new solvent composition to reach the head of the LC column. The precolumn can also serve to saturate aqueous mobile phases with dissolved silica and thereby increase the lifetime of the analytical column materials (*21*), most of which are on a silica matrix.

Guard columns are protection devices placed between the injector and the analytical column. Their main job is to protect the analytical column from sample impurities, such as irreversibly retained compounds or particulate matter. It is a good practice to filter samples through a membrane filter to prevent particulates from getting into the column or column terminator frit. Particulates may lodge at the column head or on the frit, causing high backpressures. Since the sample passes through the guard column, it is imperative that all fittings and connections be of very low dead volume. Otherwise the band broadening of sample peak, which may occur from these extra column effects, cannot be recovered as it passes through the analytical column.

Guard columns are of 3–5 cm in length with the same internal diameter as the analytical column. Because the packing in the guard column becomes contaminated, it must be replaced. The frequency of replacement depends on a number of factors such as sample cleanliness, number and type of retained components and capacity of the packing, and is a trial-and-error procedure. Both pellicular and microparticulate packings are used in guard columns. Compared to microparticulates, the advantage of pellicular packings for guard columns is that they can be easily repacked by dry packing techniques and they cost less. Their disadvantages are that they can contribute to band broadening, especially for the higher efficiency 5 $\mu$m analytical columns, and that because of a low volume of stationary phase, they have a low capacity and must be replaced more often. Microparticulate guard-column packings have the advantage of high capacity and cause little band broadening since their particle size (and hence efficiency) is the same as the analytical column. They have the disadvantage that they must be packed by high-pressure slurry techniques for which most laboratories are not equipped. Recently, prepacked microparticulate guard columns have become available from several suppliers. Although they cost a bit more, their convenience, especially the replaceable ones that fit into finger-tight holders, are worth the investment.

## VII. Modes of Liquid Chromatography

The power in HPLC is the wide variety of modes available to the chromatographers. In gas chromatography only gas–liquid (GLC) or gas–solid (GSC) chromatography are available. The mobile phase in GC—a gas—has almost no interaction with the solute or stationary phase. The gas merely serves as a carrier to assist in this transport of the solute down the column. In LC, threefold interactions occur, as depicted in Fig. 2. The mobile phase not only moves the solute down the column but also interacts with both the stationary phase and the solute.

The LC mode is designated by the nature of the predominant interaction that occurs between the sample solute and the stationary phase. There are four modes in HPLC:

### A. Liquid–Solid (Adsorption) Chromatography (LSC)

LSC uses an adsorbent, usually silica gel or alumina as the stationary phase. The adsorbent contains active sites, usually —OH groups, which interact with the polar portions of the molecules. It is the oldest technique, having been first practiced in 1906 by the founder of chromatography, Prof. M. Tswett, a Russian botanist.

### B. Bonded-Phase Chromatography (BPC)

BPC utilizes various phases chemically bonded to a silica gel base. The mechanism for solute interaction can be partition, where molecules actually penetrate the bulk of a thick bonded phase, or adsorption, when the polar or nonpolar molecules are attracted to the polar or

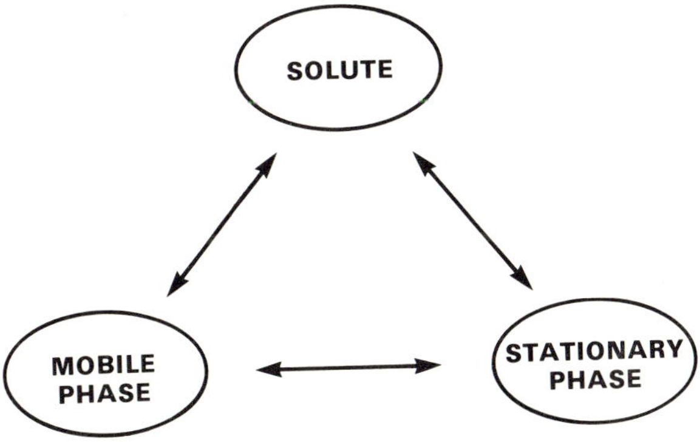


FIG. 2. Interactions in liquid chromatography.

nonpolar bonded-phase functional groups, respectively. It is the most popular form of HPLC today.

### C. Ion Exchange Chromatography (IEC)

IEC utilizes either resins or bonded silicas having ionic groups on their surfaces. The ionic portions of the solute molecule are attracted to the stationary phase ionic group of opposite charge. The technique has been used for many years for the separation of water-soluble biological substances.

### D. Exclusion Chromatography (EC)

EC separates on the basis of molecular size. The mechanism involves the selection diffusion of solute molecules into and out of mobile-phase-filled pores of a porous, three-dimensional matrix. Retention depends on the size of the solute relative to the size of the pore. The larger molecules that are excluded will elute first, while the small molecules that can diffuse into all pores will elute last. Molecules with sizes between these two extremes will permeate part of the pores and will elute in decreasing molecular size. Exclusion chromatography is frequently used to characterize industrial organic polymers, and biopolymers such as proteins or nucleic acids.

## VIII. Selection of the LC Mode

It has often been stated that "the column is the heart of the chromatograph." The selection of the LC mode along with the correct column packing is the most vital step in the development of an analytical method. No functional theory exists for correctly predicting the proper mode. The process of mode and column selection is largely empirical. Some useful guidelines in mode choice will be presented in this section. Once the mode and stationary phase are selected, the proper column and column packing materials are chosen. The column is packed by the most efficient technique. Initial operating parameters—mobile phase, mobile phase composition, flow rate, sample concentration and volume, and so forth—are then selected. Based on the outcome of the initial chromatograms, the separation is then optimized by modification of the chromatographic conditions.

First, one must select the LC mode that offers one the best chance to separate the compound, or compounds, of interest. Figure 3 presents a very general schematic for mode classification based on molecular weight, solubility, and ionic character. The choice of a

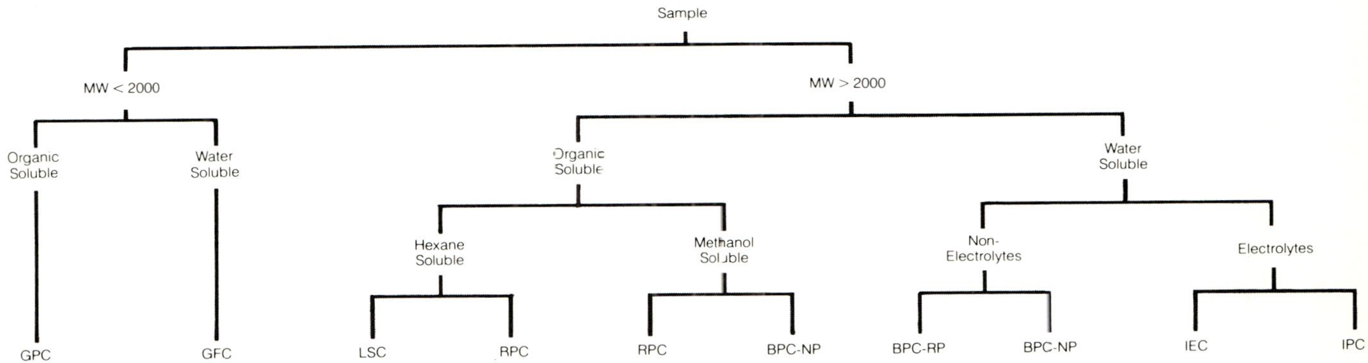


FIG. 3. Guide to LC mode selection (reprinted by permission of Varian Associates).

molecular weight of 2000 is arbitrary, since some newer exclusion packings can be used for small molecules while some large molecules (e.g., polypeptides) can be handled by reverse phase columns. For high molecular weight (mw) molecules soluble in an organic solvent, exclusion chromatography (also knows as gel permeation chromatography or GPC) is the method of choice.

If the molecule has a high mw, but is water soluble, the technique of gel filtration chromatography, GFC, is used. If the molecule is a biopolymer (e.g., proteins, nucleic acids, polysaccharides), or water soluble organic polymer (e.g., polyethylene glycol, polyvinyl alcohols, polyvinylpyrollidone), this is the method of choice. Occasionally, small molecules that are water soluble can be separated by virtue of their molecular size in solution. In fact, all exclusion chromatographic techniques separate on the basis of molecular size and thus are useful for characterization of molecular weight distribution for the separation of polymers and their lower molecular weight additives, and for separations of oligomers.

Samples with mw less than 2000 that are organic and soluble in hexane or methanol, but water insoluble, can usually be handled by reverse bonded-phase chromatography (RPC) as noted in Fig. 3.

If hexane soluble, and the samples are geometric or aromatic isomers or are water-sensitive, then adsorption chromatography on silica should be tried.

If the samples are part of a homologous series, then RPC is the first choice since in this mode samples are separated by virtue of the hydrophobicity.

If the sample is insoluble in both hexane and water, but soluble in methanol, it is moderately polar, and again, the first choice would be RPC. A second choice would be a normal phase (NP) separation using a —CM or—$NH_2$ column with an intermediate polarity solvent.

Note that normal phase chromatography refers to LC practiced with polar packings such as silica gel and nonpolar solvents such as hexane. In contrast, the technique of reverse phase chromatography uses nonpolar column packings and polar solvents such as methanol and water.

Samples with molecular weights lower than 2000 that are water soluble can be further classified as electrolytes (conductors of electric current) or non-electrolytes (nonconductors of electric current). If the sample is a nonelectrolyte, e.g., fatty acids, sugars, or polyols, a reverse phase column would be a first choice, or possibly a specialty column (to be discussed later) developed for sugars.

If the sample is an electrolyte, we must choose between ion exchange (IEC) and ion pair chromatography (IPC). Compounds that are anionic in their ionized state are generally separated on an anion exchange column of the resin or bonded silica type. Compounds that are cationic are separated on a cation exchange column. Ion pair chromatography is a technique in which ionized samples form a neutral ion pair by addition of a counterion, of opposite charge to that of the sample, to the aqueous mobile phase. This ion pair complex then is chromatographed on a monolayer reverse-phase column. A related technique is ion suppression chromatography, in which a weak acid or base is entirely converted to its unioinized form during the chromatography. In fact, ionization control is a useful technique in both IEC and IPC.

## IX. Selection of Type of Column Packing

Once the LC mode is selected, next comes the selection of the column packing or the packed column. Microparticles of 5 and 10 μm average particle diameter are the preferred packing for most high-performance analytical separations. The 10 μm particles provide adequate resolution for most analytical separations, while 5 μm particles are recommended for the most demanding separations. In general, the 5 μm particle size gives 2–3 times more theoretical plates than a 10 μm size for a given column length. However, for a given set of LC conditions, column backpressure rises a factor of 3–4 for 5 compared to 10 μm particles. Therefore, lower flow rates are recommended for the smaller particle sizes. For a 4 mm id column packed with 10 μm particle, 2 mL/min is a typical flow rate, while for the same column packed with a 5 μm particle, the flow rate would be 0.5–1 mL/min. Since microparticulate columns require special, high-pressure slurry-packing techniques, as mentioned earlier, the purchase of prepacked, pretested columns is recommended. However, should the chromatographer wish personally to fill these microparticulate columns, bulk packings for most of the LC modes are available.

## X. Columns for Bonded-Phase Chromatography

The development of stable, reproducible chemically bonded phases in the early 1970s has made this type of column the most widely used. For reverse phase chromatography alone, recent estimates ranging from 60 to 70% of HPLC applications used either $C_{18}$ (octadecyl) or $C_8$ (octyl) hydrocarbon phases bonded to silica microparticles. The earlier technique of liquid–liquid (partition) chromatography (LLC) is virtually extinct.

## A. Preparation of Bonded Phases

Understanding the methods for synthesis of chemically bonded phases will help to later explain some of their important surface characteristics. Most commercially available bonded phases are of the siloxane type having the Si—O—Si—C bond. They are prepared by the reaction of microparticulate silica gel, which possesses reactive silanol groups, with organochloro- or organoalkoxy-silanes, as depicted in Fig. 4. The reaction may be carried out with mono-, di-, or trichloroorganosilanes (or the corresponding alkoxy compounds). The nature of the R— group (which may be alkyl or aryl, with and without other functional groups such as —CN) will dictate the final property of the bonded phase. If the R— group is hydrocarbonaceous, such as an octadecylsilane moiety, the packing will be hydrophobic and will be used for reverse phase chromatography. If the R— groups contain polar functionality, such as —CN or —$NH_2$, then its main use will be in normal phase chromatography.

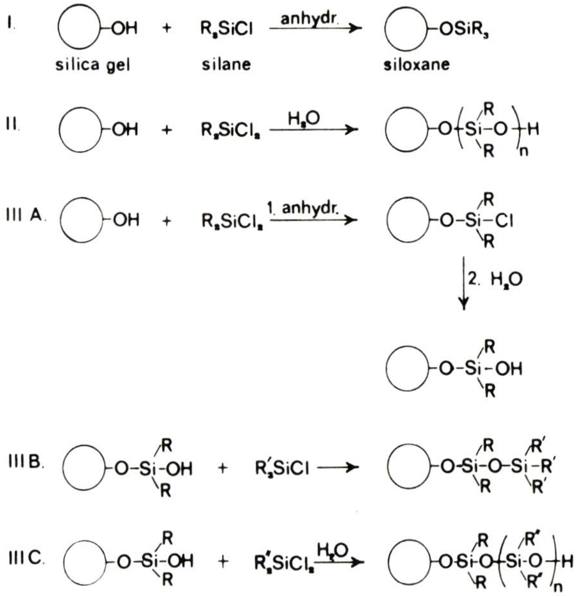


FIG. 4. Reaction for bonding organosilanes to silica gel [reprinted by permission of Preston Publications; From R. E. Majors and M. J. Hopper, *J. Chromatogr. Sci.* **12**, 768 (1974)]

Monolayers of bonded phase will be formed if the reaction is carried out with monochlorosilanes (also alkoxysilanes) or di- and trichlorosilanes under anhydrous conditions (Fig. 4, Reaction I). If controlled amounts of water are introduced, either on purpose or accidentally during the reaction of di- or trisilanes, then polymerization or crosslinking reactions may occur and the bonded-phase coverage will be increased (Fig. 4, Reaction II). Di- and trisilanes are more reactive than monosilanes. Consequently they may react with more silanols than do the monosilanes. It is doubtful whether all Si—Cl (or Si—OR) bonds of a di- or trisilane will react with the silica surface owing to steric limitations. These unreacted bonds will be converted to Si—OH when the packing is exposed to the atmosphere or aqueous solutions. Furthermore, for steric reasons it is doubtful whether all silica gel silanols will react with the organosilane reactant. Some silanols may remain unreacted on the silica backbone. Also, additional silanols may be created on the bonded phase. Both of these types of silanols may give rise to mixed mechanisms, undesired adsorption, tailing with polar solutes, especially basic compounds, or chemical reaction with some sample or mobile phase components. Some of these exposed silanols may be removed by treatment of the bonded phase packing with a small silylating reactant such as trimethylchlorosilane (TMCS), as depicted in Fig. 4, Reaction IIIB. This procedure is sometimes called "end capping."

## B. Bonded-Phase Coverage and Stability

Monolayer coverage of chemically bonded phases give rise to more rapid solute mass transfer, thereby giving better chromatographic efficiency. Polymeric or crosslinked phases show a higher bonded-phase coverage and hence more stability. Their thicker films do result in lower efficiency owing to slow solute mass transfer. Polymeric layers show greater selectivity for very nonpolar solute molecules, such as polynuclear aromatic hydrocarbons. Bonded phases are attacked by the action of hydroxide ion or water on the silica base material or the siloxane bond. Therefore, siloxane bonded phases or silica gels should not be used at high pHs (pH $>$ 8) and, since the Si—C bond can be hydrolyzed at pH values of around 1, very acidic mobile phases should be avoided. The usual recommended pH range for chemically bonded phases is 2–8.

In general, chromatographic retention increases with the degree of phase coverage (*23*) and chain length of the R— group, especially in reverse phase chromatography (*24*). There is some experimental evidence (*25*) to suggest that very short chain phases, such as —$C_2$, show less stability in aqueous solution than the long chain phases such as $—C_{18}$.

### C. Columns for Reverse-Phase Chromatograpy

In RPC, the stationary phase is usually a hydrophobic bonded phase, such as octadecylsilane or octylsilane, and the mobile phases are usually polar solvents, such as water, or mixtures of water and water-miscible organic solvents, such as methanol, acetonitrile, or isopropanol. Water is considered to be the weak solvent (solvent A), and the organic modifier the strong solvent (solvent B).

Presently, reverse-phase chromatography as practiced in its various forms is dominating the application of HPLC since:

1. Often, nonionic, ionic, and ionizable compounds can be separated using a single column and mobile phase with various additives, such as buffer salts and ion pair reagents.
2. The bonded-phase columns are fairly reproducible and relatively stable provided certain precautions are taken.
3. The predominant mobile phase, water, is inexpensive and plentiful; aqueous samples can be directly injected into the aqueous mobile phases.
4. The most frequently used modifier, methyl alcohol, can be obtained at a reasonable price and suitably pure in most places in the world. Acetonitrile is more expensive and harder to get in suitable purity (especially for sub-200nm UV detection) in some parts of the world.
5. The elution order is often predictable based on the degree of hydrophobicity of the solute molecule. The more hydrophobic parts of a molecule or the less polar the molecule can be made, the greater its chromatographic retention.

The exact mechanism of reverse-phase separation is still a matter of debate. Polar solutes tend to prefer the polar mobile phase and elute before nonpolar components, which, in a polar medium of very high cohesive energy density (arising from a three-dimensional hydrogen bonding network), are forced into the hydrocarbon stationary phase (*26–28*). Any polar functionality that may be present on a solute opposes the repulsion from the polar mobile phase. The degree of retention is based primarily on the hydrocarbon moieties of a solute. Likewise, the more hydrophobic the bonded phase, the greater the attraction of the nonpolar molecule. Thus, in general, $C_{18} > C_8 >$ phenyl $> C_2$ in terms of the degree of retention of a given solute. Table 3 lists the various bonded phases that have found use in RPC. No

Table 3
Typical Bonded Phases for RPC

| Functionality | Estimate of usage | Type | General use |
|---|---|---|---|
| $—Si—_nC_{18}H_{37}$ | 85% | Octadecylsilane (ODS) | Best general reverse phase, polymeric separations of very nonpolar molecules; monomeric; best for polar and semipolar molecules, ion pair, ion suppression. |
| $—Si—_nC_8H_{17}$ | 8% | Octylsilane | Lower retention than $C_{18+}$ good for moderately polar compounds, ion pair |
| $—Si—(CN_3)_2$ | 4% | Dimethylsilane | Least retentive; very polar compounds, beware of bonded phase stability. |
| —Si— | 2% | Phenyl | Selectivity for aromatic compounds; useful for peptides. |
| $—Si—C_3H_7CN$ | 1% | Cyano | Used mostly in normal phase work. Has found some use in analysis of tricyclic antidepressants. |

doubt additional specialty phases will be developed, but those listed in Table 3 will cover most application problems likely to be encountered.

If ionic (e.g., $—SO_3$, $—N\overset{+}{R_3}$) or ionizable (e.g., $—NH_2$, —COOH) compounds can be rendered less ionic by suitable mobile phase additives such as counterions, buffers, chelates, or certain organic solvents, they, too, can be forced to be retained on a reverse-phase packing. The use of such mobile phase additives has opened up a number of subcategories of RPC: regular, ion suppression, ionization control, ion pairing, and metal chelation.

Regular RPC is the technique referred to when simple mixtures of water and a water-miscible organic solvent are used as the mobile phase. At present, there is little quantitative data published on the relative strengths of a variety of organic modifiers when used in mixtures with water (as the Eluotropic Series in LSC), but Table 4 lists several solvents of increasing strength gathered from references *29* and *30*. In addition to the primary effect of fixing the eluent power of an

Table 4
Organic Modifiers Used in RPC[a]

| Solvent | |
|---|---|
| Ethylene glycol | |
| Methanol | ↓ |
| DMSO | Strength increases |
| Ethanol | |
| Acetonitrile | ↓ |
| DMF | |
| Dioxane | |
| Isopropanol | |
| Tetrahydrofuran | |

[a]Listed in order of increasing strength. Thus, at a fixed composition, say 50% organic modifier in water, a solute would elute earlier as the organic solvent was replaced by a solvent below it in the table.

aqueous mobile phase system, solvent selectivity effects (*26, 30, 31*) can occur. Thus, one could choose several eluents having a fixed solvent strength in a water mixture, but owing to other solvent properties solutes may show differences in retention. Such solvent selectivity effects can be exploited in reverse phase chromatography by the use of ternary solvent systems (*30, 31*).

An important example of regular RPC depicted in Fig. 5 is the separation of polynuclear aromatic hydrocarbons (PNA), many of which are highly carcinogenic. Those shown are from the Environmental Protection Agency's list of Priority Pollutants and have been found to be present in atmospheric air and water. Note that the greater the number of fused rings, the greater the chromatographic retention. Because of the great range in degree of retention, gradient elution was used to achieve the separation. In RPC, a solvent gradient consists of increasing the amount of organic modifier added to the weaker solvent water as a function of time. In Fig. 5 the initial composition was zero percent since the column was being used to concentrate traces of PNA from water. The strength was quickly increased to 40% methanol, followed by a gradual increase to 100% methanol over the course of 30 min. The column used was a MicroPak®-CH, a polymeric octadecylsilane phase bonded to 10-$\mu$m silica gel with a relatively high loading of 22% carbon.

A more complex example is illustrated in Fig. 6 which is a chromatogram of several synthetic estrogen steroids used as oral

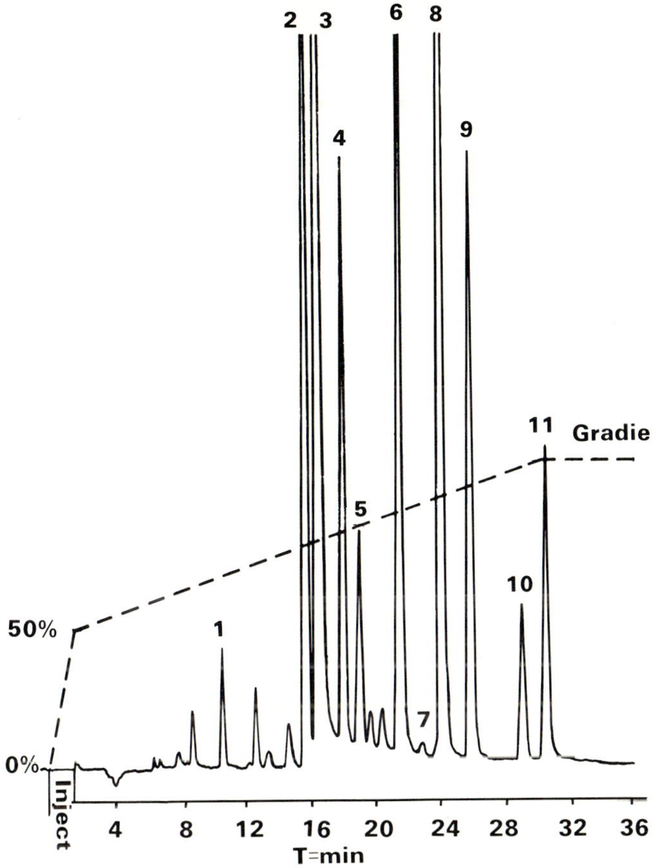


FIG. 5. Separation of polynuclear aromatic hydrocarbons (*32*). Column: MicroPak CH-10 (Varian); mobile phase: water–acetonitrile gradient (see chromatogram); flow rate: 2 mL/min; detector: 254-nm, 0.65 Aufs; sample size: 1 $\mu$g each.

contraceptives. Here the basic five-ring steroid structure is the same (biphenyl was used as an internal standard), but substituent groups are different. The basic steroid structure being a hydrocarbon skeleton would be expected to control retention of all six estrogens, but because of the unique ability of RPC to differentiate on the basis of only slight differences in organic character, there is an excellent separation. As might be expected, ethynylestradiol containing the two —OH moieties eluted first, and ethynodiol diacetate with its diacetate functionality, being the least polar of all steroids, eluted last. Reverse phase

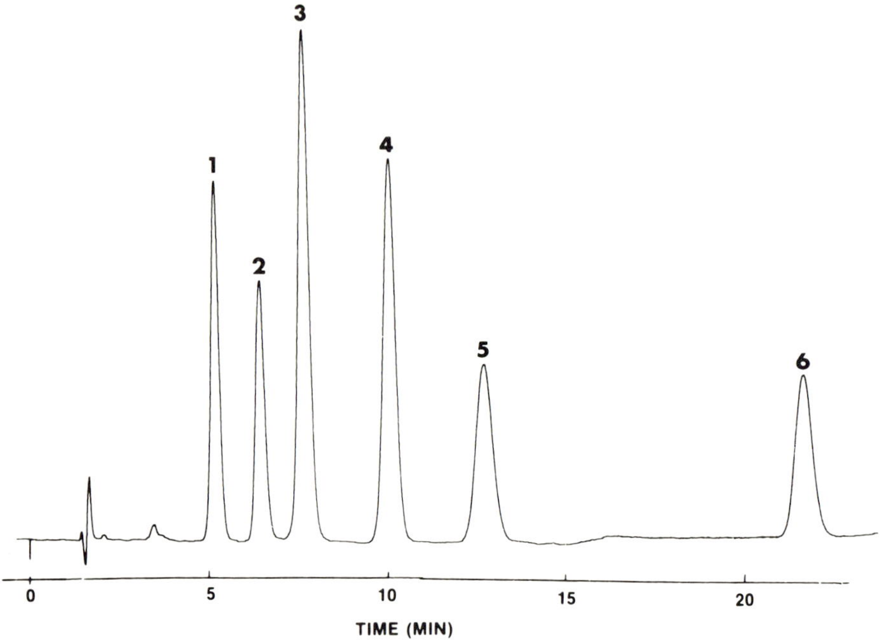


FIG. 6. Separation of synthetic estrogen standards (*57*). Column: MicroPak MCH (Varian); mobile phase: 60% acetonitrile in water; flow program: 0–15 min, 1 mL/min; 15–16 min, 1–2.5 mL/min, 16–25 min, 2.5 mL/min; detector: λ= 230 nm.

chromatography can also differentiate between members of a homologous series differing only in a single methylene group.

Ion suppression and other forms of *ionization control* utilize selective chemical equilibria in achieving better separation of polar compounds. In regular RPC, very polar compounds and especially ionizable compounds such as weak acids and bases will often elute very quickly or give poorly defined peaks, frequently with tailing. Retention and peak shape cannot be improved merely by changing the mobile phase composition. Consider the equilibrium that exists for a carboxylic acid in solution:

$$RCOOH \rightleftharpoons RCOO^- + H^+$$

If the pH of the solution is such that the acid is injected in a partially ionized state, the resulting peak would be poorly retained and skewed. By adjusting the pH sufficiently below the p$K_a$ of the acid, one can suppress ionization and chromatograph the carboxylic acid in its unionized state. Controlling the pH is one way of extending the scope of RPC for weak acids and bases. The technique can be carried out

FIG. 7. Chromatogram of plasma of patient on aspirin therapy (*37*). Column: $\mu$Bondapak $C_{18}$ (ODS), 30 cm × 3.9 mm), 10 $\mu$m (Waters Associates); mobile phase: water–methanol–acetic acid (19:4.4:1 v/v); flow rate: 2.6 mL/min; sample: plasma found to contain 0.96 mg/L of gentisic acid (GA), 3.49 mg/L of salicyluric acid (SUA), 20.46 mg/L (added) of internal standard, *o*-methoxybenzoic acid, and 120.61 mg/L of salicylic acid (SA). (Reprinted by permission of the American Association for Clinical Chemistry.)

within the range of siloxane-bonded phase stability, pH values of 2–8. For weak acids, a pH of 2–3 is sufficient to suppress ionization. An example of the ion suppression of weak acids is shown in Fig. 7, which is the injection of plasma from a patient on aspirin therapy (*33*). A reverse phase column was used and the mobile phase consisted of a mixture of water–methanol–acetic acid (19:4.4:1 v/v). The acidic compounds, all metabolites of acetylsalicyclic acid, were chromatographed in their unionized state.

Strong acids and bases cannot be handled by controlling the degree of ionization by pH adjustment since many are completely ionized in the pH 2–8 range. However, by forming an *ion pair* with a suitable counterion of opposite charge to the solute, strong acids (e.g., $—SO_3H$) or bases ($—NMe_3^+$) can be converted into electrically neutral compounds and thereby be retained on a reverse phase column. Reverse-phase ion pair chromatography (RP–IPC) is an attractive alternative to ion exchange chromatography in that many of the same types of samples can be handled by the two techniques. However, RP–IPC uses reverse phase columns that are readily available and more understood by liquid chromatographers, the columns are more efficient and thus can be used at ambient temperature, and columns tend to be more quickly regenerated than after a gradient.

The basic equations of RP–IPC have been covered elsewhere (*34–36*), but for a brief review let us consider the simplistic mechanisms. Mobile phases and columns are the same as other reverse phase techniques. In RP–IPC a counterion, which itself often contains an organic moiety, is added to the aqueous mobile phase. The counterion, being of opposite charge, forms a neutral ion pair with the solute of interest. This neutral ion pair then partitions (or associates) with the hydrophobic packing according to the following equilibria:

$$R^+_{aq} + X^-_{aq} \rightleftarrows (RX)_{aq} \rightleftarrows (RX)_{org}$$

where $R^+_{aq}$ and $X^-_{aq}$ are ions in solution, $(RX)_{aq}$ is the ion pair formed in aqueous solution, and $(RX)_{org}$ is the ion pair that has partitioned into the bonded phase.

A second postulated mechanism (*37*) is that the counterion hydrophobic portion partitions into the hydrophobic stationary phase and its ionic portion is oriented towards the more favorable aqueous media. Thus, an *in situ* dynamic ion exchanger is created. The charged solute is then attracted by electrostatic force to the ionic group at the packing surface as suggested in the following equation:

$$R^+_{aq} + X^-_{org} \rightleftarrows (RX)_{org}$$

where $X^-_{org}$ is the counterion associated with the hydrophobic bonded phase. In support of this mechanism, there is evidence that a high equilibrium concentration of counterion added to the mobile phase ends up on the column (*38–40*). Probably, the actual mechanism is more complex and involves both suggested mechanisms as well as other solution and stationary phase equilibria (*41*). Nevertheless, the

Table 5
Counterions Used in RP–IPC

| Solute | Typical solutes | Counterion |
|---|---|---|
| Cationic | Protonated amines, tetraalkylammonium salts, benzalkonium salts | Alkyl ($C_1$, $C_4$, $C_5$, $C_7$) or aryl (benzene, naphthalene) sulfonates, or alkyl (octyl or lauryl) sulfonates |
| Anionic | Carboxylic acid salts, sulfonic acids, sulfonic dyes | Quaternary ammonium (tetramethyl, $—NEt_4^+$, $NBu_4^+$) compounds or amines (mono-, di-, or trioctyl) at pH $< 4$ |

ion pair technique is a useful technique for handling ionic and ionizable substances by RPC.

Typical counterions used for cationic and ionic species are listed in Table 5. Many are available from chemical suppliers in the solid form. Chlorides or perchlorates are recommended owing to their low UV absorbance. The normal concentration range is 0.005–0.02 *M*. Retention is proportional to counterion concentration until micelle formation occurs or the effective concentration of counterion in the eluent is lowered (*35, 38*). Above a certain optimum concentration, retention actually decreases with further addition of counterion. As might be expected, the more hydrophobic the organic portion of the counterion, the greater the degree of retention (e.g., $CH_3SO_3^- < C_4H_9SO_3^- < C_6H_{13}SO_3^- < C_7H_{15}SO_3^- < C_{12}H_{25}SO_3$, etc.) The longer the alkyl chain, however, the more time required for the column to regenerate back to the initial conditions after running a gradient (*35*).

Many clinical samples, both therapeutic drugs as well as endogenous compounds, can be handled by the RP–IPC technique. Such is the case for species such as amines, which can be protonated in acidic solution and ion paired with a negatively charged counterion, such as a sulfonic acid salt. Figure 8 shows the ion pair separations of catecholamines and indoles, important chemical markers in the certain types of carcinomas. The separation shown was for urine from a patient with a malignant melanoma.

Many popular drugs contain functional groups amenable to ion pair complexation. The separation of disopyramide (Norpace®, a product of Searle Lab), an antiarrhythmic drug, from its internal standard, *p*-chlorodisopyramide, in a plasma extract is presented in

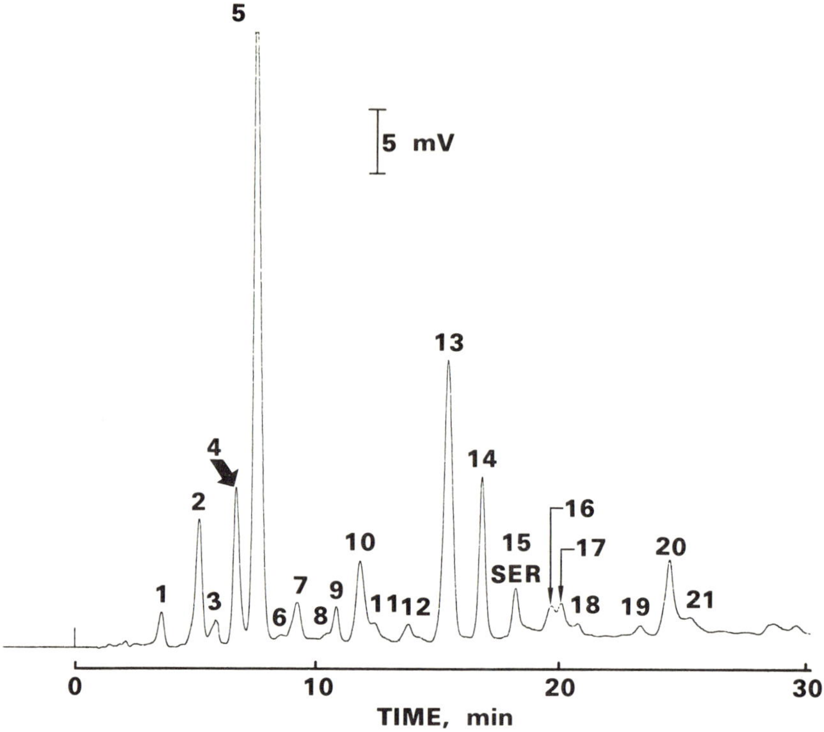


FIG. 8. Reverse phase separation of urine of patient with metastatic melanoma (*42*). Column: MicroPak MCH-10 (Varian); mobile phase: A = 0.005 *M* heptane sulfonic acid, B = 0.005 *M* heptane sulfonic acid in acetonitrile; flow rate: 2 mL/min; Detector: Fluorichrom, fluorescence, Emission filter = Corning 7-60, excitation filter = 220 nm Interference.

Fig. 9 (*43*). At acidic pH values, the amine group of the drug is protonated and capable of ion pair formation with heptane sulfonic acid. The drug and its internal standard were well separated from other extractables from plasma.

For certain species of compounds, the addition of *metal ions* and *metal chelates* to the mobile phase gives unique selectivity by the formation of selective complexes. The technique of argentation chromatography, where $Ag^+$ is added to the mobile phase, has been used to effect the selectivity of olefinic materials such as unsaturated $C_{18}$ esters (*44*) and capsaicins (*45*). The silver ion forms a charge transfer complex and renders the olefinic compound more hydrophilic and it elutes prior to its saturated analog.

Another approach involves the addition of a hydrophobic metal chelate to the aqueous mobile phase (*46, 47*). The $C_{12}$—dien—$Zn_2^+$

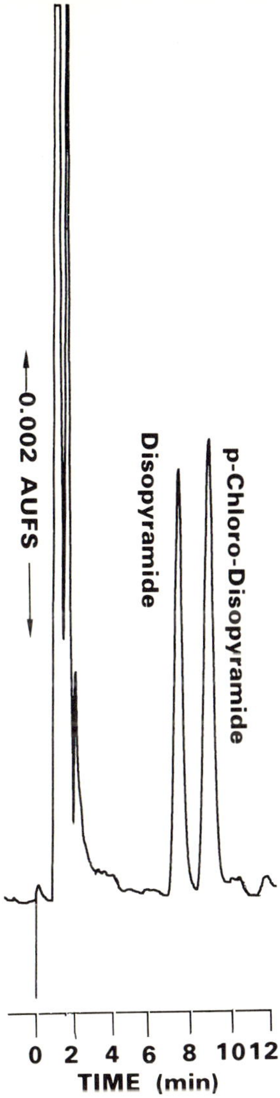


FIG. 9. Determination of diisopyramide in plasma (*43*). Column: MicroPak MCH-10 (Varian), 30 cm × 4 mm; mobile phase: 30% 0.025 *M* sodium acetate containing 0.005 *M* heptane sulfonate, 70% acetonitrile; flow rate: 2 mL/min; detector: $\lambda$ = 254 nm; sample: 20 $\mu$L extract from 100 $\mu$L plasma containing 5 $\mu$g/mL diisopyramide and internal standard.

complex can form an ion pair with a negatively charged species, such as a sulfa drug or the carboxyl group of an acid. Besides ion pairing, other effects such as steric or hydrogen bonding interactions may occur. The separation of D- and L-dansyl amino acids by the complexation with an optically active metal chelate was said to involve the formation of an ion pair (*48*).

## XI. Columns for Adsorption and Normal Bonded-Phase Chromatography

Normal phase chromatography is referred to when the stationary phase is more polar than the predominant solvents of the mobile phase. For example, a silica gel column containing the polar Si—OH groups when used with a hexane eluent would be considered a normal phase application. The term "normal" is historical in that early liquid chromatography was carried out in this manner. When a technique developed where the stationary phase was less polar than the mobile phase, it was termed "reverse phase" chromatography, as it was the opposite of normal.

### A. Liquid-Solid (Adsorption) Chromatography (LSC)

LSC uses silica gel or alumina packings and a nonpolar mobile phase, such as a hydrocarbon, mixed with a more polar solvent (e.g., chlorinated hydrocarbons, alcohols, esters, or ethers). In normal phase, the hydrocarbon (e.g., hexane, heptane, or isooctane) is considered to be the weak solvent (solvent A), while the more polar component (e.g., dichloromethane, isopropanol, or diethyl ether) is considered to be the strong solvent (solvent B). In the adsorption mode, the polar sample components are attracted to the polar Si—OH group of the silica (or Al—OH of the alumina) by hydrogen bonds and other molecular interactions. Nonpolar sample components prefer the nonpolar mobile phase and are eluted prior to the polar components.

Basically, the simplistic mechanism of adsorption considers that the competitive phenomena in the mobile phase molecule (S) and the solute molecule (X) are in competition for the surface adsorption site as expressed by the following equation:

$$X_m + nS_{ads} \rightleftharpoons X_{ads} + nS_m$$

where $X_m$ and $X_{ads}$ represent the solute molecule in the mobile phase and adsorbed state, respectively; $S_{ads}$ represents the mobile phase molecule adsorbed on the surface site while $S_m$ represents the solvent molecule in the mobile phase. The $n$ is the number of adsorbed solvent molecules that must be displaced by the adsorption of X. Thus, a solvent more polar than the solute may displace it from an adsorption site.

Adsorption chromatography is an ideal technique for class separations, that is, separations based on the type of functional groups. Functional groups can be classified according to their attraction for the silica surface (e.g., hydrocarbons $<$ halogenated hydrocarbons $<$ ethers $<$ esters $<$ ketones $\approx$ aldehydes $\approx$ alcohols $<$ amines $<$ acids). A more quantitative representation is given by the Eluotropic Series (*49*) that assigns a parameter $\epsilon^0$ representing the strength that a certain solvent has in LSC. In such a series, solvents that are more polar (i.e., higher in the Eluotropic Series) will displace solvents (or solutes) that are less polar.

Adsorption chromatography will also separate multifunctional compounds and isomers as occurs in the separations of mono-, di-, and triglycerides or the estrogen steroids depicted in Fig. 10. The steroids have one (estrone), two (estradiol), or three (estriol) hydroxyl functionalities, and elute in proportion to the number of polar groups. Owing to the topographical distribution of silanol adsorption sites, the separation of ortho-, meta-, and para-substituted aromatic compounds

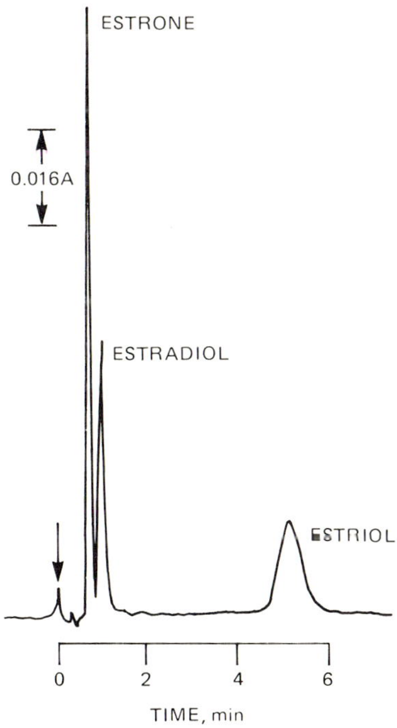


FIG. 10. Separation of estrogen steroids. Column: MicroPak Si-10 (silica); mobile phase: 5% isopropanol, 5% methylene chloride, 90% hexane (v/v/v); flow rate: 2.1 mL/min; detector: UV 254 nm; sample size: 4 μg each. (Reprinted by permission of Varian Associates.)

is possible. Likewise, the separation of geometrical isomers (e.g., *cis-trans*) can be accomplished by LSC.

One drawback of LSC for clinical samples is that, in general, the nonpolar mobile phases used in the technique are not compatible with injections of aqueous-based samples. Also, water, being strongly adsorbed by silica gel, will cause an adsorbent to be deactivated, and consequently the column may change its adsorption characteristics. Such biological samples must be extracted into a compatible solvent or the aqueous phase removed by evaporation or freeze-drying and the sample reconstituted in an organic solvent.

In adsorption chromatography, gradient elution is carried out by increasing the more polar solvent strength as a function of time. In a typical solvent system, such as hexane (solvent A) and dichloromethane (solvent B), the dichloromethane concentration would be increased usually continuously, but sometimes stepwise. At the conclusion of the gradient, the silica column must be returned to its original condition in order to maintain reproducible retention times. It has been found that adsorbents respond slowly to changes in solvent composition, mainly owing to the slow kinetics of equilibration with the trace amounts of water found in most solvents.

## B. Normal Bonded Phases

The use of *normal bonded phases* negates some of the problems experienced with silica gel columns. The chemically bonded phases will respond much more rapidly to mobile phase composition changes, especially the changes that occur in column regeneration going from the polar mobile phase to a much less polar mobile phase. Packing possessing polar functional groups have been slowly replacing the classical adsorbent silica gel in normal phase operation. The surfaces of these packings are "milder" and give rise to fewer chemisorption, tailing, and catalytic activity problems. The reason for lower surface activity is that the so-called reactive silanols responsible for strong adsorption are eliminated by the reaction with organosilane during the bonding. These surface silanol groups are replaced by functional groups such as —CN, —$NH_2$, or diol. Separations on some of these polar bonded phase materials often resemble those obtained on silica gel, but retention is usually reduced and selectivity is sometimes altered. Separations formerly done on extremely dry silica and dry nonpolar mobile phases, such as separations of polynuclear aromatic hydrocarbons, are better done by RPC.

The functionality of the most widely used bonded phases of normal phase work are amino, cyano (nitrile), and diol. Several other phases are commercially available, such as nitro, dimethylamino, and ester, but these phases have not found much use.

The amino phases are particularly useful in that, being basic, they impart quite a different chromatographic selectivity when compared to the slightly acidic surface of silica gel. It may function as a Brønsted acid or base depending on the solute. Its strong hydrogen bonding properties result in excellent separation of polyfunctional compounds. The most successful application of $—NH_2$ phases has been in the separation of carbohydrates using water–acetonitrile as a mobile phase, as depicted in Fig. 11. To elute mono-, di-, and trisaccharides in a single run in a reasonable time, a solvent gradient was required. Since a refractive index detector, often used for saccharide detection, is not compatible with gradient elution, a variable wavelength detector set at a detection wavelength of 192 nm was used. At such a low UV wavelength the water, being the stronger solvent in the gradient, is more transparent. Hence, as the gradient proceeded, there was a slight baseline decrease owing to the differential absorbance of the acetonitrile and water.

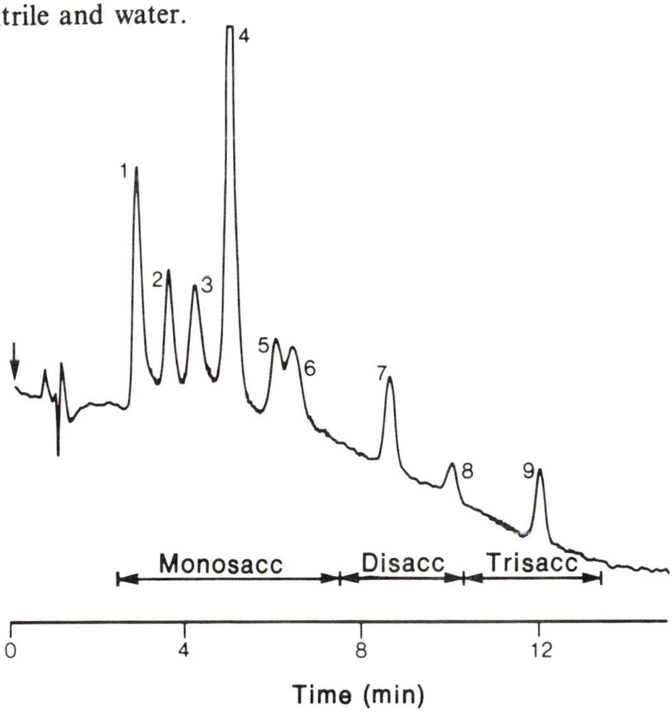


FIG. 11. Gradient elution separation of saccharides (*50*). Column: MicroPak $NH_2$; gradient: 10% $H_2O$ to 50% $H_2O$ in acetonitrile at 2%/min; flow rate: 1 mL/min; detector: Varichrom, $\lambda$ = 192 nm.

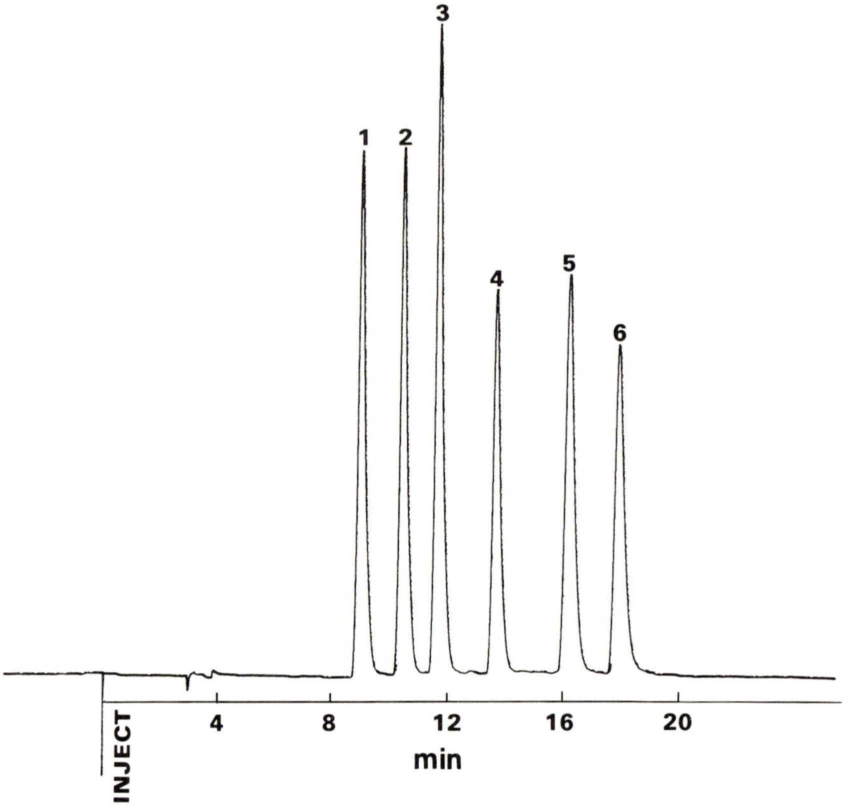


FIG. 12. Separation of ketosteroids by normal bonded phase chromatography. Column: MicroPak CN-10 (Varian); mobile phase: A = hexane, B = 33% isopropanol in dichloromethane; gradient: 15–75%B in 20 min; flow rate: 1 mL/min; detector: λ = 254 nm, 0.32 Aufs.

The alkyl–nitrile phase is of intermediate polarity and is less retentive than silica gel, but displays similar selectivity. Compared to silica, weaker mobile phases are used with these columns and they can frequently be used for the normal phase chromatography of more polar compounds with less tailing. The separation of several polar ketosteroids on a —CN column, depicted in Fig. 12, shows very little tailing when chromatographed with the hexane–methylene chloride–isopropanol mobile phase. The nitrile triple bond confers excellent selectivity in the separation of double bond isomers and ring compounds differing in double bond content.

The diol-type of phase, prepared by the hydrolysis of a bonded epoxysilane phase, has found use in the exclusion chromatography of proteins using aqueous mobile phases (*17*). Proteins, which are sometimes irreversibly adsorbed on silica silanol groups, interact much less with the carbinol bonds of the diol phase.

Sometimes, in working with very polar compounds (e.g., carboxylic acids, phenols, polyamines) in normal phase chromatography, poor peak shape or tailing may occur. Such behavior may owe to the presence of unreacted silanols or nonlinear isotherms, but often these effects may be reduced or eliminated by the inclusion of a small amount of acid or base in the nonpolar mobile phase. For acidic compounds (e.g., carboxylic acids or phenols), addition of 0.5–1% by volume of acetic or phosphoric acid may be used. For basic compounds propylamine or ammonia may be used. Such modifiers are miscible with the normal polar mobile phases.

## XII. Columns for Ion Exchange Chromatography

Ion exchange chromatography is well suited to the separation of water soluble substances in biological fluids, since the technique uses aqueous mobile phases. Ion exchange chromatography is generally applicable to ionic compounds, to ionizable compounds such as organic acids or bases, and to compounds (such as chelates and ligands) that can interact with ionic groups. The packing may be a polystyrene–divinylbenzene resin or silica gel to which has been bonded an ionogenic group. A typical structure of an anion exchange resin is illustrated in Fig. 13. For a resin, the amount of divinylbenzene (usually 4–12%) incorporated into the polymer determines its porosity and the structural rigidity of the packing beads. The higher the crosslinking the more rigid the packing, but the smaller the pore size.

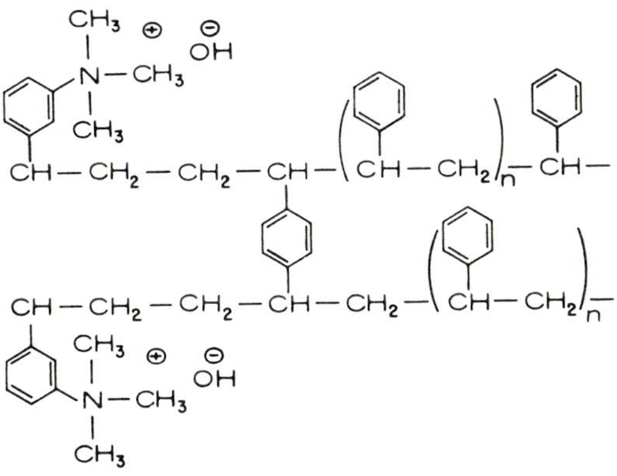


FIG. 13. Structure of an anion exchange resin.

Table 6
Classification of Ion Exchange Resins

| Type | Strength | Functional group | Typical forms |
|---|---|---|---|
| Anion | Strong | $—N(CH_3)_3^+$ | Chloride, hydroxide |
| Anion | Weak | $—NH_2$ or $—NH(CH_3)_2^+$ | Phosphate, chloride |
| Cation | Strong | $—SO_3^-$ | Sodium, ammonium |
| Cation | Weak | $—COO^-$ | Hydrogen |

Ion exchangers are characterized by the type, number, and strength of ionogenic functional groups. Table 6 lists the most popular types of cation and anion exchangers. Strongly acidic or basic packings are the most widely used. They are ionic at all pH values, whereas the weak packings operate over a much smaller pH range, dependent upon the $pK_a$ (or $pK_b$) of the functional group. A high degree of selectivity can be achieved near the $pK_a$ by slight variations in pH. Ion exchangers are further classified by their ion exchange capacity. The capacity is defined as the number of available functional groups for ion exchange, and is usually expressed as milliequivalents per gram of dry resin. Pellicular packings, which have the thin layer of stationary phase polymerized or bonded to a glass bead, show very low ion exchange capacities (in the range of 0.01 meq/g), while silica-bonded-phase exchangers and porous resins show capacities in the range of 0.5–5 meq/g.

Mobile phases used in ion exchange chromatography are salt buffers whose pH is adjusted to optimize ionic interactions with the packing functional groups. Similar to the competition between solute and mobile phase for the adsorption site in LSC, there is a competition between the charged solute and the buffer counterion of like charge for the oppositely charged site on the packing. Equilibria are established as follows:

Cation exchange: $X^+ + R^-Y^+ \rightleftharpoons Y^+ + R^-X^+$
Anion exchange: $X^- + R^+Y^- \rightleftharpoons Y^- + R^+X^-$

where X = the sample ion, Y = the mobile phase counterion, and R = the ionic site on the exchanger. Thus, to decrease retention in ion exchange, the buffer ionic strength is increased. The most common type of gradient elution is carried out by increasing buffer strength as a function of time.

Ion exchange involves more variables than other forms of chromatography. Distribution coefficients and selectivities are not only functions of ionic strength, but also of pH, solute charge and radius, packing porosity, type of buffer ion present, type of solvent (if any), temperature, backbone of packing (silica or polystyrene resin) and so forth. The number of experimental variables makes ion exchange a very versatile technique since each may be used to effect a better separation, but nevertheless a difficult technique, because of the time needed to optimize a separation. In addition to "pure" ion exchange, other interactions may also govern solute retention, especially when using resin packings. For example, because of the "solvent" effects of the aromatic polystyrene resin matrix, phenols are more strongly retained in anion exchange than their weak ionization would suggest. Even nonionic compounds may be separated on resins, probably by a partition mechanism. In these cases, the presence of a buffer decreases compound solubility in the mobile phase, therefore increasing its affinity for the resin. Electrically neutral species that can complex with ions can be separated by the exchange process. A well-known example is the separation of sugars through the adducts formed with the borate buffer used to elute them. Ligands can be separated through their interaction with metallic ions sorbed by the resin.

Besides being the predominant technique for the separation of metallic cations, ion exchange is widely applied to the separation of amino acids, nucleic acid constituents, proteins, vitamins, pharmaceuticals, and endogenous compounds in body fluids. The most widespread use of ion exchange resins in biochemistry is in amino acid analysis. Figure 14 presents such a separation using a modern microparticulate resin. A citrate buffer step gradient was used to effect a separation of 21 amino acids on a cation exchange resin.

As can be seen in Fig. 15, using a unique silica-based weak anion exchanger, the separation of bases, nucleosides, and nucleotides can be carried out in a single run using a three solvent gradient (*51*). Prior separations were carried out by two separate experiments using reverse phase for the nucleosides and ion exchange for the nucleotides. A relatively high phosphate buffer strength was required to elute the di- and triphosphonucleotides from the column. Bonded phase columns are more rapid to regenerate at the conclusion of the gradient; thus, overall time between samples is reduced.

Proteins can be separated by ion exchange or by exclusion chromatography (as will be seen later). For the ion exchange separation of proteins, wide pore packings are required so that the proteins can diffuse into the pores to interact with the ionogenic groups. The separation of oligonucleotides is shown in Fig. 16 on a polyethyleneimine–glycidyl ethyl bonded phase whose average pore size was 300 Å (*52*).

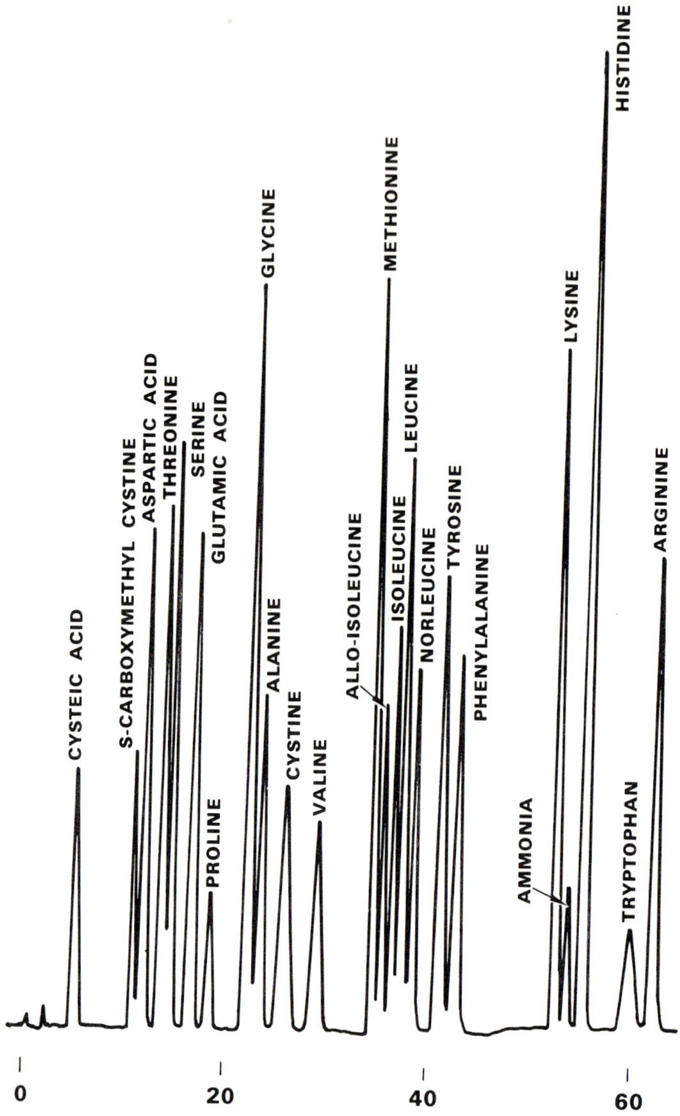


FIG. 14. Separation of amino acids by ion exchange chromatography. Column: Beckman AA-10, 9 ± 0.5 m, 2.8 × 300 mm; mobile phase: first buffer, pH 3.28, 0.20 *N* $Na^+$ with 2% isopropanol; second buffer, pH 3.90, 0.35 *N* $Na^+$ with 2% isopropanol (35 min); third buffer, pH 4.95, 1.40 *N* $Na^+$ with 4% isopropanol (47 min); flow rate: 0.17 mL/min; detector: ninhydrin post column system, λ = 570 nm; column temperature: 50° C, stepped to 65° C at 35 min; sample size: 12.5 nmol/component. (Reprinted with permission of Beckman Instruments.)

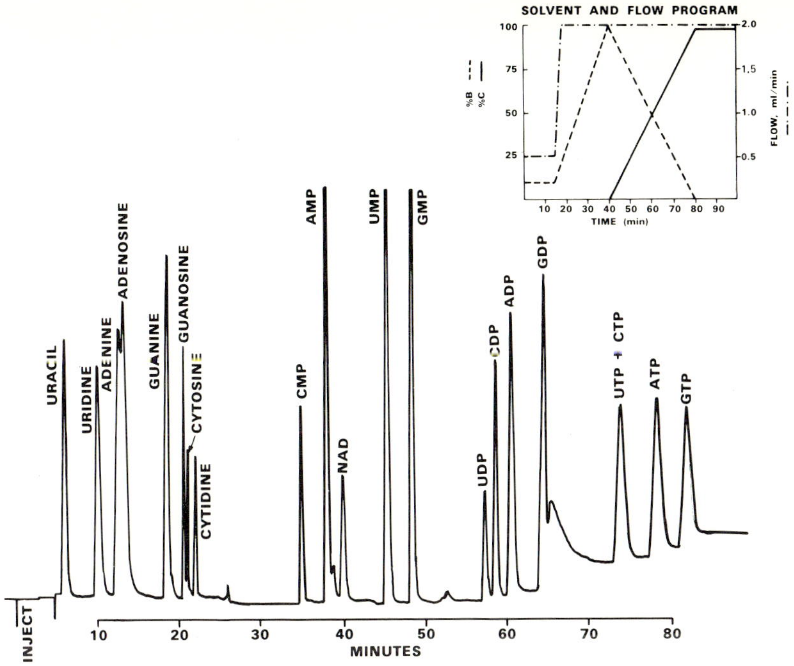


FIG. 15. Separation of nucleotides, nucleosides, and bases (*82*). Column: MicroPak AX-10 (Varian); mobile phase: A = 5 m*M* $NH_4H_2PO_4$ (pH 2.85)/acetonitrile (13/87), B = 5 m*M* $NH_4H_2PO_4$ (pH 2.85), C = 0.75 *M* $NH_4H_2PO_4$ (pH 4.5); gradient: see graph for program; flow rate: see graph for program; detector: λ = 254 nm, 0.08 Aufs.

FIG. 16. Chromatogram of oligonucleotides (*52*). Column: polyethyleneimine-pentaerythritol tetraglycidyl ether crosslinked on 10μm Lichrosorb Si500 (500 Å pore size); mobile phase: A = 0.05 *M* $KH_2PO_4$ (pH 7.0) with 30% methanol, B = 1.0 *M* $NH_4Cl$ in solvent A, linear gradient: 5%B to 100%B in 25 min; flow rate: 1.5 mL/min; detection: λ = 254nm.

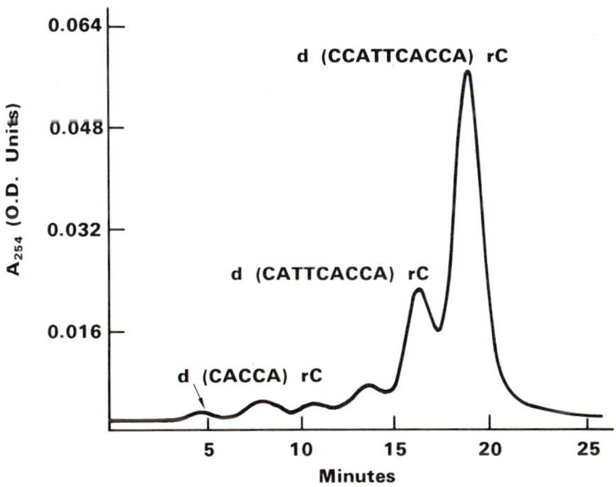

## XIII. Columns for Exclusion Chromatography

Unlike the other chromatographic techniques, exclusion chromatography does not involve the interaction between the solute and the stationary phases. The mechanism is one of solute diffusion into and out of the pores in a porous matrix. All molecules larger than the pores are excluded and elute in the exclusion volume. All molecules smaller than the smallest pore elute at the total permeation volume. Molecules that permeate part of the pores but are excluded from others can be separated. All separation occurs between the exclusion volume and the permeation volume. The elution order is in terms of decreasing molecular size. Column selection involves matching of the pore size of the packing to the molecular size of the solute molecules. In this mode of HPLC, several columns with packings of varying pore sizes may be required to separate a sample consisting of species of widely different molecular sizes.

Exclusion chromatography has also been known by other names, such as gel chromatography, size or steric exclusion chromatography, molecular sieve chromatography, but the most common were gel permeation chromatography (GPC) and gel filtration chromatography (GFC). The GPC technique is referred to when separations are carried out in organic solvents. Its main use is in the characterization of industrial organic polymers, oligomers, and polymer additives. The GFC technique is referred to when separations are performed in aqueous solution. Since most biological and physiological substances are water-soluble, we will confine our coverage to exclusion chromatography carried out in aqueous solutions, or GFC.

In the past, GFC used very soft gels that could not withstand the pressures generated by HPLC pumps. Since only a few psi of pressure was permissible, low flow rates had to be used. Therefore, separation times were very long: hours, and in extreme cases, days. The recent advent of microparticles for aqueous exclusion work has revolutionized the application of this technique to biopolymers. The packings in use now are rigid silica gel base materials possessing chemically bonded phases with slightly hydrophilic functional groups such as diol [i.e., —$CH_2(OH)$—$CH_2(OH)$], as discussed earlier. Some semirigid packings based on hydrophilic polymeric material have also become available.

In applying the exclusion technique we must have some idea of the separation range of a particular pore size packing. A calibration curve defines the operating range of a particular packing. To construct such a curve, standard substances of known molecular size are injected into the exclusion column and a plot of log molecular size (or molecular weight for a homologous series of standards) versus peak elution volume is obtained. Figure 17 gives such a plot for a series of polydextrans and polyethylene glycol (PEG) standards on 60 cm × 7.5 mm MicroPak TSK Type SW columns. As can be seen, there are three different pore size columns available covering the range of molecular weight from 500 to 600,000. Since PEG standards were not available to cover the entire molecular weight range, polydextran standards were used for the upper ends of the curves. Note the slight offset of the curves for the PEG and dextrans standards of similar molecular weight. The reason for the offset is that the molecular sizes of solvated PEG and dextrans molecules are different in solution.

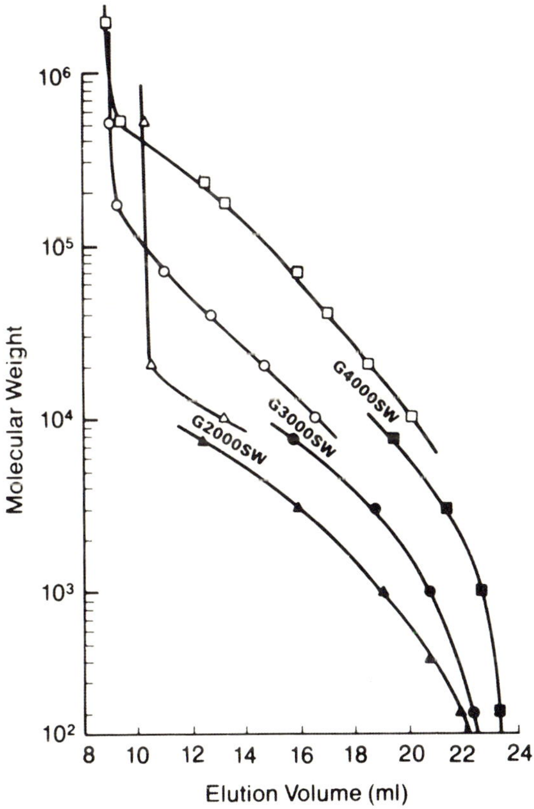


FIG. 17. Calibration curves for MicroPak TSK type SW columns. Columns: MicroPak TSK 2000SW, 3000SW, and 4000SW, 7.5 mm 30 cm × 2 mm; mobile phase, $H_2O$; flow rate: 1 mL/min; sample: dextran (△, ○, □), polyethylene glycol (▲, ●, ■), detection: refractive index.

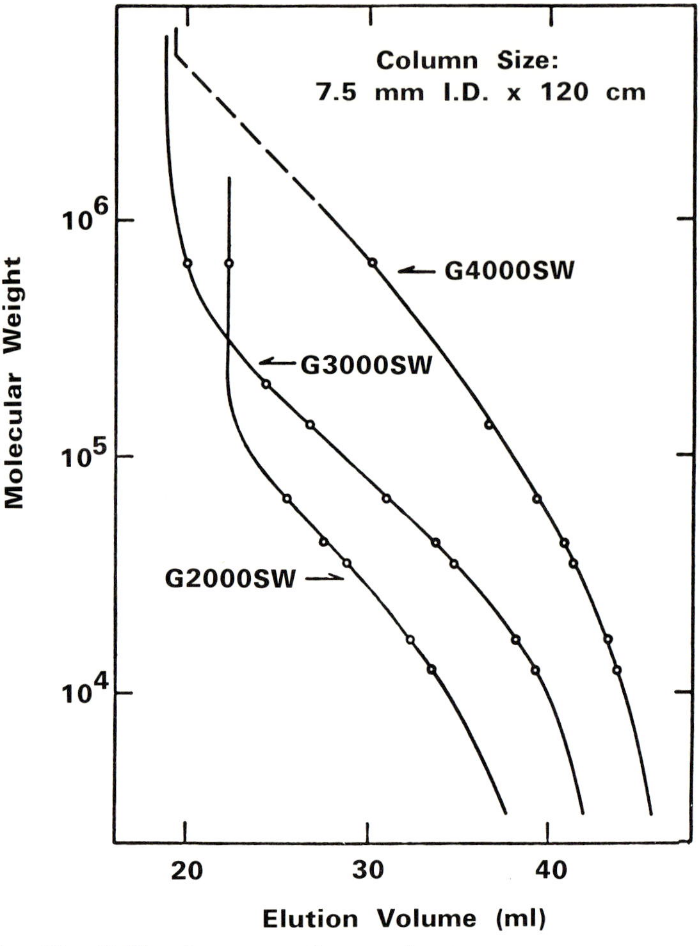


FIG. 18. Calibration curves for proteins for MicroPak TSK type SW columns.

In Fig. 17 the exclusion volume for the MicroPak TSK 2000SW column (60 cm × 7.5 mm) would be the extrapolated volume where the curve turns sharply upward on the upper end, or 10.5 mL. The total permeation volume would be the extrapolated volume where the curve turns downward on the lower end, or 22 mL. The separation range lies between these two extremes, or a total of 11.5 mL.

Similar calibration curves for the three columns are shown for various protein standards in Fig. 18. Often, owing to shape factors, the effective size (hydrodynamic volume) of a biopolymer may differ from other molecules of similar molecular weight. Table 7, taken in part

Table 7
Size of Common Biopolymers

| Biopolymer | Molecular weight | Size, A × A |
|---|---|---|
| Fibrinogen (human) | 580,000 | 700 × 38 |
| Edestin | 310,000 | 237 × 55 |
| Meromyosin (H) | 232,000 | 435 × 29 |
| $\alpha_1$-Lipoprotein | 200,000 | 300 × 50 |
| Gamma-globulin | 156,000 | 235 × 44 |
| Serum albumin (human) | 69,000 | 150 × 38 |
| Hemoglobin (human) | 68,000 | 57 × 34 |
| Prothrombin | 62,700 | 119 × 34 |
| Cytochrome C | 15,600 | 98 × 18 |
| Lysozyme | 14,100 | 60 × 24 |

from reference *53,* gives the molecular weights, as well as the major and minor axes, of selected proteins that would fall inside the molecular weight ranges of the MicroPak TSK Type SW columns shown in Fig. 18. A case in point is the actual chromatogram of protein standards illustrated in Fig. 19 (*54*), which shows that bovine serum albumin (mw 69,000), elutes at an earlier volume than hemoglobin (mw 69,000), which, according to Table 7, would probably exhibit a different molecular size.

To illustrate the advances made in microparticulate exclusion packings, one can compare the 8-min separation of human serum (*55*) shown in Fig. 20 with a similar separation of human serum (not shown here) done on Sephadex G200 in reference *56*. In the latter case, the time required to elute the serum sample from a 140 cm column of Sephadex was 3.8 h and the resolution was much poorer. Other examples of the use of these columns include the separation of many other proteins and enzymes, liver extracts, insulin, heparin-sodium, nucleic acids, gelatins, and oligosaccharides.

## XIV. Future Developments in Columns and Column Technology

Reverse phase chromatography will continue to be developed by further manipulation of the mobile phase through ternary systems, ionization control, and selective, complex-forming additives. Such molecular manipulations will force chromatographers to remember their chemistry. Further understanding of the role of the type chain

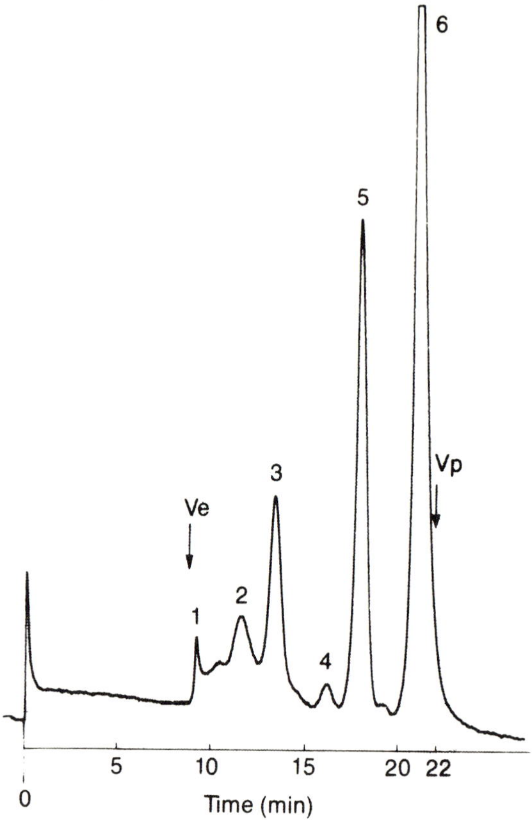


FIG. 19. Exclusion separation of proteins (*54*). Column: MicroPak TSK 3000SW, 8 mm × 30 cm (× 2) (Varian); mobile phase: 1/15 *M* Sorensen phosphate buffer, pH 6.8 + 0.1 *M* NaCl; flow rate: 1 mL/min; peaks: 1, bovine α-globulin, mw 200,000; 2, 3, bovine serum albumin, mw 69,000; 4, hemoglobin, mw 69,000; 5, myoglobin, mw 17,500; 6, lysozyme, mw 14,000.

length and coverage of hydrophobic of bonded phases and the role of unreacted silanols on the separation of polar substances is required. Already there is a tendency for chromatographers to use moderately polar normal bonded-phase materials, such as cyanopropyl, with reverse phase mobile phases to gain some additional stationary phase selectivity. The introduction of ternary pumping systems will allow the easier manipulation of mobile phase composition and the further exploitation of solvent selectivity effects. More work needs to be done on criteria for the optimization of ternary gradients (binary gradients are frequently difficult enough!). Reverse phase columns that can truely deliver 100,000 theoretical plates in a meter, rather than those

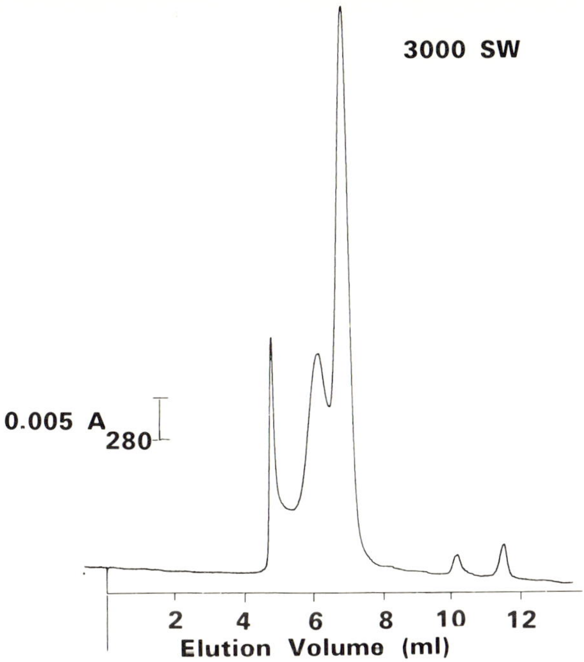


FIG. 20. Separation of human serum (*35*). Column: MicroPak TSK 3000SW, 30 cm × 7.5 mm (Varian); mobile phase: 0.067 *M* $KH_2PO_4$ + 0.1 *M* KCl + 6 × $10^{-4}$ *M* $NaN_3$, pH 6.8; flow rate: 1 mL/min; detection: λ = 280 nm; temperature: 30° C.

calculated to have 100,000 plates from 10 cm lengths, would be a boon to the handling of the more complex separations. A bonded-phase matrix other than silica gel that can be used in basic solution is needed. Perhaps high efficiency microparticulate polymeric gels with polar and nonpolar functionality incorporated into the polymer would serve this purpose.

The recent introduction of microparticulate exclusion packings will cause many biochemists using the older soft gel materials to switch to these high performance, high speed columns, which unfortunately are also of high cost. Further developments in rigid large pore ion exchange packings will allow even more sophisticated separations of biopolymers to occur.

As the analyses of complex samples becomes more and more demanding, the use of a single chromatographic mode will be supplanted by the application of multidimensional approaches where two or more LC techniques are used to further fractionate a sample. Already, the on-line coupling of exclusion chromatography and RPC has been applied to complex natural products, and the technique of coupling two RPC systems has been used to resolve components of physiological fluids.

The success of capillary gas chromatography has spurred research into capillary HPLC. The advantages would be very low solvent consumption, possibly better sensitivity, and the easier application of LC–mass spectrometry coupling. A new generation of instrumentation (pumps, injectors, detectors) would be required to handle the small samples, small peak volumes, and low flow rates required.

## XV. Conclusions

We have attempted to provide an overview of HPLC columns and column technology. A detailed discussion of all possible LC modes was beyond the scope of this chapter. For further reading on column technology, several all-inclusive texts are recommended (*2, 3, 58–61*).

## Acknowledgment

The author would like to thank Ms. Sally Bird for her preparation of the manuscript and several of his colleagues in the LC Applications Laboratory at Varian who generated some of the chromatograms.

## References


*1.* Majors, R. E., ed., *J. Chromatogr. Sci.* **15,** 333–439 (1977); ibid., in press.
*2.* Bristow, P. A., *LC in Practice,* hetp Ltd., Cheshire, U. K., 1976, pp. 32–33.
*3.* Johnson, E. L., and Stevenson, R., Basic Liquid Chromatography, Varian, Palo Alto, California 1978, pp. 70–71.
*4.* Martin, M., and Guiochon, G., *Chromatographia* **10,** 194 (1977).
*5.* Majors, R. E., *Anal. Chem.* **44,** 1722, 1726 (1972).
*6.* Cassidy, R. M., Legal, D. S., and Frei, R. W., *Anal. Chem.* **46,** 340, (1974).
*7.* Strubert, W.,*Chromatographia* **6,** 50 (1973).
*8.* Kirkland, J. J., *J. Chromatogr. Sci.* **10,** 593 (1972).
*9.* Endele, R., Halasz, I., and Unger, K., *J. Chromatogr.* **99,** 377 (1974).
*10.* Coq, B., Gonnett, C., and Rocca, J. L., *J. Chromatog.* **106,** 249 (1975).
*11.* Webber, T. J. N., and McKerrell, E. H., *J. Chromatogr.* **122,** 243 (1976).
*12.* Bather, J. M., and Gray, R. A. C., *J. Chromatog.* **122,** 159 (1976).
*13.* Knox, J. H., *J. Chromatogr. Sci.* **15,** 352 (1977).

*14.* Cox, G. B., Luscombe, C. R., Slucutt, M. J., Sudgen, K., and Upfield, J. A., *J. Chromatog.* **117,** 269 (1976).
*15.* Linder, H. R., Keller, H. P., and Frei, R. W., *J. Chromatogr. Sci.* **14,** 234 (1976).
*16.* Little, C. J. Dale, A. D., Ord, D. A., and Marten, T. R., *Anal. Chem.* **49,** 1311 (1977).
*17.* Chang, S. H., Gooding, K. M., and Regnier, F. E., *J. Chromatog.* **125,** 103 (1976).
*18.* Kirkland, J. J., and Antle, P. E., *J. Chromatogr. Sci.* **15,** 137 (1977).
*19.* Waters Associates, Milford, Mass. 01757, Product Bulletin D99, January, 1979.
*20.* Knox, J. H., and Parcher, J. F., *Anal. Chem.* **41,** 1599 (1969).
*21.* Atwood, J. G., Schmidt, G. H., and Slavin, W., "The Use of High pH Mobile Phase in HPLC Methods Using Silica Guard Columns," paper delivered at the 1979 Pittsburgh Conference on Analytical Chemistry and Spectroscopy, March 5–9, 1979, Cleveland, Ohio, paper 209.
*22.* Majors, R. E., and Hopper, M. J., *J. Chromatogr. Sci.* **12,** 767 (1974).
*23.* Cox, G. B., *J. Chromatogr. Sci.* **15,** 385 (1977).
*24.* Karger, B. L., and Giese, R. W., *Anal. Chem.* **50,** 1048A (1978).
*25.* Wahlund, K. G., and Lund, U., *J. Chromatog.* **122,** 269 (1976).
*26.* Karger, B. L., Gant, J. R., Hartkopf, A., and Weiner, P. H., *J. Chromatog.* **128,** 65 (1976).
*27.* Horvath, C., and Melander, W., *J. Chromatogr. Sci.* **15,** 393 (1977).
*28.* Horvath, C., and Melander, W., *International Lab.* Nov/Dec, 11 (1978).
*29.* Schmit, J. A., Henry, R. A., Williams, R. C., and Dieckman, J. F., *J. Chromatogr. Sci.* **9,** 645 (1971).
*30.* Bakalyar, S. R., McIlwrick, R., and Roggendorf, E., *J. Chromatog.* **142,** 353 (1977).
*31.* Bakalyar, S., *Amer. Lab.,* June 43 (1978).
*32.* Realini, P., "Determination of PNA in Waste Water, Liquid Chromatography At Work," No. 91, Varian Instrument Division, Walnut Creek, CA, 1979.
*33.* Cham, B. E., Johns, D., Bochner, F., Imhoff, D. M., and Rowland, M., *Clin. Chem.* **25,** 1420 (1979).
*34.* Schill, G., in *Assay of Drugs and Other Trace Compounds in Biological Fluids,* E. Reid, ed., North Holland, New York, 1976.
*35.* Gloor, R., and Johnson, E. L., *J. Chromatogr. Sci.* **15,** 413 (1977).
*36.* Johansson, I. M., Wahlund, K. G., and Schill, G., *J. Chromatog.* **149,** 281 (1978).
*37.* Kissinger, P. T., *Anal. Chem.* **49,** 883 (1977).
*38.* Knox, J. H., and Laird, J. R., *J. Chromatog.* **122,** 17 (1976).
*39.* Knox, J. H., and Jurand, J., *J. Chromatog.* **149,** 297 (1978).
*40.* Konijnendijk, A. P., and Van De Venne, J. L. M., in *Advances in Chromatography 1979,* A. Zlatkis, ed., Chromatography Symposium, University of Houston, Houston, Texas, 1979, pp. 451–462.
*41.* Bidlingmeyer, B. A., Deming, S. N., Price, W. P., Sachok, B., and Petrusek, M., *ibid,* pp. 435–450.

*42.* Abbott, S. R., and Banda, P. W., *HPLC-Fluorescence Analysis of Urinary Indoles,* Liquid Chromatography At Work No. 103, Varian Instrument Division, Walnut Creek, CA, 1980.
*43.* Larson, L., *Determination of Norpace® in Biological Fluids,* Liquid Chromatography At Work No. 71, Varian Instrument Division, Walnut Creek, CA 94598 (1979).
*44.* Vonach, B., and Schomburg, G., *J. Chromatog.* **149,** 417 (1978).
*45.* Johnson, E. L., Majors, R. E., Werum, L., Reiche, P., in *Analysis of Food and Beverages,* Vol. 1, New York, G. Charalombous, ed., Academic Press, New York, 1979, pp. 17–29.
*46.* Cooke, N. H. C., Viarattene, R. L., Eksteen, R., Wong, W. S., Davies, G., and Karger, B. L., *J. Chromatog.* **149,** 391 (1978).
*47.* Karger, B. L., Wong, W. S., Viarattene, R. L., LePage, J. N., and Davies, G., in *Chromatog.,* G. Schomburg and L. Rohrschneider, eds., Elsevier, Amsterdam, 1978, pp. 73–92.
*48.* LePage, J. N., Lindner, W., Davies, G., Seitz, D. E., and Karger, B. L., *Anal. Chem.* **51,** 533 (1979).
*49.* Snyder, L. R., *Principles of Adsorption Chromatography,* Dekker, New York, 1968, 194–198.
*50.* Hettinger, J., and Majors, R. E., *Varian Instrument Applications* **10,** 6 (1976).
*51.* Wehr, C. T., Liquid Chromatography At Work No. 82, Varian Instrument Division, Walnut Creek, CA, 1979.
*52.* Alpert, A. J., and Regnier, F. E., *J. Chromatography,* to be published, 1980.
*53.* Florkin, M., and Stotz, E. H., eds., *Comprehensive Biochemistry,* Vol. 7, Elsevier, New York, 1963, p. 50.
*54.* Abbott, S., Varian, unpublished data, 1979.
*55.* Wehr, C. T., *High Speed Exclusion Chromatography of Proteins,* Liquid Chromatography At Work No. 100, Varian, Walnut Creek, CA, 1979.
*56.* Flodin, P., and Killander, J., *Biochem. Biophys. Acta* **63,** 403 (1962).
*57.* Burce, G. L., Determination of Synthetic Estrogens in Oral Contraceptive Dosage Forms, Liquid Chromatography At Work No. 79, Varian Instrument Division, Walnut Creek, CA, 1979.
*58.* Snyder, L. R., and Kirkland, J. J., *Introduction to Modern Liquid Chromatography,* Wiley-Interscience, New York, 1974, 534 pp.
*59.* Parris, N. A., *Instrumental Liquid Chromatography,* Elsevier, Amsterdam, 1976.
*60.* Simpson, C. F., *Practical High Performance Liquid Chromatography,* Heyden, London, 1976.
*61.* Hamilton, R. J., and Sewell, P. A., *Introduction to High Performance Liquid Chromatography,* Chapman & Hall, London, 1977.

# Part II

# Therapeutic Drug Monitoring and Toxicology

# Chapter 4

# Why Measure Drug Levels?

Lewis B. Sheiner

*Department of Laboratory Medicine & Division of Clinical Pharmacology*
*Department of Medicine, University of California, San Francisco, California*

## I. Introduction

In this chapter, I will discuss briefly some traditional uses of measurements of drug levels, and then discuss in greater detail the justification for, and certain features of, the use of drug level determinations for a newer purpose: therapeutic monitoring.

Before doing this, however, a definition of drug level is required: A drug level is a measurement of the concentration of drug or drug products in a body fluid. Because it is not my purpose to discuss the utility of drug levels for scientific inquiry, I restrict the body fluid in question to one derived from a patient receiving the drug. "Drug level" may thus refer to the concentration of drug or metabolite in blood, sweat, or tears; or in such less poetic fluids as urine, saliva, or blood fractions, notably plasma.

## II. Well-Accepted Uses of Drug Levels

### A. Overdosage

Drug level measurements have traditionally been made in cases of drug overdose. Clearly, when dealing with drug overdose, it is important to know what agent has been ingested. A "toxicologic screen," performed on gastric contents, blood or urine can sometimes provide crucial information upon which to base therapeutic choices. Likewise, knowledge of the actual magnitude of the drug concentration in blood may prove prognostically or therapeutically useful if certain other information also is known, such as time of overdose ingestion. In some cases, certain measures (such as hemoperfusion) may be indicated only when concentrations are quite high, or normal removal processes are proceeding too slowly. However, even in such cases, there is some controversy over whether other than supportive measures should be undertaken (*1*), so that only the qualitative use of drug levels in overdosage is unequivocally accepted.

Also related to toxicology is the use of quantitative measurements of drug levels to diagnose toxicity occurring, not from intentional overdosage, but in the course of therapy with an agent. The rationale here is that if clinical signs suggesting toxicity are present and the level is high, toxicity is likely, and if the level is not high, toxicity is unlikely. This use has problems that will be discussed more fully below, after a conceptual scheme relating drug levels to both doses and effects has been outlined.

## B. Failure of Regimen

Other accepted uses of drug level determinations occur when the therapeutic regimen is failing for no apparent reason.

When faced with a therapeutic regimen that appears to be ineffective, the reason may sometimes be that drug is poorly absorbed, or eliminated more rapidly than usual. Drug levels, determined during a carefully supervised and planned "experimental" dosage regimen designed to elucidate absorption and elimination (*4*), can be definitive in resolving the issue. The appropriate therapeutic decision obviously depends on whether, in effect, inadequate drug has been given, or whether an "adequate" drug is nonetheless ineffective.

Although drug-level work-up of a therapeutic problem may sometimes be complex and require multiple quantitative drug level determinations, there is one common cause of failure of therapeutic regimen that requires only semiquantitative drug levels to "diagnose." This is lack of patient compliance with the therapeutic regimen. Drug levels in blood or urine can often provide definitive evidence in this matter.

For the toxicologic uses of drug levels, then, and for at least one of the causes of an ineffective regimen, only qualitative or semi-quantitative drug levels are needed. The drugs for which such levels may be useful are virtually any, since almost any drug can be suspected of causing an overdosage, and certainly any regimen can fail to produce the desired results. For toxicology, the turnaround time of the drug level assay must be rapid, often minutes or only a few hours, whereas to work-up a therapeutic problem, delays of a day are often quite acceptable.

## III. A Conceptual Model for Drug Use

Before turning to a discussion of the use of drug levels for therapeutic monitoring, a description of a conceptual model that relates dosage to drug effects will prove useful. A convenient conceptual model for drug use states that after choosing a drug, the pharmacologic action of which is, in principle, likely to benefit a given patient, its benefits can be maximized only if the route of drug administration and amount and timing of drug application are properly chosen. Let all of these features be termed "dosage." Dosage will determine ultimate benefit because doses of drug (via any route) give rise to concentrations of drug in tissues, including the tissue where the drug is intended to act. Drug at its site of action presumably interacts with "receptors," ultimately to give rise to pharmacologic effects. These, in turn, interact with normal or disease-altered physiologic processes that augment or oppose the drug's effects to finally give rise to net clinical effects. These result in some net benefit (efficacy) or harm (toxicity) to the patient. Under this model, at the crudest level,the greater the dosage (i.e., the greater the intensity of application of drug) the higher the blood concentration; hence, the higher the tissue and active site concentration, the more receptors "occupied," and the more intense the drug's effects. Thus, because the intensity of drug effect relates directly to efficacy and toxicity, dosage relates directly to patient benefit. It does so, however, through the multistage "series" system I have described. Anything that causes variability or uncertainty in any stage of the system can increase variability or uncertainty in the entire system. We may thus be uncertain of the effect of a given dosage of drug on a given patient (a) because we cannot predict the blood concentration (the drug may not be absorbed completely, or it may be eliminated extra-rapidly, or extra-slowly); (b) because we cannot predict the tissue concentration (blood supply to the site of action may be altered); (c) because we cannot predict receptor occupancy (receptor affinity for drug may be unusual); (d) because we cannot predict the pharmacologic effects of a given degree of receptor occupancy (unknown factors may alter the quantitative nature of the linkage); and/or (e) because we cannot predict the resultant clinical effects (variable degrees of disease-induced alterations in normal physiologic response may be present).

## IV. Drug Levels for Therapeutic Monitoring

Accepting the above model, it can be proved that anything that reduces uncorrelated variability or uncertainty in any part of the overall system necessarily reduces uncertainty in the entire system. In particular, knowing the drug concentration in blood reduces the uncertainty in the first step of the dosage-to-clinical effect relationship. This first step is denoted pharmacokinetics, a term encompassing the quantitative and temporal aspects of drug absorption, distribution, and elimination. Since knowledge of a patient's drug levels can reduce pharmacokinetic uncertainty, overall uncertainty can also be reduced. And because such uncertainty can complicate therapeutic decision making, knowing drug levels can aid in therapeutic decision making. Stripped of some embellishments and particulars to follow, this is the essential justification for use of drug levels for therapeutic monitoring. It remains to discuss under what circumstances uncertainty about the dosage–effects relationship in fact confuses dosage decisions, and in which such cases those drug level measurements actually available to us can clarify them.

First, however, let me return to the use of drug levels to diagnose toxicity (or efficacy) in the therapeutic (vs overdose) context, a use I have already indicated has problems.

### A. Diagnosing Toxicity or Efficacy

Unquestionably, it is important to know whether a therapeutic drug is causing toxicity, if that toxicity is potentially harmful to the patient. It is tempting to try to measure drug levels in order to resolve the issue. Assuming that some clinical signs suggest, but do not prove, drug toxicity, it seems reasonable to diagnose definite toxicity if the level is in a range usually associated with toxic effects. Alternatively, if the drug level is in a range usually unassociated with toxicity it seems reasonable to exclude toxicity and to conclude that the drug may be safely continued. Although it is true that finding a "toxic" or "nontoxic" drug level may influence one's *tendency* to diagnose or exclude drug toxicity, it can rarely be definitive. Prudence usually dictates withholding a drug when toxicity is suspected, no matter what the drug level, except for the most life-saving and irreplaceable of agents. I have discussed the reasons for this elsewhere (*2*). In essence, the variability inherent in the level-effect portion of the dose–effect relationship means that "toxic" and "therapeutic" ranges of drug levels must overlap. The drug level can therefore almost never be "pathognomonic" of either toxicity or lack of it, and so it must be regarded in the same way as any other laboratory test, historical item, or physical sign: it may add to the totality of evidence upon which a rational clinician bases a reasoned judgment, but it cannot substitute for that judgment.

This may seem obvious, and it should be. For those who doubt it, the general lack of utility of drug levels for diagnostic purposes need not be accepted on theoretical grounds: it has been empirically verified for at least one drug (*3*).

The other side of the same coin is attempting to use drug levels to determine whether a drug is optimally effective, and for the same reason as above, one can no more conclude with certainty that inefficacy is present when the drug level is "low" than one can that toxicity is present when it is high. Again, it is true that the probability of inefficacy is greater when the drug level is relatively low, but many other factors influence efficacy as they do toxicity. Thus, if the patient's clinical status is satisfactory, it makes little sense to raise dosage because of a low drug level alone. An exception to this statement concerns certain drugs used purely for prophylaxis of dangerous events.

## B. Rationale for Target Level Strategy

The discussion immediately above reduces to the following: for deciding about both drug efficacy and toxicity, if readily observable clinical endpoints are available, drug levels will usually contribute little. But what if clinical endpoints are not available? This happens more often than might be thought. When we treat pulmonary emboli with anticoagulants, the relevant therapeutic endpoint is the nonoccurrence of another embolus, and the relevant toxic endpoint is abnormal bleeding. The first can only be "observed" by the *absence* of an event, and the second is too dangerous to permit it to occur. We therefore must and do substitute an intermediate endpoint, the prothrombin time, and attempt to control it. This is rational because we believe that the prothrombin time correlates, albeit imperfectly, with both efficacy and toxicity. Its advantage relative to the true clinical endpoints is that it is easy to observe, it is continuous rather than all-or-none, and its observation carries little risk for the patient.

At the risk of belaboring the point, consider another example. In the treatment of hypertension the therapeutic endpoint is prevention of stroke, kidney failure, blindness, heart failure and other ravages of the disease. The single toxic endpoint common to most antihypertensive drugs is excessive hypotension signalled by fainting, and so forth. Whereas the toxic endpoint can be directly observed, the therapeutic one is again the absence of a catastrophe. It does not provide any ongoing feedback concerning the efficacy of treatment. For hypertension, then, we do just as we do for pulmonary emboli, we choose an intermediate endpoint, the blood pressure, and control it. To do so is rational because the blood pressure correlates with the true endpoint, although, again, only imperfectly. If we may so use some laboratory test (the prothrombin time) or physical sign (the blood pressure), why not the drug level? According to the conceptual scheme above, the drug level *should* correlate with clinical effect, and in many instances empirical evidence suggests that it does (*4*). Like the prothrombin time, the drug level is relatively easy to measure and provides a continuous and graded intermediate endpoint. When used as an intermediate endpoint, drug levels are being used prospectively, to increase the probability of benefit and reduce the probability of harm. This prospective use is quite distinct from the retrospective use of drug levels for diagnosis, as discussed above. The prospective use is rational, when certain criteria are met, although the retrospective use often is not.

Consider the criteria for the rational use of drug levels as intermediate therapeutic endpoints. All of the following must simultaneously hold (*4*).

1. A target "effect" strategy is not possible, as was mentioned above. A target effect strategy is difficult or impossible in a number of circumstances: (a) when the true therapeutic endpoint is long delayed relative to the time available for accomplishing therapeutic regulation (as in hypertension); (b) when lack of efficacy can be confused with drug toxicity, as when a new arrhythmia appears during anti-arrhythmic therapy; (c) when the true endpoint is not sufficiently quantitative or graded to be useful for subtle, but necessary dosage adjustments; or (d) when measurement of the true objective is either difficult, risky, or excessively costly.

2. Another established intermediate endpoint (such as the prothrombin time) is not available.

3. The drug has a small therapeutic index. That is, in any given individual, the dosage that will produce toxicity is likely to be only moderately more than that required for efficacy. Moreover, the toxicity that may appear must be of some consequence. Thus, to keep within the narrow range between efficacy and dangerous

toxicity it is important to have fine control over dosage and a graded endpoint is required. In contrast, for a drug with a wide gap between therapeutic dosage and that causing consequential toxicity (for example, penicillin, in the absence of allergy to it, when used to treat streptococcal pharyngitis), we usually simply give somewhat more drug than the least responsive case would require, and do not carefully monitor either efficacy or toxicity.

4. Drug levels must directly relate to drug effect at most times. The relation between drug level and effect need not be perfect, however, just as the relation between blood pressure and stroke is not perfect. Some relations here, however, are so imperfect as to be useless. An example of such a relation between level and effect is that for a "hit and run" drug, where the drug irreversibly alters cellular function, as, for example, some anticancer agents do. For such a drug, the only drug level that might relate to drug effect would be the peak level; drug levels at other times would be irrelevant. Additionally, such drugs might have cumulative effects if cell recovery time were slow, so that even if the peak level were known, the drug's "effect" would be somehow dependent on all previous peak levels, and not only on the current one. These complications usually preclude prospective use of drug levels for such drugs. Thus, the only drugs for which levels have proven useful as an intermediate endpoint are those for which the levels relate to current effects, relatively independently of past effects. Moreover, for such drugs this relation holds most of the time during therapy, so that the *exact* time of drawing the level relative to the last previous dose is not crucial (but see below, for qualifications to this).

5. Interindividual variability, not intraindividual variability must account for the majority of pharmacokinetic uncertainty (see below).

6. The physician must know how to use the drug level properly to regulate dosage, and the laboratory must provide timely, accurate, and specific assay values to him.

The above criteria are quite restrictive and only a handful of drugs fulfill them: a number of antibiotics, antiseizure drugs, cardiac glycosides, antiarrhythmics, theophylline, and some psychoactive agents, anti-inflammatory and anticancer drugs. Yet these few drugs are quite important: they are responsible for the majority of dose-related adverse effects of drugs seen in hospital (*5*). It is, however, worth stressing that the target level strategy applies to very few drugs: if the strategy is inappropriately applied to drugs that do not fulfill the above criteria, drug levels will appear at best, useless, and at worst, dangerous.

Although the last requirement (No. 6) above seems obvious, many physicians do not use drug levels optimally. Inappropriate use of drug levels, even for "appropriate" drugs, can also render drug levels useless, or actually harmful. This subject will receive additional comment below. First, however, criterion 5 needs further explanation and demonstration.

## C. Sources of Pharmacokinetic Variability

Pharmacokinetics, recall,is the name for the portion of the overall dose-to-response relationship that describes the relationship between dosage and drug level. Pharmacodynamics describes the relationship between drug level and effect. I have already commented on the wide interindividual variability in pharmacodynamics. The difficulties caused by this variability are not reduced by knowledge of drug levels. There is also wide interindividual variability in pharmacokinetics. Drug levels can help here if, at the same time, intra-individual pharmacokinetic variability; that is, variability within a patient, day-to-day, is considerably smaller than variability between patients. If intraindividual variability is small, one can reasonably hope to learn about an individual's kinetics from measuring his drug levels, and to be able to use this knowledge to adjust dosage. This is because one may reasonably expect that future drug levels in the individual will bear the same proportionality to dosage as past ones have.

That this expectation, which is one of the bases of the use of drug levels for therapeutic monitoring (criterion number 5), is true is supported by considering the sources of interindividual variability in pharmacokinetics.

The determinants of the average drug level that will ultimately be obtained in an individual when he is put on a regular dosage schedule are two: absorption and elimination. The first can be measured as the fraction of each dose that ultimately reaches the systemic circulation, and the latter can be measured as drug clearance; the volume of plasma from which all the drug it contains would have to be removed to account for the actual loss rate of drug.

The fluctuation about the average drug level during an interdose interval will be proportional to the rate of fall of the drug level during the interval. This is measured by a drug's half-life, the time it takes for the concentration to fall by half. A drug's half-life is inversely proportional to the ratio of clearance to the apparent volume in which the drug is distributed. This volume is called the volume of distribution of the drug. Therefore this quantity, too, influences dosage decisions.

Variability in fraction absorbed, clearance and volume of distribution all contribute to interindividual variability in pharmacokinetics. Consider fraction absorbed: although this may be influenced by differences in tablet manufacture, most of this variability can be eliminated by using reliable preparations. Variability that cannot be eliminated so easily comes from the patient: some individuals have different gastrointestinal function from others. Some, for example, have achlorhydria. A drug such as penicillin, which is inactivated by gastric acid, will pass into the intestine intact in such a patient, but not in another.

Clearance will vary markedly from individual to individual. Genetic influences on the ability to metabolize drugs exist, as in fast and slow acetylators of isoniazid. One drug may induce the metabolism of another; an example is phenobarbital, which induces the metabolism of many drugs. Diseases of the drug clearing organs, such as cirrhosis of the liver or kidney failure, can cause major changes in the ability to clear various drugs. Patients vary in size, and eliminating capacity will also vary in parallel to this.

Finally, volume of distribution will also be a function of patient size. It will vary with the degree of binding of drug to plasma proteins, and may be influenced as well by certain pathologic states, such as uremia.

There are methods that allow approximate prediction of a patient's clearance or volume of distribution to be made as a function of patient size (weight), or renal function (as reflected by creatinine clearance). There are no commonly available or accepted tests that can be used to predict the influence of gastrointestinal function on absorption, or of liver disease on clearance to name only two examples.

Yet, most of the influences on drug kinetics I have mentioned are ones that, although they may change in a patient, can often be expected to remain relatively constant or change only slowly: liver or kidney failure often progresses slowly; other drugs are continued for protracted periods; chronic conditions such as achlorhydria are stable; and genetics never change. Thus, the premise above is generally supported: individuals vary from one another unpredictably with respect to pharmacokinetics, but often vary only slowly if at all, with respect to themselves. Criterion 5 can, therefore, often be satisfied, and drug levels can prove useful in therapeutic monitoring.

## D. Use and Misuse of Drug Levels

I return now to a fuller discussion of criterion number 6, knowledge of how to use drug levels. Before discussing some pitfalls, a brief indication of how they are generally used will be helpful.

Having chosen initial dosage, drug is administered for a period of time long enough so that the drug level can rise to its steady-state value. Normally, this is at least 4 or 5 half-lives of the drug in the patient. Although one may sample a drug level sooner than this—and indeed one is advised to do so in certain circumstances (*4*)—interpretation of these "early" drug levels is difficult, and beyond the scope of this chapter.

When the drug level is sampled at steady-state, it is a simple matter to use this level to adjust future dosage. The new dosage rate, in most cases, is simply chosen so that its ratio to the old dosage rate bears the same ratio as the target drug level (perhaps the center of the usual therapeutic range) bears to the measured drug level. The adjustment of dosage rate can be accomplished either by changing the dose amount in direct proportion to the ratio of target to observed level or by changing the dosing interval in inverse proportion to that ratio. The former adjustment method is almost always preferable. The proportional adjustment represents a logical way to adjust for the individual peculiarities in absorption and clearance that are present in the patient and that cause the drug level on the original dosage to differ unpredictably from the target level. Turnaround time for this use of drug levels need never be sooner than ca. 2 half-lives of the drug (see above). Since most drugs in clinical use have half-lives of at least 6 h, half-day turn around times are almost always quite acceptable, and full day times are usually so.

I may now point out some of the pitfalls in the use of drug levels when they are used in this manner. I do so to point out that often, when drug levels appear to mislead us, it may be that we have inadequate knowledge of how to use them rather than there being any intrinsic irrationality in their use.

The most major, and most obvious, pitfall in the use of drug levels concerns inaccuracy in their measurement or in the associated data necessary for their interpretation. Typical therapeutic plasma concentrations of drugs are usually less than 10 mg/L, and often less than 100 μg/L. Such low concentrations of almost anything are difficult to measure. Therefore, poorly equipped laboratories, or ones with less than meticulous quality control, or ones that rarely perform drug level assays are especially likely to perform them poorly. Obviously one can make no sense of a grossly inaccurate assay, and if one then acts on its basis, great harm can result.

Even if a drug assay is accurate, in order to make sense of it, past dosage must be known. If the dosage history is inaccurate, a wrong interpretation must result, and the drug level will appear to be misleading.

In addition to the assay and the dosage history, attention must be paid to the time of drawing the level in relation to the last previous dose. Drug levels drawn shortly after a dose tend to be higher than those drawn midway between doses or just prior to a dose, not only because drug is eliminated during a dosage interval, but because drugs are often rapidly absorbed, reach high concentrations in the blood, and then more slowly diffuse (distribute) out of the blood and into the tissues. Drug levels obtained before distribution is complete are difficult to interpret and usually cannot be used to assess how extensively a patient absorbs or eliminates a drug. This, however, is the most crucial information to be obtained from a drug level: if we know this, we can predict future levels resulting from various choices of future dosage, and may therefore rationally choose that dosage likely to result in the target level. Drug levels drawn too soon after a dose are therefore almost useless for dosage adjustment, and if they are used as if they had been drawn later, they will be misleading. This is not to say that drug level measurement soon after a dose may not correlate with drug effect at that time: it may or may not, depending on how rapidly drug equilibrates between blood and active site. Whichever is the case, such early levels almost never tell us very much about the important aspects of individual pharmacokinetics. Thus, although the *exact* time of drawing the level is not crucial, it should not be drawn too soon after a dose.

Somewhat less prevalent as a source of confusion in interpreting drug levels are certain pharmacokinetic peculiarities exhibited by some drugs. For example, certain drugs are extensively bound to serum proteins, often albumin. Drug level assays almost always measure total drug in blood or plasma, not unbound drug. However, it is usually the unbound drug that is free to diffuse to its active site, so that the unbound drug concentration in blood or plasma correlates with effect. Moreover, often only unbound drug is free to diffuse to its elimination site (usually in the liver or kidney) and so drug elimination rate is proportional to unbound drug concentration rather than total drug concentration. Now, in certain metabolic states, or when albumin is low, the extent of binding of certain drugs may be decreased. When this happens, and dosage is kept constant, drug elimination for these drugs will rise until unbound levels fall to their former value. The net effect is that the same unbound drug levels will be present as were present with normal binding, but at a lower total drug level. Since unbound drug levels are the same as usual, so is drug effect. An example of such a drug is phenytoin. Its binding is usually about 90%, but falls to 80% or less in uremia (*6*). If a physician were to measure a phenytoin plasma level in a uremic patient on a standard dose and find it to be one-third to one-half of the usual "therapeutic" value (10–20 mg/L), and raise the phenytoin dose for this reason alone, he might cause drug toxicity. The original, apparently subtherapeutic, drug level might well have been associated with a "therapeutic" unbound level.

The above is not the only example of pharmacokinetic complexity that must be appreciated in order to intelligently use drug level measurements. Some drugs exhibit nonlinear kinetics, meaning that changes in drug levels are not proportional to associated changes in dosage. The proportional adjustment approach previously outlined can have disastrous consequences if used for such a drug. Moreover, such drugs reach steady-state very slowly so one may falsely believe that the level is no longer rising (say 3 weeks after beginning or changing a regimen) when, in fact, it is, and will only reach toxic levels in the next few weeks, when vigilance has waned.

Some drug-level assays measure inactive metabolites along with active parent compound and some assays fail to measure active metabolites. Both situations obscure the relationship between level and effect.

The answer to such problems is to know about their possibility: ignorance misleads, accurate drug level measurements, properly interpreted, do not. If a patient receiving digitalis developed paroxysmal atrial tachycardia with 2:1 block, every alternate P-wave on the EKG could be hidden in the T-wave; or the physician could misinterpret the EKG for another reason. In either event, if the physician then continued giving digitalis and the patient died from digitalis toxicity, would we reject electrocardiograms in general as useless or harmful? The drug level can only be rejected as useless for therapeutic regulation in the context of a target level strategy if empirical evidence concerning intelligent and informed use in such a context reveals no benefit. In fact, available empirical evidence confirms theoretical expectation and indicates the contrary: drug levels can be of considerable benefit in this context.

### E. Empirical Results of Using Drug Levels for Therapeutic Monitoring

Some evidence is available for phenytoin (*7*), and for some other drugs, but the best studied example is digoxin. Initial studies suggested that the incidence of toxicity with this drug, using then current methods of administration, approached 20% (*8*). When a partial target level approach was used (initial dosage was adjusted as far as possible to compensate for expected pharmacokinetic differences between individuals, for example, by lowering dosage in the presence of renal disease), toxicity was decreased by about half, to ca. 12% (*9*). When a full target level strategy was used; that is, when initial doses were adjusted for expected individual pharmacokinetic peculiarities, and then subsequent drug levels were measured (and presumably used to ensure achievement of the target), toxicity fell by half again, to ca. 6% (*10*). These empirical results concerning drug effect confirm others that indicate that if drug levels are measured and dosage adjusted accordingly, target concentrations can be achieved reasonably accurately (*11*). Thus, theory and practice combine to support the use of drug level values as therapeutic targets, in appropriate circumstances. Greater use can be expected as more sensitive and precise drug assays become available (as discussed in this volume), and as drugs with narrower therapeutic indices are marshalled to combat the major intractible medical problems, such as cancer, against which we are now beginning to make encouraging headway.

## References

*1.* Lorch, J. A., and Garella, S., *Ann. Int. Med.* **91,** 301 (1979).
*2.* Holford, N., and Sheiner, L. B., *Amer. Heart J.* **94,** 529 (1977).
*3.* Ingelfinger, J. A., and Goldman, P. N., *Engl. J. Med.* **294,** 867 (1976).
*4.* Sheiner, L. B., and Tozer, T. N., "Clinical pharmacokinetics—the use of plasma concentrations of drugs," In Melmon, K. L., and Morrelli, H. F., eds., *Clinical Pharmacology*, 2nd ed., Macmillan, New York, 1978.
*5.* Melmon, K. L., *N. Engl. J. Med.* **284,** 1361 (1971).
*6.* Reidenberg, M. M., Odar-Cederlof, I., Von Bahr, C., Borga, O., and Sjoqvist, F., *N. Engl. J. Med.* **285,** 254 (1971).
*7.* Lund, L., *Arch. Neurol.* **31,** 289 (1974).
*8.* Beller, G. A., Smith, T. W., Abelmann, W. H., Haber, E., and Hood, Jr., W. B., *N. Engl. J. Med.* **284,** 989 (1971).
*9.* Ogilvie, R. I., and Ruedy, J., *JAMA* **222,** 50 (1972).
*10.* Koch-Weser, J., Duhme, D. W., and Greenblatt, D. J., *Clin. Pharmacol. Ther.* **16,** 284 (1974).
*11.* Sheiner, L. B., Halkin, H., Peck, C., Rosenberg, R., and Melmon, K. L., *Ann. Int. Med.* **82,** 614 (1975).

# Chapter 5

# Anticonvulsants

Pokar M. Kabra, Brian E. Stafford,
Donna M. McDonald, and Laurence J. Marton

*Department of Laboratory Medicine*
*School of Medicine*
*University of California*
*San Francisco, California*

## I. Introduction

Until recently, it was difficult to explain why an identical drug dosage may exert a toxic effect in one patient and a therapeutic, or no, response in another patient. It has now been demonstrated that the concentration of drug at the tissue receptor site is the most important parameter for adjusting drug dosage. However, the concentration of drug at the receptor site cannot be measured directly, and must therefore be correlated with the concentration of drug in body fluids in contact with these tissue receptors. The ability to correlate plasma drug concentration with tissue concentrations and with therapeutic or toxic effects has enabled those interested in optimizing drug dosages to generate much of the information needed to make useful therapeutic decisions. The exact nature of drug–receptor interactions is not firmly established. It is, however, known that the concentration of free drug correlates best with therapeutic effect. The free drug in plasma is in equilibrium with protein-bound drug and, because of this relationship, it is important to understand to what extent a particular drug is protein-bound. Various attempts to measure free drug concentration in saliva (*1*) and cerebrospinal fluid (*2*) have been reported.

Numerous clinical studies have adequately demonstrated the importance of total plasma or serum concentrations of a drug in relation to its efficacy. Measurement of plasma levels gives the clinician a better handle on drug dosage and scheduling in the presence of such complicating factors as age, sex, compliance, metabolism, absorption, and drug–drug interactions.

Anticonvulsants are no exception to these rationales for therapeutic drug monitoring. Modern analytical techniques have been valuable in identifying noncompliance, individual variations in drug disposition, altered utilization owing to the disease state, and altered physiological condition in patients receiving anticonvulsant drugs. With the advent of such techniques as gas–liquid chromatography (GLC), radioimmunoassay (RIA), enzyme multiplied immunochemical technique (EMIT), and liquid chromatography (LC), plasma levels of anticonvulsant drugs can be measured routinely, resulting in improved patient care.

Figure 1 shows the structures of the majority of anticonvulsant drugs. It is evident from this figure that most anticonvulsant drugs belong to five main classes. These include the following:

1. Barbiturates: phenobarbital, mephobarbital, primidone (desoxybarbiturate).
2. Hydantoins: phenytoin, mephenytoin.
3. Succinimides: ethosuximide, methsuximide, phensuximide.

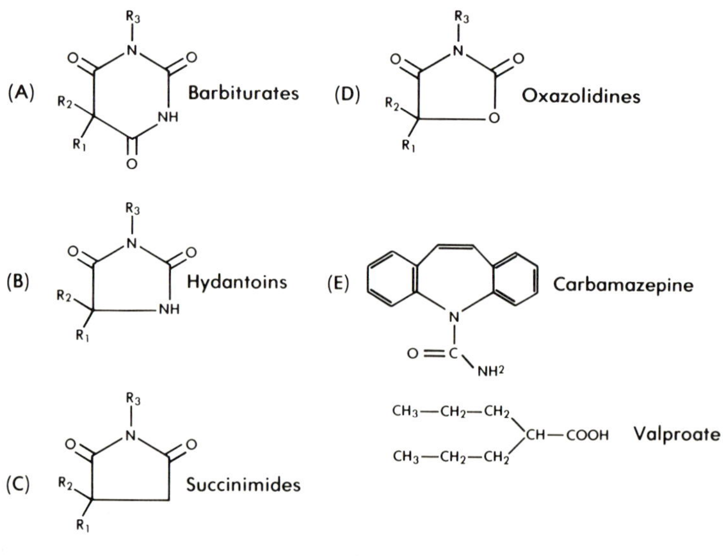


FIG. 1. Molecular structure of common anticonvulsant classes.

4. Oxazolidines: trimethadione, paramethadione.
5. Miscellaneous: carbamazepine, valproate.

Table 1 lists the principal uses of many of these anticonvulsants in different epileptic disorders, their serum half-lives, the usual dose required to obtain therapeutic levels in plasma, and the levels at which

Table 1
Properties of the Anticonvulsant Drugs

| Drug | Use | Serum half-life, h | Dose, mg/day | Therapeutic range, mg/L | Toxic level, mg/L |
|---|---|---|---|---|---|
| Carbamazepine | Trigeminal neuralgia grand mal, psychomotor | 12 (± 3) | 800–1200 | 2–10 | 12 |
| Ethosuximide | Petit mal | 45 (± 15) | 500–1000 | 40–100 | 100 |
| Mephenytoin | Grand mal and focal seizures | 34 | 200–600 | | |
| Metabolite: 5-phenyl-5-ethylhydantoin (Nirvanol) | Sedative (abandoned in 1920 owing to its toxicity) | 72 | | 15–40 (combined) | 40 |
| Mephobarbital | Grand mal | 24 (demethylated to phenobarbital) | 200–400 | Relative to phenobarbital level | |
| Valproic acid | Petit mal | 8–15 | 1400 | 60–100 | |
| Methsuximide | Psychomotor, petit mal | 6 (± 2) | 600–1200 | 10–40 (combined) | 40 |
| Metabolite: α-methyl-α-phenylsuccinimide (*N*-desmethyl methsuximide) | | | After ½ hour the concentration is much higher (as much as 700 times the concentration of methsuximide) | | |
| Phenobarbital | Grand mal | 96 (± 12) | 10/kg load | 10–40 | 40 |
| Primidone | Grand mal | 7.5 (± 2.5) | 750 | 2–10 | 12 |
| Metabolite: phenylethylmalonamide (PEMA) | 36 (± 12) | | | Same or higher than primidone | |
| Phenytoin diphenylhydantoin) | Grand mal, psychomotor | 24 (± 12) | 300 | 10–20 | 20 |

toxic symptoms are observed. A number of bioactive metabolites are also listed in this table.

A variety of analytical techniques have been employed for the analysis of the anticonvulsants. Many of the initial studies on the therapeutic or toxic effects of these drugs were facilitated by the use of ultraviolet spectroscopy. The 5,5-disubstituted barbiturates in particular have been analyzed by spectrophotometric methods, such as the one developed by Goldbaum (*3*). Some of the problems associated with this technique are the inability to distinguish one barbiturate from another, the requirement for a large amount of sample, and the potential interference of other compounds that may co-extract.

Thin layer chromatography (TLC) possesses adequate resolution for identifying many of the anticonvulsants, but suffers from the inability to accurately and precisely quantitate these drugs. It is a laborious and time-consuming technique with inadequate sensitivity for therapeutic monitoring. TLC, nevertheless, remains a very useful technique in the toxicology laboratory.

Radioimmunoassay is sensitive and reasonably precise, but requires the use of radionuclides. Cross-reactivity with other closely related drugs and metabolites is a potential problem with this technique.

EMIT offers some advantages over RIA, in that no radioactive tracer is required and there is no need to separate the bound and unbound fractions. The technique can also be automated (e.g., on a centrifugal analyzer), thus increasing throughput. However, the potential for cross-reactivity still exists. For example, 5-(4-hydroxyphenyl)-5-phenylhydantoin, a major metabolite of phenytoin, can interfere with the EMIT analysis of phenytoin in patients with renal failure (*4*). Another potential pitfall in this technique is enzyme inhibition occurring secondarily to the presence of inhibitors in the sample. Nevertheless, good correlation between EMIT and chromatographic techniques has been noted (*5, 6*), and EMIT assays for anticonvulsants have found their place in many laboratories throughout the world.

GLC is a very popular technique for the analysis of anticonvulsants in clinical laboratories. GLC packings have been specifically developed for anticonvulsants, and it is possible to analyze a number of drugs simultaneously without derivatization (*7*). With the development of the sensitive nitrogen–phosphorus detector, it is also possible to analyze microsamples.

Gas–liquid chromatography coupled with mass spectrometry is the most specific technique for monitoring anticonvulsants. However, the complexity and cost of available instrumentation is rather prohibitive of its routine use in a clinical laboratory.

High-performance liquid chromatography offers several advantages over the other techniques. Sample manipulation prior to chromatography is minimal, and several classes of compounds, including their metabolites, can be analyzed simultaneously with good specificity, precision, and accuracy. A distinct advantage of LC is its nondestructive nature. Compounds may be collected from the column effluent for further analysis and characterization.

## II. Collection and Preparation of Samples

Current analytical methods used to routinely monitor anticonvulsant drugs in serum or plasma do not distinguish between bound and unbound drugs. As indicated previously, it is generally accepted that only the unbound (free) drug freely diffuses into tissues and actually manifests control of seizures. It would therefore seem more reasonable to correlate free drug concentration with clinical efficacy or toxicity. Unfortunately, the techniques commonly employed for the measurement of free drug in plasma require ultrafiltration, dialysis, or ultracentrifugation. These techniques are laborious, time consuming, and require specialized apparatus, rendering them impractical for routine use.

Free levels for some of these drugs may be estimated by the analysis of saliva or CSF. Lumbar puncture, required for the collection of CSF, is associated with some risks and hence is not a routine procedure. Although collection of saliva is non-invasive, saliva does not appear to be a suitable fluid for the determination of drugs that are ionized at the pH of saliva (e.g., phenobarbital) (*8*). For some anticonvulsants, such as phenytoin, primidone, and ethosuximide, measurement of saliva concentration can be used as an alternative to the measurement of total drug in plasma (*1*).

It should also be recognized that certain drugs, for instance phenytoin, may bind to red cells in addition to plasma proteins. In certain pathological conditions, such as uremia, alcoholic cirrhosis, and salicylate therapy, the amount of phenytoin bound to red blood cells versus that in plasma may increase substantially, further complicating the interpretation of total plasma or serum concentrations of the drug (*9*).

Extraction methods for the isolation and concentration of anticonvulsants from the specimen may vary from a simple one-step solvent extraction, to complicated back extractions, column extractions, and charcoal adsorption techniques. Generally, the type of extraction and the amount of sample cleanup is dictated by the efficiency and the selectivity of the chromatographic technique used for analysis. The more specific and efficient the chromatographic system, the less sample extraction and cleanup are necessary to obtain the desired results. Sometimes extraction steps are also necessary to improve the sensitivity of the assay method by concentrating the analyte.

In the reported micro extraction procedure of Adams et al. (*10*), 50 μL of serum is extracted with 0.5 mL of chloroform containing the internal standard at pH 8.0. After phase separation, the chloroform is evaporated at 40°C and the residue reconstituted with 20 μL of methanol before chromatography. This simple extraction eliminates most of the proteins, and acidic and basic compounds. Excellent recoveries were reported for several anticonvulsant drugs.

We have reported (*11, 12*) a single-step chloroform extraction method for both macro (2 mL) and micro (100 μL) samples at acidic pH for phenobarbital, primidone, phenytoin, and carbamazepine. Recoveries of 95–110% were obtained for these drugs.

Atwell et al. (*13*) reported a single-step methylene dichloride extraction, from 0.5 mL of plasma at neutral pH (6.8) for phenytoin and phenobarbital. Most single-step solvent extraction methods are adequate for normal phase or reverse-phase liquid chromatographic analysis.

A charcoal adsorption method was utilized by Adams et al. (*14*) to isolate several anticonvulsants and their metabolites from serum. The adsorbed drugs were removed from the charcoal surface by a ternary solvent system composed of dichloromethane, isopropanol, and diethylether. Though the reported precision of their overall liquid chromatographic analysis, utilizing an octadecyl stationary phase, was adequate for routine use, analytical recoveries ranged from 12 to 92%. The method was sensitive enough to detect 0.5 mg/L of anticonvulsant drugs when 0.5 mL of serum was extracted.

Disposable extraction columns, such as Clin Elut (Analytichem International) (*15*) or XAD-2 (Rohm-Haas) (*16*) can be used to extract a number of anticonvulsants from serum or plasma. The Clin Elut extraction method is based upon the principle of liquid–liquid extraction. The sample is poured onto the top of a column that has been prepacked with an inert matrix of large surface area. A suitable organic solvent is used to elute the anticonvulsants, while the polar components, particulate matter, and other impurities are retained on the column. These columns are very versatile and easy to use for the extraction of several classes of drugs.

Protein precipitation is probably the simplest and most rapid method of sample preparation prior to LC analysis. The basic principle of this technique is to precipitate serum proteins in the sample using a water-miscible organic solvent such as acetonitrile, methanol, or ethanol. Though the proteins are precipitated, the drugs remain freely solubilized in the supernatant. Probably the best solvent for this purpose is acetonitrile, because it yields a clear supernatant in a short period of time. It is recommended that the sample be centrifuged at high speed, preferably greater than 10,000*g* to pack the protein into a tight pellet at the bottom of the sample tube and to assure a clear supernatant. The procedure involves pipeting 15–500 μL of serum into a tube to which an equal, or double, the volume of acetonitrile is added; the sample is vortexed and centrifuged to obtain a clear supernatant. The addition of internal standard to the acetonitrile facilitates quantitation. The major advantage of this sample preparation technique is its simplicity and speed. It is, however, advisable to use a pre-column packed with macrosize packing material to assure that unprecipitated proteins and polypeptides do not enter the analytical column.

An automated sample extraction system (Dupont Prep I) can process several samples simultaneously. This automated sample processor utilizes a solid-phase column extraction technique using an inert hydrophobic sorption resin to extract the materials of interest. The sample matrix containing the analyte(s) is pipetted into the extraction column reservoir, buffered, and then placed into the rotor cup. The instrument automatically dispenses water onto the extraction column, washing away interfering and contaminating compounds into an effluent cup. Acetone is then automatically dispensed to elute the drugs of interest into a recovery cup. Acetone and residual water from the recovery cup are evaporated, and the extracted drugs reconstituted and chromatographed.

## III. Chromatography

Since normal phase columns (packed with silicic acid) were the first to be introduced in the early 1970s, most early investigators utilized these columns for the analysis of anticonvulsant drugs. And because silicic acid is a polar packing material that has active hydroxyl groups on its surface, it generally retards polar compounds, while the nonpolar compounds are eluted with ease. Polar organic solvents such as methanol, isopropanol, or water are used, either alone or in combination, to elute the polar compounds. Acids or bases may be used in the mobile phase to affect possible secondary equilibria, such as ionization of the solute, thereby making the solute more or less polar to increase or decrease its retention.

Normal phase methods require more extensive sample cleanup prior to chromatography than reverse phase methods because the aqueous nature of biofluids and the normal presence of polar constituents are not compatible with the silicic acid packing material. These polar constituents, including some of the metabolites, are retarded on the column and may interfere with the subsequent analysis. Therefore, all the analytes must be isolated from the aqueous matrix prior to chromatography.

Atwell et al. (*13*) used a silicic acid column to analyze phenobarbital and phenytoin utilizing 0.5 mL of serum. These drugs were isolated prior to chromatography by a simple solvent extraction. The mobile phase, consisting of chloroform, dioxane, isopropanol, and acetic acid (310/9/7/0.1 by volume) was pumped at 1.5 mL/min to elute these drugs. Total chromatographic time was approximately 7 min.

We (*11*) analyzed primidone, phenytoin, and phenobarbital on a silicic acid column utilizing a mobile phase consisting of chloroform, methanol, and ammonia (95/4.5/5/0.5 by volume) at a flow rate of 1.0 mL/min. These drugs were isolated by a simple one-step chloroform extraction. Total chromatographic time was less than 10 min for all three anticonvulsants (Fig. 2).

A multistep extraction method to isolate the anticonvulsants phenytoin and phenobarbital prior to chromatography on silicic acid column was reported by Evans (*17*). This multistep extraction was employed to eliminate interfering neutral and basic constituents present in serum.

Among all of the factors contributing to the surge of interest in clinical analysis by LC, none stands out as prominently as the development of reversed phase chromatography using *N*-alkyl chemically bonded phases. In reversed phase chromatography, the stationary phase (typically hydrocarbonaceous) is less polar than the mobile phase (typically water/methanol or water/acetonitrile mixtures). Substances thus elute in a general order of decreasing polarity. Mobile phase strength increases with decreasing polarity (e.g., increasing acetonitrile concentration in mobile phase).

Bonded phases are typically made by reacting the appropriate chloro- or alkoxysilane with a fully hydroxylated silica gel. The role of

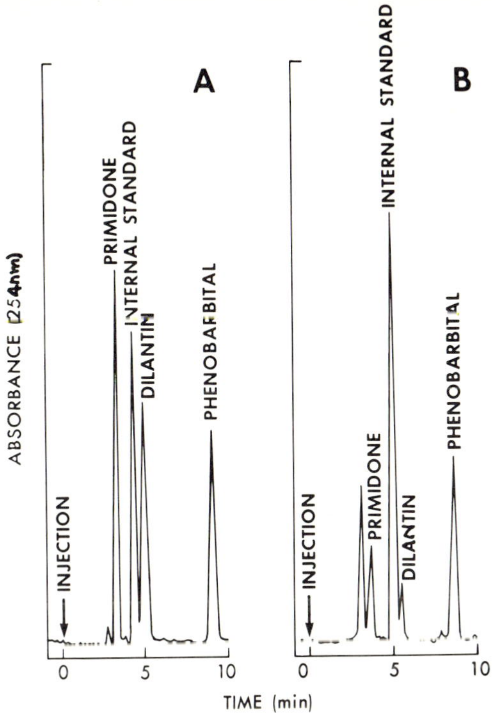


FIG. 2. Adsorption liquid chromatography of primidone, diphenylhydantoin (dilantin), and phenobarbital. Trace A shows a chromatogram from whole blood of a patient. Reprinted from reference *11* by courtesy of American Association for Clinical Chemistry.

the alkyl chain length in separation is not fully understood at the present time. However, it has been observed that retention of solute increases with chain length, and that column capacity also increases with longer carbon chains.

There are a number of favorable aspects of reversed phase chromatography in clinical analysis. The most important being the aqueous nature of the mobile phase, which is compatible with biofluids. The ability to separate a wide range of substances of varying hydrophobicities is particularly useful in the complex mixture of compounds found in biofluids. Other useful features of reversed phase chromatography are rapid analysis, and rapid equilibration following gradient elution or mobile phase alteration. These features are a consequence of the weak surface energies of the bonded alkyl phases.

Adams et al. (*14*) reported the simultaneous analysis of several anticonvulsant drugs in serum using an octadecylsilane reversed phase column. Their mobile phase, consisting of acetonitrile/water (17/83 by volume), separated phenobarbital, phenytoin, primidone, ethosuximide, methsuximide, and carbamazepine in about 14 min at 65°C. These drugs were detected at 195 nm. The charcoal adsorption technique utilized for the extraction of these drugs resulted in poor recovery; subsequently, they used a simple chloroform extraction for improved recovery. Later the method was expanded to include other minor anticonvulsants (*29*).

Soldin et al. (*18*) described the use of reversed phase LC for the microanalysis of five anticonvulsant drugs (ethosuximide, primidone, phenobarbital, phenytoin, and carbamazepine) in 25-$\mu$L serum samples. They also used an octadecylsilane column, which was eluted with an equivolume mixture of phosphate buffer (10 mM, pH 8.0) and acetonitrile at a flow rate of 0.8 mL/min. The serum supernatant prepared by the acetonitrile precipitation method was injected into the chromatograph and the effluent monitored at 200 nm. However, baseline resolution was not obtained, and the method did not separate phenylethylmalonamide (a bioactive metabolite of primidone) from the parent drug, resulting in an overestimation of primidone. In addition, the use of a pH 8.0 mobile phase is detrimental to silica-based columns.

In our laboratory, a reversed-phase LC method for the analysis of five major anticonvulsant drugs was developed in 1976 (*19*). The method incorporated the simple acetonitrile precipitation step, followed by injection of the serum supernatant. The five anticonvulsant drugs were resolved with baseline separation in 15 min using an octadecylsilane reversed-phase column with a mobile phase consisting of phosphate buffer, pH 4.4, and acetonitrile (19/81 by volume), at a flow rate of 3.0 mL/min (Fig. 4). The use of an acidic pH buffer, and an elevated temperature of 50°C, facilitated the chromatographic resolution.

The method was accurate, precise, and reasonably fast. No significant interference from endogenous or exogenous components was noted. We subsequently extended out method to include a total of twelve anticonvulsant drugs and their major bioactive metabolites (*20*) (Fig. 3).

Kitazawa et al. (*21*) reported an anion exchange chromatographic method for the analysis of phenytoin, phenobarbital, and carbamazepine. These drugs were extracted from 0.5 mL of serum utilizing a multistep extraction method. The chromatography was not selective and several unknown peaks eluted in the regions of interest.

The use of elevated temperature has a marked effect on the retention time and resolution of anticonvulsants on octadecylsilane

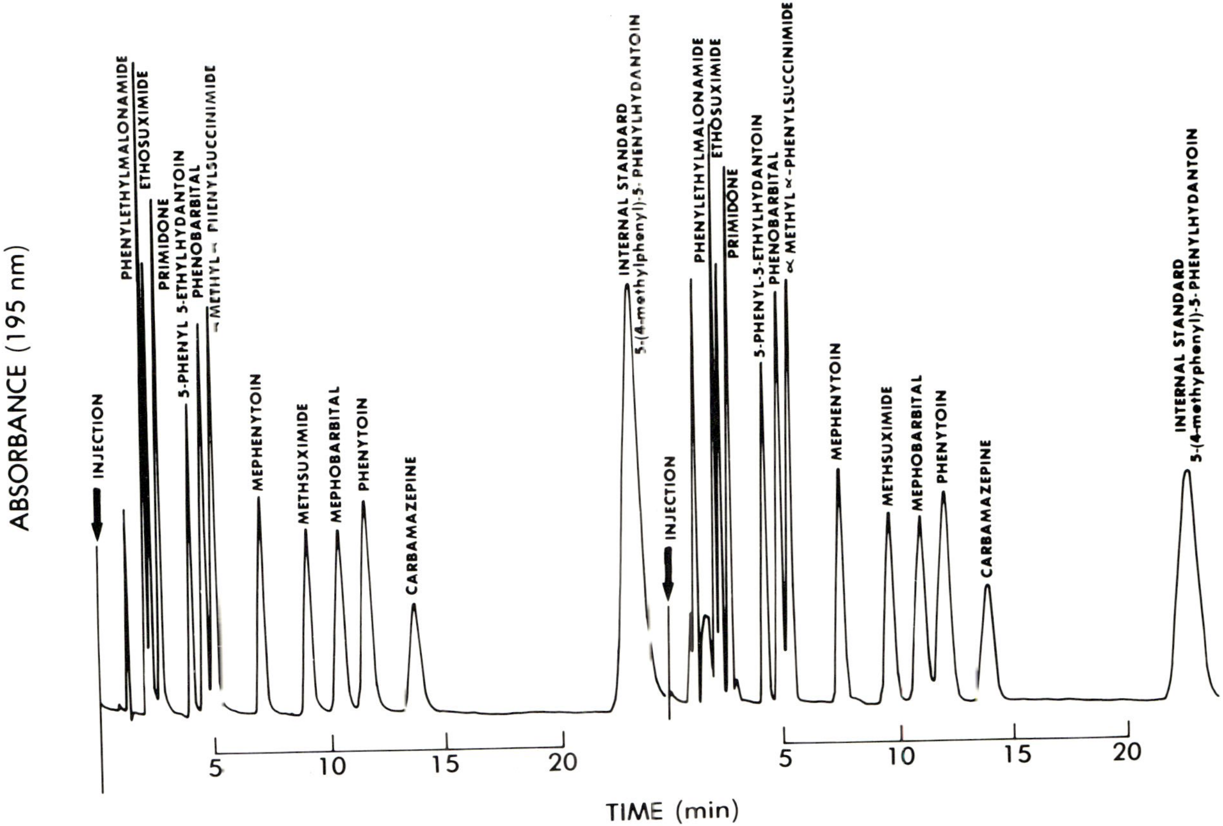


FIG. 3. Left: Chromatogram of a standard mixture of drugs. Quantity chromatographed was 100 ng of each drug, except the internal standard, which was 500 ng. Right: Chromatogram of serum supplemented with 25 mg/L of each drug, except internal standard, which was 50 mg/L. Reprinted from reference *20* by courtesy of *Journal of Analytical Toxicology*.

reversed phase columns. There are several advantages to the use of elevated temperature in reverse phase LC. Elevated temperature helps reduce the viscosity of the mobile phase, resulting in a lower pressure at a given flow rate. Elevated temperature also facilitates the mass transfer properties of the analyte between the mobile phase and the stationary phase. This usually results in an increase in chromatographic efficiency. Figure 4 illustrates the analysis of five anticonvulsant drugs at 50° C using the method of Kabra et al. (*19*). The same analysis, when performed at ambient temperature, requires 19 min (Fig. 5). It is evident from comparing these two chromatograms that though all five major anticonvulsants are resolved with baseline separation at 50° C, phenytoin and carbamazepine coelute at ambient temperature. Thus we see that elevated temperature can effect selectivity. Besides shortening analysis time, elevated temperature results in better reproducibility of retention times. Since drugs usually

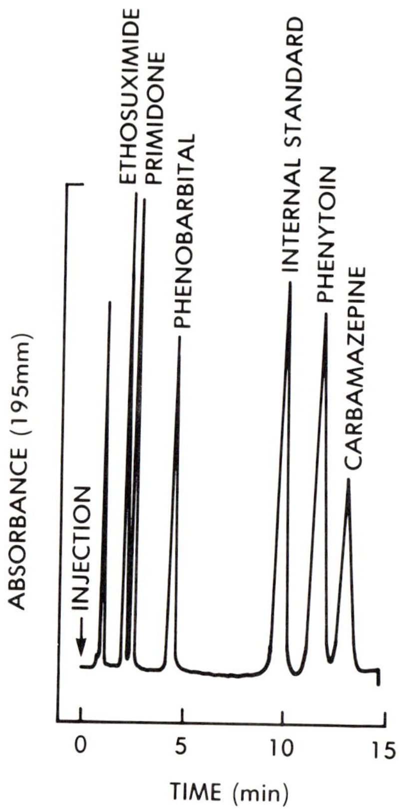


FIG. 4. Reversed-phase chromatographic analysis of anticonvulsants at 50° C.

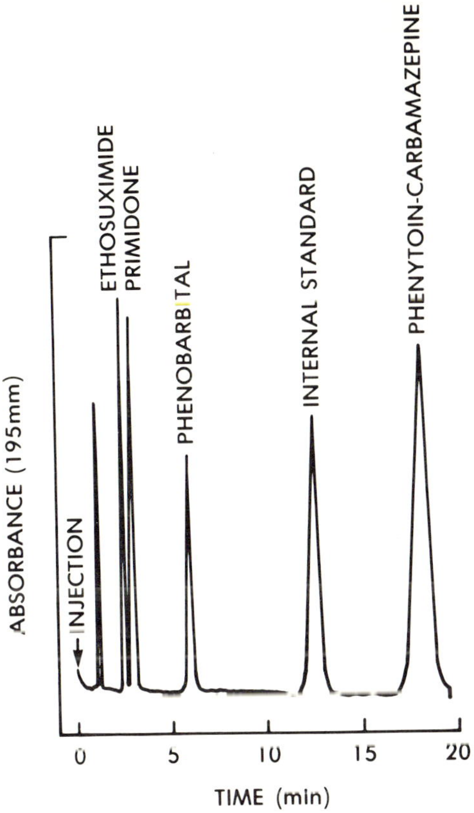


FIG. 5. Reversed phase analysis of anticonvulsants at ambient temperature.

are identified on the basis of their relative or absolute retention times, reproducibility of retention times is critical, especially when data handling is performed by an integrator or computer. Table 2 summarizes different LC methods for the analysis of anticonvulsants.

## IV. Detection and Quantitation

Virtually all LC methods for the analysis of anticonvulsants employ ultraviolet detectors to monitor the effluent. Many anticonvulsant drugs can be detected at 254 nm if they are concentrated and isolated from biofluids prior to chromatography. Barbiturates, as a class, possess very poor UV absorption at 254 nm when eluted with a neutral or an acidic mobile phase. This creates a problem in the LC analysis, because, as previously mentioned, a basic mobile phase is frequently associated with the dissolution of silica-based stationary phases. Since there is no concentration step involved in the protein precipitation technique, a fixed-wavelength 254-nm detector cannot be used. In addition, anticonvulsants, such as ethosuximide and primidone, have very poor UV absorption at 254 nm.

It is well known that many chemical compounds having little or no absorption in the near UV region may possess fairly strong absorption in the region below 200 nm. This phenomenon is commonly referred to as "end absorption" and generally results from the $n$–$\sigma^*$ and $\pi$–$\pi^*$ energy transitions. Thus, compounds containing an isolated double bond as a chromophore have strong absorption in the region near 200 nm, but are transparent at higher wavelengths. In addition, the presence of auxochromes, such as —OH, —$NH_2$, —S, or halogen, in a molecule may lead to $n$–$\sigma^*$ transitions that permit detection in the region at or below 200 nm. Though most aromatic compounds possess good absorption at 254 nm, the sensitivity can be greatly enhanced if the detector is set at 220 nm or lower. The detection of anticonvulsants at 195 nm allowed us to analyze these drugs without the usual sample concentration steps. Samples as low as 10 ng of anticonvulsant drug could be detected at 195 nm. This enables one to detect 1 mg/L of anticonvulsant drugs when 10 $\mu$L of serum supernatant is injected into the chromatograph.

Although sensitivity is greatly enhanced at 195 nm, there are several limitations with detection in this region. First, the mobile phase must be transparent at these wavelengths. A mobile phase that usually conforms to these criteria of transparency is a mixture of phosphate buffer and glass-distilled acetonitrile (UV cutoff $<$ 190 nm). Other solvents, such as methanol, ethanol, tetrahydrofuran, and acetate buffer, cannot be used as the mobile phase below 200 nm. Additionally, the detection wavelength is fairly nonspecific; hence, the resolution must be adequate to separate potentially interfering substances from the analyte.

Derivatization is a powerful technique to enhance the detectability of compounds that otherwise could not be detected by UV detectors. This technique was applied for the analysis of valproic acid in serum by Sutheimer et al. (*22*). Valproic acid cannot be detected without derivatization even in the far UV range. Phenacyl ester derivatization provides a suitable UV chromophore for detection at 254 nm.

Quantitation can be accomplished either by peak height or peak area measurements, provided the chromatographic peaks are Gaussian or symmetrically shaped. Internal standardization is the preferred method whether extraction, derivatization, or simple protein precipitation methods are used in sample preparation. An internal standard will usually compensate for a number of variables in the methodology. The injection of a reference standard containing the analytes and internal standard in known concentrations, and the subsequent normalization of the peaks to obtain a response factor, is a suitable means of calibrating a method. A number of data systems that automatically accomplish this task are available.

## V. Stability of Columns

Most LC analytical columns cost are relatively costly (approximately $300). A properly packed reversed-phase LC column should last through several months of routine use, if proper care is exercised in its maintenance. The pH of the mobile phase should not deviate from the range of 2.0–7.5, to prevent dissolution of the silica. The mobile phase should be suitably filtered to prevent the accumulation of particulate material on the column. If a protein precipitation technique is employed, it is advisable to use a precolumn packed with the same type of packing material that is in the analytical column. A 50 × 2.1 mm precolumn dry-packed with 25–40 μm octadecylsilane material has been used with good success in our laboratory. The precolumn increases the useful life-span of the analytical column. The precolumn should be changed every 4–5 weeks.

## VI. Metabolites

A number of anticonvulsant drugs are transformed into bioactive metabolites. For instance, carbamazepine is metabolized into carbamazepine-10,11-epoxide, which was found to be equipotent to carbamazepine in rats (*23*). It was also reported that the dosage of carbamazepine is better correlated with the levels of the epoxide metabolite than with the parent drug. Mephobarbital is almost completely metabolized into phenobarbital. Primidone is converted into phenobarbital and phenylethylmalonamide (PEMA), both of which exhibit anticonvulsant activity (*24*). Phenytoin is metabolized to 5-(4-hydroxyphenyl)-5-(phenylhydantoin (HPPH) in the liver. In cases of malabsorption or unusual metabolism of phenytoin, measurement of urinary HPPH can provide useful information. An LC method for the determination of urinary HPPH was reported from our laboratory (*25*).

Table 2
Review of LC Procedures That Can Monitor Anticonvulsant Drugs at Therapeutic Concentrations

| Drug | | Sample preparation | Column[a] | Detection | Remarks | Ref. |
|---|---|---|---|---|---|---|
| Carbamazepine | (a) | 200 μL serum + 200 μL acetonitrile | $C_{18}$RP | UV 195 nm | 50° C (oven temp.) | *19, 20* |
| | (b) | 300 μL supernatant extracted with chloroform | $C_{18}$RP | UV 195 nm | 50° C (oven temp.) | *20* |
| | (c) | Automated extraction of 150 μL serum | $C_8$RP | UV 200 nm | Technicon Autoanalyzer interfaced with LC | *26* |
| | (d) | 25 μL serum + 25 μL of acetonitrile | $C_{18}$RP | UV 200, UV 254 nm | pH 8.0 mobile phase 30° C oven temp. | *18* |
| | (e) | 500 μL plasma extracted with chloroform | AE | UV 254 | Two analytical columns in series | *21* |
| | (f) | 200 μL serum extracted with butyl chloride and isobutanol | $C_{18}$RP | UV 254 | Ion pair with hexane sulfonic acid | *27* |
| | (g) | 0.5–1.0 mL plasma extracted with chloroform | SG | UV 250 | | *28* |
| | (h) | 2 mL plasma extracted with chloroform | $C_{18}$RP | UV 254 | | *12* |

| | | | | | | |
|---|---|---|---|---|---|---|
| | (i) | 1.0 mL plasma extracted with dichloromethane and back extraction with 0.1 *M* NaOH | SG | UV 254 | | *23* |
| | (j) | 50–100 μL serum extracted with chloroform | $C_{18}$RP | UV 195 | 65° C (oven temp.) | *29* |
| | (k) | Charcoal adsorption, 0.5 mL serum | $C_{18}$RP | UV 195 nm | 40° C (oven temp.) | *14* |
| | (l) | 200 μL serum + 400 μL acetonitrile | $C_{18}$RP | UV 200 nm | Methanolic mobile phase, 55° C (oven temp.), pH 6.8 mobile phase | *30* |
| Carbamazepine-10,11-epoxide | (a) | Automated extraction of 150 μL serum | $C_3$RP | UV 200 nm | Technicon autoanalyzer interfaced with LC | *26* |
| | (b) | 0.5–1.0 mL plasma extracted with dichloromethane | SG | UV 250 nm | | *28* |
| | (c) | 1.0 mL plasma extracted with dichloromethane and back extraction with 0.1 m NaOH | SG | UV 254 nm | | *23* |

*(continued)*

Table 2 continued

| Drug | | Sample preparation | Column[a] | Detection | Remarks | Ref. |
|---|---|---|---|---|---|---|
| | (d) | 50–100 μL serum extracted with chloroform | $C_{18}$RP | UV 195 nm | 65° C oven temp. | *29* |
| Ethosuximide | (a) | 200 μL serum + 200 μL acetonitrile | $C_{18}$RP | UV 195 nm | 50° C oven temp. | *19, 20* |
| | (b) | 300 μL supernate extracted from (a) with chloroform | $C_{18}$RP | UV 195 nm | 50° C oven temp. | *20* |
| | (c) | 50 μL serum extracted with chloroform | $C_{18}$RP | UV 195 nm | 65° C oven temp. | *29* |
| | (d) | 25 μL serum + 25 μL acetonitrile | $C_{18}$RP | UV 200 nm, UV 254 oven temp. | pH 8.0 mobile phase; 30° C | *18* |
| | (e) | 0.5 mL serum charcoal adsorption | $C_{18}$RP | UV 195 nm | 40° C oven temp. | *14* |
| | (f) | 200 μL serum + 400 μL acetonitrile | $C_{18}$RP | UV 200 nm 55° C oven temp., pH 6.8 mobile phase | Methanolic mobile phase, | *30* |
| Mephenytoin | (a) | 200 μL serum + 200 μL acetonitrile | $C_{18}$RP | UV 195 nm | 50° C oven temp. | *20* |

| | | | | | | |
|---|---|---|---|---|---|---|
| | (b) | 300 μL supernatant from (a) extracted with chloroform | $C_8$RP | UV 195 | 50° C oven temp. | *20* |
| | (c) | 50 μL serum extracted with chloroform | $C_8$RP | UV 195 nm | 65° C oven temp. | *29* |
| 5-Ethyl-5-phenyl hydantoin (Nirvanol) | (a) | 200 μL serum + 200 μL acetonitrile | $C_8$RP | UV 195 nm | 50° C oven temp. | *20* |
| | (b) | 300 μL supernatant from (a) extracted with chloroform | $C_8$RP | UV 195 nm | 50° C oven temp. | *20* |
| Methsuximide | (a) | 200 μL serum + 200 μL acetonitrile | $C_8$RP | UV 195 nm | 50° C oven temp. | *20* |
| | (b) | 300 μL supernatant from (a) extracted with chloroform | $C_8$RP | UV 195 nm | 50° C oven temp. | *20* |
| | (c) | 0.5 mL serum, charcoal adsorption | $C_8$RP | UV 195 nm | 40° C oven temp. | *14* |
| | (d) | 50 μL serum extracted with chloroform | $C_8$RP | UV 195 nm | 65° C oven temp. | *29* |
| 2-Methyl-2-phenyl succinimide | (a) | 200 μL serum + 200 μL acetonitrile | $C_8$RP | UV 195 nm | 50° C oven temp. | *20* |
| | (b) | 300 μL supernatant from (a) extracted with chloroform | $C_{18}$RP | UV 195 nm | 50° C oven temp. | *20* |

*(continued)*

Table 2 continued

| Drug | | Sample preparation | Column[a] | Detection | Remarks | Ref. |
|---|---|---|---|---|---|---|
| Mephobarbital | (a) | 200 μL serum + 200 μL acetonitrile | $C_{18}$RP | UV 195 nm | 50° C oven temp. | *20* |
| | (b) | 300 μL serum supernatant from (a) extracted with chloroform | $C_{18}$RP | UV 195 nm | 50° C oven temp. | *20* |
| | (c) | 50 μL serum extracted with chloroform | $C_{18}$RP | UV 195 nm | 65° C oven temp. | *29* |
| Phenobarbital | (a) | 200 μL serum + 200 μL acetonitrile | $C_{18}$RP | UV 195 nm | 50° C oven temp. | *19, 20* |
| | (b) | 300 μL serum supernate extracted from (a) with chloroform | $C_{18}$RP | UV 195 nm | 50° C oven temp. | *20* |
| | (c) | 50 μL serum extracted with chloroform | $C_{18}$RP | UV 195 nm | 65° C oven temp. | *29* |
| | (d) | automated chloroform extraction 150 μL of serum | $C_8$RP | UV 200 nm | Technicon Autoanalyzer interfaced | *26* |
| | (e) | 25 μL serum + 25 μL acetonitrile | $C_{18}$RP | UV 200 nm + 254 30° C oven temp. | Basic pH, mobile phase, | *18* |

| | | | | | | |
|---|---|---|---|---|---|---|
| | (f) | 2 mL blood or serum extracted with chloroform | SG | UV 254 nm | | *11* |
| | (g) | 40–200 μL serum extracted with ether, back extraction with alkali | SG | UV 254 nm | | *17* |
| | (h) | 0.5 mL serum extracted with methylene chloride | SG | UV 254 nm | | *13* |
| | (i) | 0.5 mL serum extracted with chloroform | AE | UV 254 nm | 2 columns in series | *21* |
| | (j) | 200 μL serum extracted with butylchloride and isobutanol | $C_{18}$RP | UV 254 nm | Ion-pair with hexane sulfonic acid | *27* |
| | (k) | 0.5 mL serum, charcoal adsorption | $C_{18}$RP | UV 195 nm | 40°C oven temp. | *14* |
| | (l) | 200 μL serum + 400 μL acetonitrile | $C_{18}$RP | UV 200 nm | 50°C oven temp. | *30* |
| Phensuximide | (a) | 50 μL serum extracted with chloroform | $C_{18}$RP | UV 195 nm | 65°C oven temp. | *29* |
| Phenytoin | (a) | 200 μL serum + 200 μL acetonitrile | $C_{18}$RP | UV 195 nm | 50°C oven temp. | *19, 20* |
| | (b) | 300 μL supernatant from (a) extracted with chloroform | $C_{18}$RP | UV 195 nm | 50°C oven temp. | *20* |

*(continued)*

Table 2 continued

| Drug | | Sample preparation | Column[a] | Detection | Remarks | Ref. |
|---|---|---|---|---|---|---|
| | (c) | automated extraction of 150 $\mu$L serum | $C_8$RP | UV 200 nm interfaced with LC | Technicon Autoanalyzer | *26* |
| | (d) | 25 $\mu$L serum + 25 $\mu$L acetonitrile | $C_{18}$RP | UV 200 nm, UV 254 nm 30°C oven temp. | pH 8.0, mobile phase, | *18* |
| | (e) | 2 mL blood or plasma extracted with chloroform | SG | UV 254 nm | | *11* |
| | (f) | 50–200 $\mu$L serum followed by alkali back extraction | SG | UV 254 nm | | *17* |
| | (g) | 0.5 mL serum extracted with methylene chloride | SG | UV 254 nm | | *13* |
| | (h) | 0.5 mL plasma extracted with chloroform | AE | UV 254 nm | | *21* |
| | (i) | 200 $\mu$L serum extracted with butylchloride isobutanol | $C_{18}$RP | UV 254 nm | Ion-paired with hexane sulfonic acid | *27* |

| | | | | | | |
|---|---|---|---|---|---|---|
| | ( ) | 0.5 mL serum, charcoal adsorption | $C_{18}$RP | UV 195 nm | 40° C oven temp. | *14* |
| | (k) | 50 μL serum extracted with chloroform | $C_{18}$RP | UV 195 nm | 65° C oven temp. | *29* |
| | ( ) | 200 μL serum + 400 μL acetonitrile | $C_{18}$RP | UV 200 nm | Methanolic mobile phase, pH 6.8, 50° C oven temp. | *30* |
| 5-(4-Hydroxyphenyl)-5-phenylhydantoin (HPPH) | (a) | 0.5 μL hydrolyzed urine extracted with ethylacetate dichloroethane | $C_{18}$RP | UV 254 nm | Acid hydrolysis to hydrolyze HPPH glucuronide | *25* |
| | (b) | 2 mL hydrolyzed urine extractd with ethylacetate | SG | UV 254 nm | Acid hydrolysis | *31* |
| Primidone | (a) | 200 μL serum + 200 μL acetonitrile | $C_{18}$RP | UV 195 nm | 50° C oven temp. | *19, 20* |
| | (b) | 300 μL supernatant from (a) extracted with chloroform | $C_{18}$RP | UV 195 nm | 50° C oven temp. | *20* |
| | (c) | 50 μL serum extracted with chloroform | $C_{18}$RP | UV 195 nm | 65° C oven temp. | *29* |

*(continued)*

Table 2 continued

| Drug | | Sample preparation | Column[a] | Detection | Remarks | Ref. |
|---|---|---|---|---|---|---|
| | (d) | automated chloroform extraction of 150 μL serum | $C_8$RP | UV 200 | Technicon Autoanalyzer interfaced with LC | *26* |
| | (e) | 25 μL serum + 25 μL acetonitrile | $C_{18}$RP | UV 200 | pH 8.0 mobile phase; 30° C oven temp.; primidone not resolved from PEMA | *18* |
| | (f) | 0.5 mL serum, charcoal adsorption | $C_{18}$RP | UV 195 nm | 40° C oven temp. | *14* |
| | (g) | 200 μL serum + 400 μL acetonitrile | $C_{18}$RP | UV 200 nm | Methanolic, pH 6.8 mobile phase; 55° C oven temp. | *30* |
| Phenylethylmelonamide (PEMA) | (a) | 200 μL serum + 200 μL acetonitrile | $C_{18}$RP | Uv 195 nm | 50° C oven temp. | *20* |
| | (b) | 300 μL supernatant from (a) extracted with chloroform | $C_{18}$RP | UV 195 nm | 50° C oven temp. | *20* |

| | | | | | | |
|---|---|---|---|---|---|---|
| | (c) | 50 $\mu$L serum extracted with chloroform | $C_{18}$RP | UV 195 | 65°C oven temp. | *29* |
| | (d) | Automated chloroform extraction of 150 $\mu$L serum | $C_8$RP | UV 200 nm | Technicon Autoanalyzer interfaced with with LC | *26* |
| Valproate | (a) | 0.5 mL acidified serum extracted with butylchloride/methylene chloride derivatization with phenacylbromide in the presence of crown ether | $C_{18}$RP | UV 247 nm | 50°C oven temp.; 25 min derivatization | *22* |
| | (b) | 250 $\mu$L plasma extracted with pentane, derivatization with $\alpha$-bromoacetophenone in the presence of triethylamine | $C_{18}$RP | UV 246 | 1-h derivatization | *32* |

[a]RP, reversed phase; SG, silica gel; AE, anion exchange.

Thus, comprehensive therapeutic monitoring of anticonvulsant drugs demands the availability of a suitable technique to monitor various metabolites in addition to parent drugs. LC provides this capability (Fig. 3).

## VII. Recent Developments and New Horizons

Promising innovations and improvements in column technology have recently been reported. A series of highly efficient (> 65,000 plates per meter) reversed-phased columns, with almost total coverage of the silica base by the hydrocarbonaceous phase, have been introduced. The development of these selective and highly efficient columns ushers in a new era in the therapeutic monitoring of anticonvulsants by liquid chromatography. Taking advantage of these developments, Technicon has recently introduced a FAST-LC method for the simultaneous analysis of phenytoin, phenobarbital, primidone, carbamazepine, and two therapeutically active metabolites of these drugs: phenylethylmalonamide (PEMA) and carbamazepine epoxide (*26*). The method is fully automated, including sample aspiration, protein precipitation, particle removal, extraction, phase separation, evaporation, and reconstitution into the "pickup reagent," prior to LC separation and detection. This total automation, with a throughput of 7.5 samples per hour, offers considerable convenience when compared with various manual LC procedures.

The development of a new generation of LC detectors, such as LC coupled to either a mass-spectrometer or a high-speed scanning ultraviolet spectrophotometer, will provide more definitive identification of drugs.

## Acknowledgments

Laurence J. Marton is the receipient of NCI Research Career Development Award CA-00112. We wish to thank Ms. Mary Stawski for her careful typing of this manuscript.

## References

*1.* Horning, M. G., Brown, L., Nowlin, J., Lertratanangkoon, K., Kellaway, P., and Zion, T., *Clin. Chem.* **23,** 157 (1977).
*2.* Kalman, S. M., and Clark, D., *DAL Newsletter* **2,** Stanford University Hospital, August, 1976.

3. Goldbaum, L. R., *Anal. Chem.* **24,** 1604 (1952).
4. McDonald, D. M., and Kabra, P. M., *Clin. Chem.,* **26,** 361 (1980).
5. Pesh-Imam, M., Fretthold, D. W., Sunshine, E., Kumar, S., Terrentine, S., and Willis, C. E., *Clin. Chem.* **25,** 1118 (1979).
6. Castro, A., Ibanez, J., DiCesare, J., Adams, R., and Malkus, H., *Clin. Chem.* **24,** 710 (1978).
7. Godolphin, W., and Thoma, J., *Clin. Chem.* **24,** 483 (1978).
8. Troupin, A. S., and Fried, P., *Epilepsia* **16,** 223 (1975).
9. Kurata, D., and Wilkinson, G. R., *Clin. Pharmacol. Therapeut.* **16,** 355 (1974).
10. Adams, R. F., Schmidt, G. J., and Vandemark, F. L., *J. Chromat.* **145,** 275 (1978).
11. Kabra, P. M., Gotelli, G., Stanfill, R., and Marton, L. J., *Clin. Chem.* **22,** 824 (1976).
12. Kabra, P. M., and Marton, L. J., *Clin. Chem.* **22,** 1070 (1976).
13. Atwell, S. H., Green, V. A., and Haney, W. G., *J. Pharm. Sci.* **64,** 806 (1975).
14. Adams, R. F., and Vandemark, F. L., *Clin. Chem.* **23,** 25 (1976).
15. Lehrer, M., *Clin. Chem.* **25,** 1090 (1979).
16. Pranistis, P. A. F., Mitzoff, J. R., *J. Forensic Sci.* **19,** 917 (1974).
17. Evans, J. E., *Anal. Chem.* **45,** 2428 (1973).
18. Soldin, S. J., and Hill, J. G., *Clin. Chem.* **22,** 856 (1976).
19. Kabra, P. M., Stafford, B. E., and Marton, L. J., *Clin. Chem.* **23,** 1284 (1977).
20. Kabra, P. M., McDonald, D. M., and Marton, L. J., *J. Anal. Tox.* **2,** 127 (1978).
21. Kitazawa, S., and Komino, T., *Clin. Chim. Acta.* **73,** 31 (1976).
22. Sutheimer, C., Fretthold, D., and Sunshine, I., *Chromat. Newsletter* (Perkin-Elmer) **7,** 1 (1979).
23. Eichelbaum, M., and Bertilsson, L., *J. Chromat.* **103,** 135 (1975).
24. Kutt, H., *Clin. Pharmacol. Ther.* **16,** 243 (1974).
25. Kabra, P. M., and Marton, L. J., *Clin. Chem.* **22,** 1672 (1976).
26. Dolan, J. W., Van der Wal, S. J., Bannister, S. J., and Snyder, L. R., *Clin. Chem.* **26,** 87 (1980).
27. Draper, P., Shapcott, D., and Lemieux, B., *Clin. Biochem.* **12,** 52 (1979).
28. Westenberg, H. G. M., and Zeeuw, R. A. D., *J. Chromat.* **118,** 217 (1976).
29. Adams, R. F., Schmidt, G. J., and Vandemark, F. L., *J. Chromat.* **145,** 275 (1978).
30. Freeman, D. J., and Rawal, N., *Clin. Chem.* **25,** 810 (1979).
31. Gupta, R. N., Keane, P. M., and Gupta, M. L., *Clin. Chem.* **25,** 1984 (1979).

# Chapter 6

# Theophylline and Antiarrhythmics

F. L. Vandemark

*Liquid Chromatography Department*
*The Perkin-Elmer Corporation*
*Norwalk, Connecticut*

## I. Introduction

Liquid chromatography has become a major analytical technique in the laboratory concerned with therapeutic drug monitoring. This acceptance arises from a number of important factors, including the unusual versatility of the technique, its potential use in the routine determination of drug substances, and the nondestructive nature of the detection systems commonly used.

My intention in this article is to review the liquid chromatographic procedures that have been developed for the determination of theophylline and some of the important cardiac drugs. Plasma concentrations of these drugs have been shown to be directly related to therapeutic effectiveness (see Chapter 4 of this volume). Therefore, it is useful to monitor them closely in order to adjust individual therapeutic dosages for optimal patient care.

## II. Analysis of Antiasthmatic Drugs

Theophylline, a bronchodilator, is a useful antiasthmatic drug. Therapeutic serum concentrations of theophylline usually fall within the range of 10–20 mg/L. Drug concentrations outside this range may yield either ineffective therapy or toxicity.

Numerous techniques have been utilized for the determination of theophylline. The first technique routinely used was based upon ultraviolet spectroscopy (*1*). These procedures required large specimen volumes and extensive sample pretreatment. An additional drawback of these procedures was interference from other xanthines, including theobromine and caffeine. This problem was to some extent alleviated utilizing detection at multiple wavelengths.

Gas chromatography (GC) has also been utilized for the determination of this drug (*2, 3*). In most gas chromatographic procedures theophylline is derivatized to minimize adsorption and thermal degradation. Gas chromatographic procedures using flame ionization detectors usually require large sample volumes, and extraction and back extraction of the drug from the sample prior to analysis. Those procedures that employ nitrogen–phosphorus detection (*4*) permit the determination of theophylline in 50 μL of serum and obviate the need for back extraction cleanup.

Liquid chromatography has become the technique of choice for the determination of theophylline. This owes in part to the hydrophilic nature of this drug, which lends itself well to reversed-phase chromatography.

### A. Sample Pretreatment

The measurement of theophylline is commonly performed on serum or plasma. Sample size varies between 50 μL and 1 mL, depending upon the sensitivity of the particular analytical method being used. Most procedures utilizing reversed-phase chromatography and ultraviolet detection require only 50 μL of serum.

In general, there are three methods of sample pretreatment for the analysis of therapeutic drugs in serum or plasma. These include the use of solvent extraction, protein precipitation, or direct sample injection.

*1. Solvent Extraction.* A commonly used procedure for isolating theophylline from the serum matrix is the use of liquid–liquid partition or solvent extraction. The pH of the serum is adjusted to approximate the p$K_a$ of the drug, which facilitates extraction by reducing ionization. The sample is then extracted with an appropriate immiscible organic solvent, the organic extractant is evaporated to dryness, and an aliquot of the reconstituted extraction residue is injected into the chromatograph.

Extraction serves two purposes: The first is the isolation of theophylline from the serum matrix constituents, thereby minimizing potential interference by endogenous compounds. Second, concentration of the drug improves sensitivity, permitting the use of small sample volumes.

Solvent extraction has been used extensively for the determination of theophylline by liquid chromatography (*5–7*). An extraction procedure optimized for pediatric samples has been previously reported by our laboratory (*8*). This procedure entails the extraction of the drug from 50 μL serum buffered to pH 6. The organic extractant consists of 200 μL of chloroform/isopropanol, 1/1 (v/v). The mixture is centrifuged, and the lower layer (organic phase) evaporated to dryness. The extraction residue is redissolved in 20 μL of methanol and a 10 μL sample is injected into the chromatograph.

*2. Protein Precipitation.* A commonly used method of sample pretreatment for theophylline determination, is the use of protein precipitation (*9*). This technique requires the mixing of a small volume of serum, approximately 50–100 μL, with a four volume excess of ethanol or a two volume excess of acetonitrile. The solution is then mixed, centrifuged, and an aliquot is injected directly into the chromatograph.

These procedures are relatively easy to perform and tend to minimize the amount of protein injected onto the column. However, the technique does have some limitations; thus, the utilization of protein precipitation results in dilution of the sample. For theophylline analysis this is not a serious problem because of the drug's relatively high therapeutic concentration range, 10–20 mg/L. However, sample dilution may be a problem for other therapeutic drug agents present at lower concentrations, such as tricyclic antidepressants, propranolol, and the benzodiazepines.

In addition to the problems of sample dilution, protein precipitation is also not selective. Since only the proteins are precipitated, other compounds present in the sample will be injected into the LC chromatograph with the theophylline and may interfere with the determination. A number of interferences have been reported with the use of protein precipitation techniques, including such compounds as some of the common antibiotics, cephalosporin, cefazolin, cephapirin, ampicillin, methacillin, and acetazolamide (*10–13*).

*3. Direct Injection.* The use of direct serum injection has been used only in isolated instances. Recently, there has been some renewed interest in using direct injection for the determination of theophylline (*14*); however, the procedure is not widely accepted. These procedures require the use of a guard column to protect the analytical column

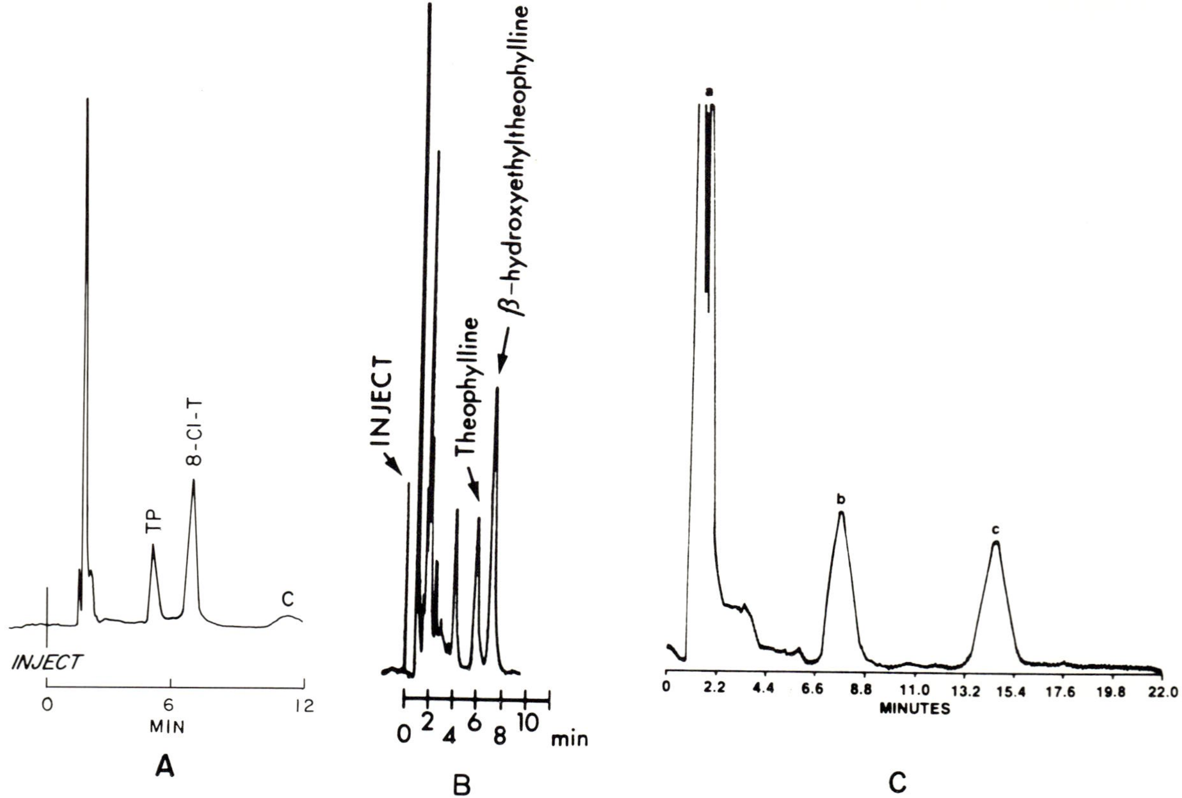


FIG. 1. Chromatograms of patient serum containing theophylline subjected to different sample pretreatments. (A) Solvent extraction, theophylline (TP), 8-chlorotheophylline (8-Cl-T), and caffeine (C); (B) protein ppt; and (C) direct injection, impurities (a), theophylline (b), and 8-chlorotheophylline (c).

from contamination by serum matrix material. After approximately 20–30 injections the guard column must be replaced.

The importance of sample preparation on the specificity and sensitivity of the analysis can be observed from a comparison of the three pretreatment procedures applied to serum samples (Fig. 1). For each pretreatment method used there is a certain amount of nondrug background material that is chromatographed. As expected, samples analyzed using solvent extraction show the least amount of nondrug background.

## B. Chromatography

There are several separation modes that have been used for the determination of theophylline. These include liquid–solid (*15*), ion exchange (*16*), and reversed-phase chromatography (*17–19*).

*1. Ion Exchange.* Ion-exchange chromatography is the preferred separation mode for the analysis of ionic compounds. The chromatographic separation may be optimized by altering the mobile phase pH or ionic strength. Most ion-exchange packings currently used in liquid chromatography are derived from high molecular weight polymers. Typically they do not have the particle size or surface area properties of silica based materials. Therefore, chromatography on ion-exchange columns usually results in low separation efficiency.

Weinberger and Chidsey (*16*) used ion-exchange liquid chromatography for the analysis of theophylline, utilizing a 0.45 *M* ammonium phosphate, pH 3.65, mobile phase at a flow rate of 0.4 mL/min. Serum was mixed with an equal volume of a saline solution containing 8-chlorotheophylline as an internal standard. An 8 $\mu$L aliquot of this mixture was injected into the chromatograph (Fig. 2). The procedure was linear over a concentration range of 1–40 mg/L. Total chromatographic time was about 30 min.

*2. Liquid–Solid Chromatography.* Historically, liquid–solid chromatography was one of the first separation modes applied to the determination of theophylline. However, this separation mode has some practical limitations. Liquid–solid chromatography uses silica gel column packings having very polar surface characteristics. Sample analytes are eluted using nonpolar mobile phase solvents, usually mixtures of chloroform and hexane. This increases the cost per test over chromatographic systems, such as reversed-phase, which use aqueous eluents. In addition, sample constituents may be adsorbed onto the silica surface and modify the chromatographic separation over time. In addition, in adsorption chromatography, nonpolar

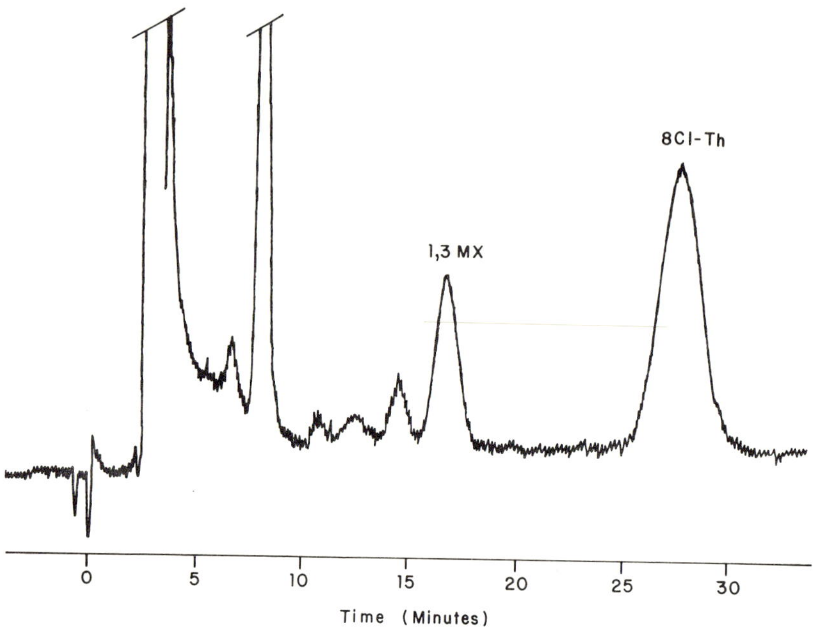


FIG. 2. Chromatogram of a serum sample from a patient receiving theophylline using ion-exchange chromatography theophylline (*1, 3* MX) 8-chlorotheophylline (8Cl-Th).

analytes elute before polar substances. Therefore sufficient time must be allowed to elute the polar matrix materials that may be present.

Sitar and coworkers (*5*) analyzed theophylline in serum using liquid–solid chromatography. The mobile phase consisted of chloroform/isopropanol/acetic acid in the proportions 84/15/1 (v/v). A half-mL serum sample was extracted with chloroform/isopropanol, 95/5 (v/v). Theophylline was detected at 273 nm. Internal standardization was not utilized.

Manion et al. (*15*) also used liquid–solid chromatography for gradient elution (hexane/isopropanol) of theophylline at 37° C. Detection was at 270 nm. One mL of serum was extracted with 30 mL of chloroform/isopropanol 95/5 (v/v). Oxazepam was used as an internal standard, and analysis time was approximately 15 min.

Evenson and Warren (*6*) developed a rapid liquid–solid chromatographic procedure that permitted the determination of theophylline in 8 min. Two hundred $\mu$L of serum was extracted with 5 mL chloroform/isopropanol, using $\beta$-hydroxyethyltheophylline as an internal standard. Theophylline was eluted from the column using a mobile phase of chloroform/heptane/acetic acid/ethanol. The drugs were detected at 254 nm with reliable quantitation of the drug over the range of 1–40 mg/L. A serum sample analyzed according to the procedure is shown in Fig. 3.

Maijub et al. (*20*) used liquid–solid chromatography for the determination of theophylline, and also monitored dyphylline and phenobarbital concentrations. One mL samples were extracted with 10 mL of chloroform/isopropanol 95/5 (v/v) using monohydroxypropyltheophylline as the internal standard. The compounds were eluted with a chloroform/heptane/methanol, 39/65/6 (v/v), mobile phase and detected at 254 nm. They found liquid–solid adsorption chromatography advantageous because the eluent from a particular fraction could be evaporated and the residue subjected to mass spectral analysis for confirmation.

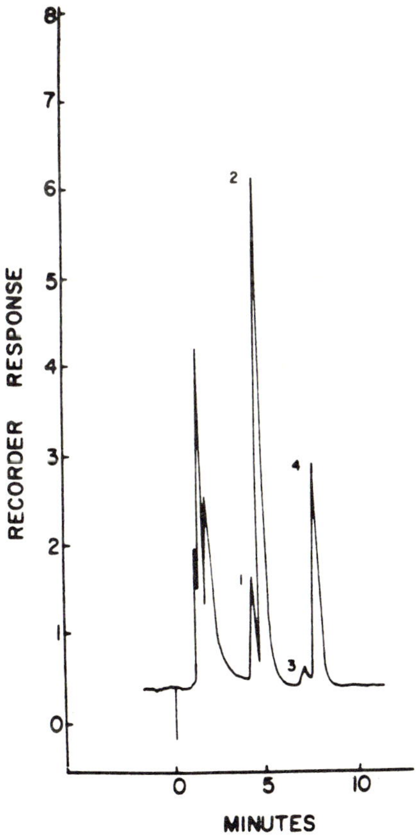


FIG. 3. Extraction of human serum sample containing caffeine (1), theophylline (2), theobromine (3), and β-HPT (4) using liquid–solid chromatography.

*3. Reversed-Phase Chromatography.* In this separation mode a stationary phase material is chemically bonded to the surface of the silica. In normal-phase partition chromatography, a polar liquid phase is bonded to the surface of the silica packing and a nonpolar organic solvent is used for elution. In reversed-phase chromatography, a nonpolar chemical species, usually a $C_{18}$ hydrocarbon, is chemically bonded to the silica surface. Elution is performed using a polar mobile phase, usually a mixture of methanol and water or acetonitrile and water.

Reversed-phase chromatography offers several significant advantages for the analysis of therapeutic drugs and other biochemicals, including the use of aqueous mobile phases that are less expensive and more optically transparent than organic solvent mobile phases. In addition, polar compounds are generally not retained on reversed-phase packings; therefore, polar matrix constituents are less prone to interfere with the less polar drugs of interest.

Franconi et al. (*17*) were the first group to use C-18 reversed-phase chromatography for the determination of theophylline. Theophylline was eluted using a 10% acetonitrile/10 m*M* acetate buffer, and detected at 254 nm. One mL serum samples were prepared prior to chromatography by ultrafiltration. Twenty μL of the ultrafiltrate was injected into the chromatograph. No internal standard was used.

A procedure developed in our laboratory (*8*) utilized reversed-phase chromatography and a micro extraction of 50 μL of serum. Theophylline was separated from other xanthine compounds using a 2% acetonitrile in water mobile phase adjusted to pH 5.5. Detection was at 273 nm. Sample preparation was as discussed in Section II.A.1. Figure 4 illustrates the analysis of a number of serum samples by this procedure. The procedure was linear over a concentration range from 1 to 30 mg/L. Total analysis time for a single sample was 18 min. The micro extraction technique provides selectivity while greatly reducing the time required to perform the solvent extraction.

Orcutt and coworkers (*9*) used a $C_{18}$ reversed-phase column and a 7% acetonitrile/10 mmol/L sodium acetate mobile phase. Detection was at 254 nm and β-hydroxyethyltheophylline was used as an internal standard. To avoid the sample manipulations required by solvent extraction, these workers utilized a protein precipitation prior to analysis. Fifty μL of serum was mixed with 100 μL of acetonitrile, the mixture was centrifuged and 20 μL injected into the chromatograph. A patient sample analyzed by this method is shown in Fig. 1B.

Popovich (*14*) reported a $C_{18}$ reversed-phase procedure for the determination of theophylline using direct serum injection. The mobile phase consisted of 35% methanol in 25 mmol/L phosphate buffer.

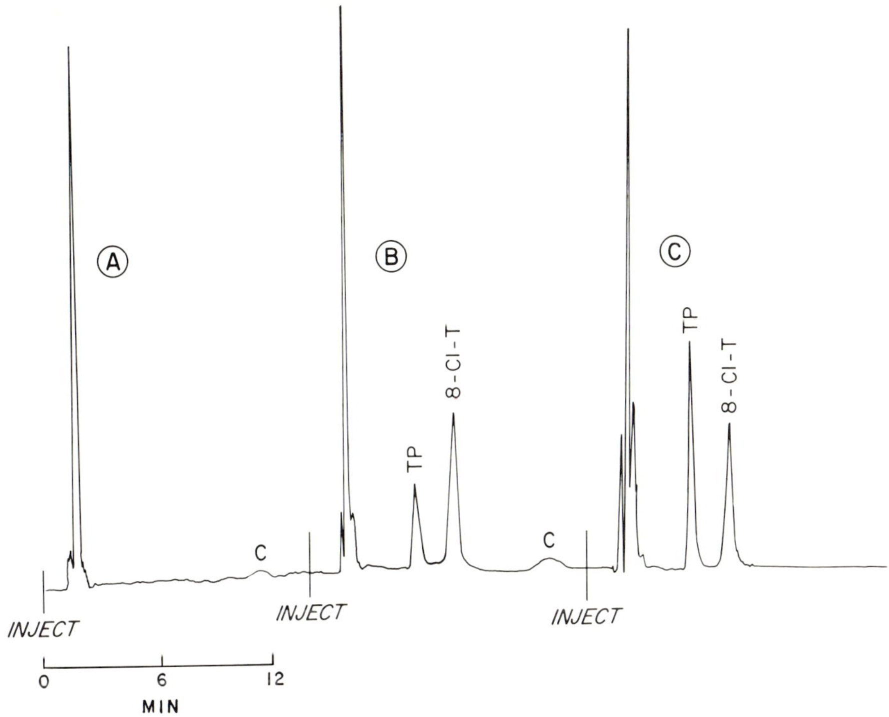


FIG. 4. Chromatograms of sera extracts: (A) theophylline-free serum without internal standard; (B) patient's serum, calculated to contain 7.6 mg/L theophylline; and (C) patient's serum containing 15.1 mg/L (TP, theophylline; 8-Cl-T, internal standard; C, caffeine).

Detection was at 254 nm. A chromatogram of a serum sample analyzed by this procedure is shown in Fig. 1C.

## III. Antiarrhythmics

### A. Lidocaine and Procainamide

Lidocaine and procainamide are used in the treatment of cardiac arrhythmias. Analogous to many other drugs, there exists a low therapeutic/toxic ratio. The therapeutic concentration ranges are 1–6 mg/L, and 4–8 mg/L for lidocaine and procainamide, respectively.

Prior to the development of liquid chromatography, the most common procedures for determining these drugs involved the use of gas chromatography (GC). GC procedures required separate

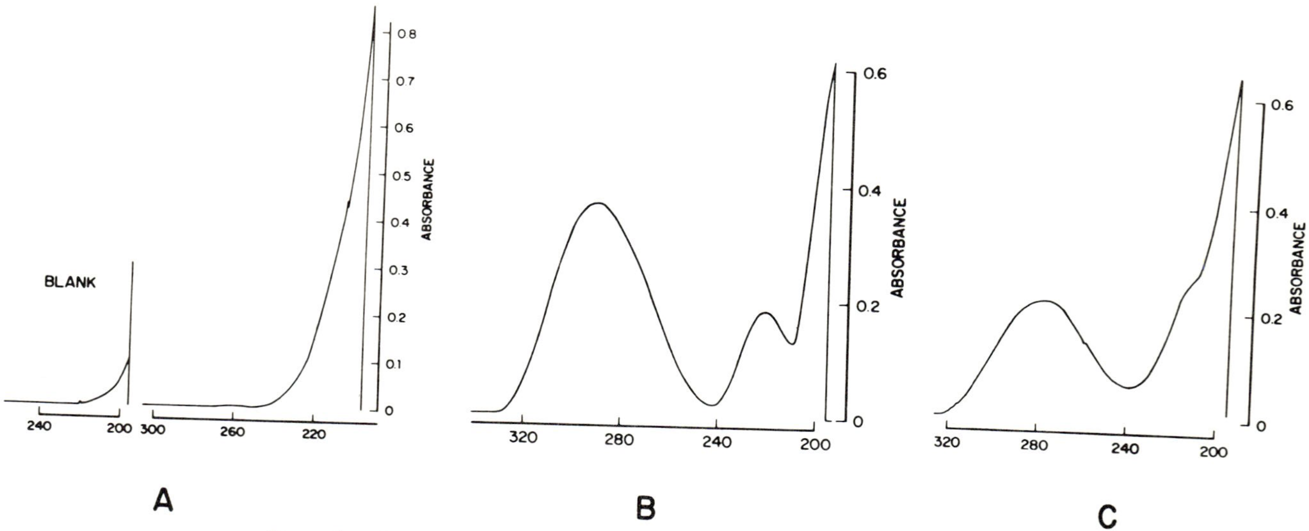


FIG. 5. UV absorption spectra for pure standards of lidocaine, procaine, and procainamide.

analytical conditions for the determination of each drug (*21, 22*) and sample preparation was time consuming.

Because of the hydrophobic nature of these drugs, reversed-phase liquid chromatography has since been widely applied to their analysis (*23, 24*). Usually, $C_{18}$ bonded-phase packings and acetonitrile and water mobile phase eluents are used to effect separations.

An important consideration for developing good chromatographic separations for these drugs is the choice of mobile phase pH and ionic strength. We have found that a mobile phase consisting of 10% acetonitrile in 100 mmol/L phosphate buffer adjusted to pH 4.0 gives good chromatographic separations in approximately 12 min.

The choice of detection wavelength for the simultaneous detection of these two drugs is an important factor that must be considered. Figure 5 illustrates the UV spectra for lidocaine, procainamide, and procaine, as an internal standard. Procaine has a maximum absorption of about 288 nm, while procainamide absorbs strongly at 278 nm. Lidocaine has no distinguishing absorbance features, although it does have a strong absorption shoulder at approximately 205 nm. Therefore, detections at 280 nm allows only two of the three compounds to be detected. Detection at 205 nm allows for the analysis of all three drugs simultaneously.

We developed a liquid–liquid extraction for these drugs. Fifty μL of serum with added procaine was extracted with 2 mL of diethyl ether at a basic pH (0.5 mL carbonate buffer). The mixture was centrifuged and the organic layer evaporated to dryness. The residue was reconstituted with a small amount of methanol and an aliquot injected into the chromatograph. A blank serum sample, and a blank serum sample with added lidocaine and procainamide, were analyzed according to this procedure, as shown in Fig. 6.

Procainamide is metabolized to *N*-acetyl procainamide. This metabolite has been shown to be as pharmacologically active as the parent drug compound (*25*). Therefore, both the parent drug compound and the metabolite should be monitored.

We have observed that *N*-acetylprocainamide coeluted with procaine. Thus, it was necessary to change the internal standard to carbocaine for the analysis of procainamide. For the analysis of lidocaine, procaine was maintained as the internal standard. The analysis of patient samples is shown in Fig. 7.

Recently, Rocoo et al. reported an assay for the determination of procainamide and *N*-acetylprocainamide (*26*). The method utilizes reversed-phase chromatography on a $C_{18}$ column with 40% methanol and water, and detection at 280 nm. Serum samples were buffered with sodium hydroxide and extracted with methylene chloride. A synthetic

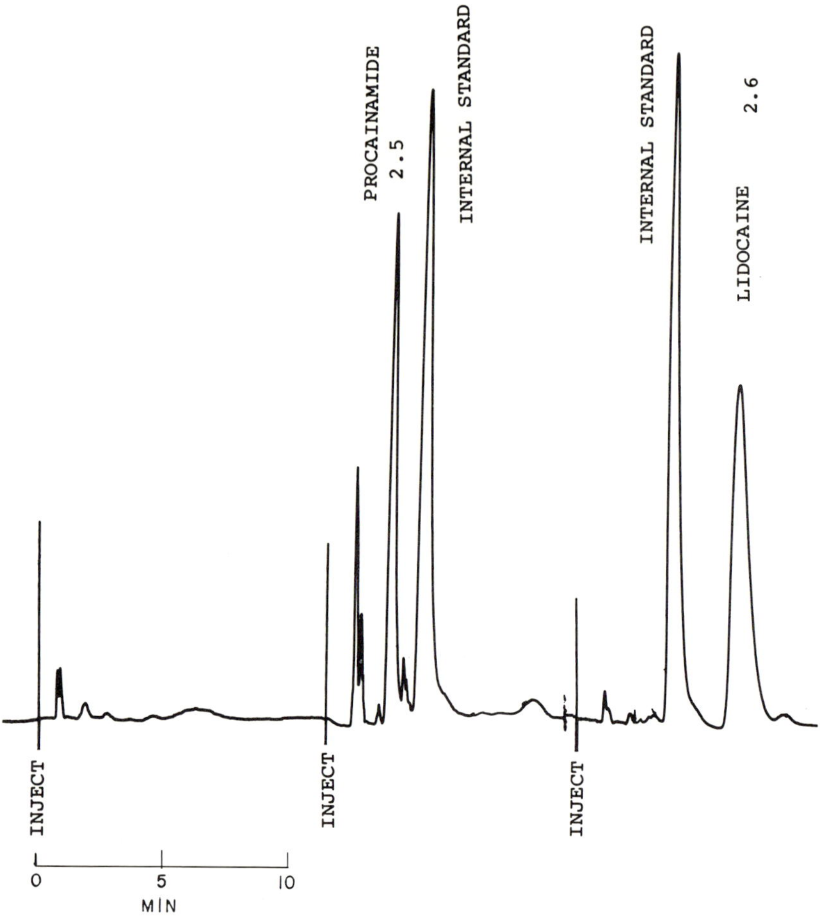


FIG. 6. Chromatograms of three serum extracts. The first chromatogram (on left) is of a known drug-free serum, and illustrates a typical serum background. The second chromatogram was from serum containing procainamide (2.5 mg/L); the third chromatogram was obtained from serum where lidocaine was calculated to be 2.6 mg/L.

analog, *N*-proprionylprocainamide, was used as an internal standard. Chromatographic time was less than five minutes per sample.

## B. Propranolol

Propranolol is used in the treatment of cardiac arrhythmias, angina pectoris, and hypertension. Although therapeutic concentrations have not been well-defined, they usually fall in the range of 50–150 μg/L.

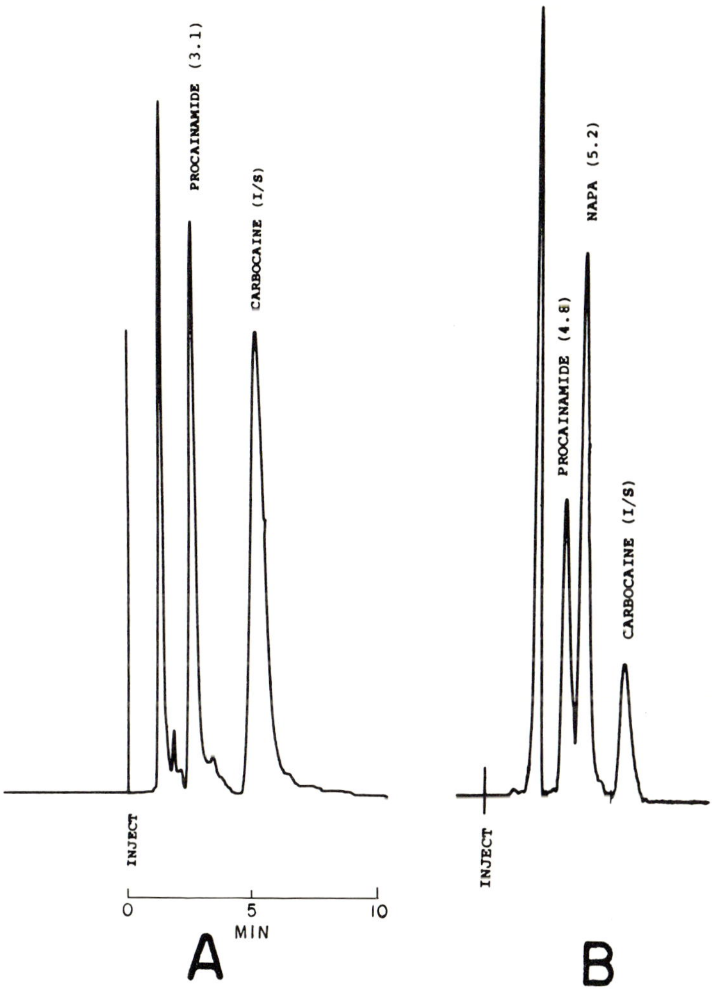


FIG. 7. Chromatograms from a patient on procainamide therapy.

Propranolol has been assayed by fluorescence, but the assay is not specific (*27*). Electron-capture gas chromatography has also been used, but the pretreatment of the sample is rather tedious (*28*). Ultraviolet detectors do not have adequate sensitivity for measuring the low concentrations of this drug in small samples. This drug does fluoresce, however, and on-line fluorescence detection has proved to be the system of choice. Solvent extraction utilizing a 1–2 mL sample is desirable to improve detection.

LC procedures for the analysis of this drug frequently utilize reversed-phase liquid chromatography with fluorescence detection. These techniques have been used successfully for the determination of propranolol in plasma (*29, 30*). One to two mL of plasma made alkaline with 2*N* sodium hydroxide were extracted with hexane/isoamyl alcohol, 97/3 (v/v). This extraction procedure resulted in the co-extraction of compounds that interfere with propranolol. For this reason a back extraction step is necessary. Jatlow et al. (*31*) have recently reported a micro-extraction method incorporating back extraction into phosphate buffer, after which injection into the chromatograph is possible. This procedure reduces the number of interfering substances. They used a $C_{18}$ reversed-phase column and a mobile phase consisting of 25% acetonitrile in water. Using a column temperature of 50°C, the chromatogram was complete in about 8 min. Serum samples analyzed by these procedures are shown in Fig. 8.

Propranolol is partly metabolized in the liver to 4-hydroxypropranolol. This metabolite has been shown to also be a pharmacologically active compound (*32*). Its measurement has been difficult because of the low concentrations present in serum and its short half-life. Analytically, the determination is further complicated by the different fluorescent characteristics it possesses from those of the parent drug.

## C. Quinidine

Quinidine is another antiarrhythmic agent. It is effective over the concentration range of 3–6 mg/L. Historically, fluorescence has been the widely used detection technique (*33*).

Drayer et al. (*34*) used reversed-phase chromatography with a 12% acetonitrile and water mobile phase and fluorescence detection for the analysis of quinidine and the 3-hydroxy metabolite. They utilized a benzene extraction of 50-$\mu$L serum samples.

However, dihydroquinidine and quinidine could not be separated. Crouthamel (*35*) used a 25% methanol and water reversed-phase system to separate quinidine from dihydroquinidine. Serum samples made alkaline with sodium hydroxide were extracted with benzene. No attempts were made to separate the 3-hydroxy metabolite.

Sved (*36*) used a silica column with a methylene chloride/hexane/methanol, 60/35/5.5 (v/v), eluent and fluorescence detection. Separation of quinidine from dihydroquinidine and other metabolites was obtained; however, there was a potential shift in the retention time of the dihydroquinidine peak from injection to injection.

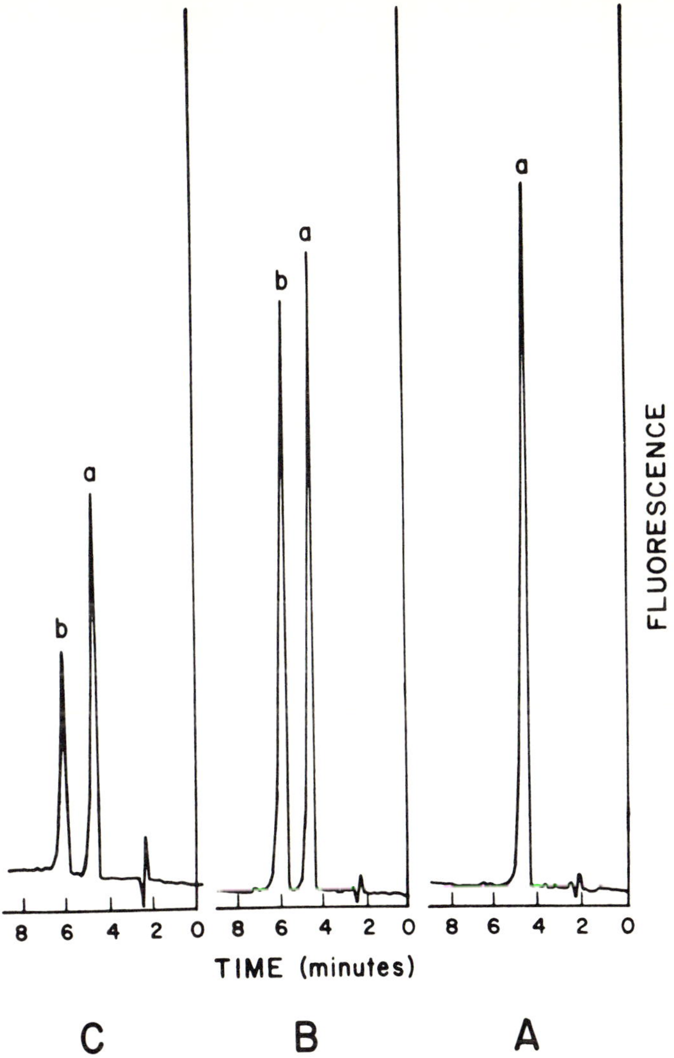


FIG. 8. Chromatograms of extracted plasma samples (A) a plasma blank, (B) a plasma standard containing 150 μg/L propranolol and, (C) a patient receiving propranolol (99 μg/L). Propranolol (a), Pronethalol (I/S), (b).

Recently, Kline (*37*) used reverse-phase chromatography with fluorescence detection for the analysis of quinidine. Utilizing a mobile phase consisting of methanol/acetonitrile/ammonium carbonate, 65/31/4 (v/v), they were able to resolve quinidine, 3-hydroxy-

Table 1
Review of LC Procedures for Determination of Theophylline and Antiarrhythmics at Therapeutic Levels

| Reference | Mobile phase | Flow rate mL/min | Detector | nm | Temp. | Trade Name | Manufacturer | Particle size, μm |
|---|---|---|---|---|---|---|---|---|
| | | | *Theophylline* | | | | | |
| *5* | $CHCL_3$/isopropanol/acetic acid, 84/15/1 (v/v) | 0.7 | UV | 273 | Amb. | Micropak SI 10 | Varian | 10 |
| *6* | 6% Ethanol; 94% chloroform/hexanol/acetic acid, 3/2/4 (v/v) | 0.4 | UV | 254 | Amb. | Zorbax SIL | Dupont | 6-8 |
| *7* | 2-Butanol/methanol/water, 25/5/1 in hexane (v/v) | 3.33 | UV | 280 | Amb. | μ-Porasil | Waters | 10 |
| *8* | 2% Acetonitrile in water containing 0.01% acetic acid, pH 5.5 | 1.5 | UV | 273 | 55°C | ODS-Sil-X-I | Perkin-Elmer | 13 |
| *9* | 7% Acetonitrile in 10 m*M* sodium acetate, pH 4.0 | 2 | UV | 254 | Amb. | μ-Bondapak C-18 | Waters | 10 |
| *10* | 10% Acetonitrile in 0.1 *M* sodium acetate, pH 4.0 | | UV | 280 | Amb. | Partisil 10 ODS | Whatman | 10 |
| *11* | 10% Acetonitrile in 20 m*M* sodium acetate, pH 4.0 | 1.8 | UV | 254 | Amb. | μ-Bondapak C-18 | Waters | |
| *12* | 7% Acetonitrile in 10 m*M* sodium acetate, pH 4.0 | 2 | UV | 254 | Amb. | μ-Bondapak C-18 | Waters | 10 |
| *13* | 7% Acetonitrile in 10 m*M* sodium acetate, pH 4.0 | 2 | UV | 254 | Amb. | Partisil-10 | Whatman | 10 |
| *14* | 35% Methanol in 25 m*M* potassium phosphate, pH 2.5 | 1.5 | UV | 254 | Amb. | Partisil-10 ODS-2 | Whatman | 10 |
| *15* | Hexane to isopropanol gradient, 14–22% at 0.6%/min. | 1.8 | UV | 270 | 37°C | Porasil C | Waters | 37 |
| *16* | 0.45 *M* ammonium phosphate, pH 3.65 | 0.4 | UV | 254 | 55°C | Aminex A-5 | Bio-Rad | |
| *17* | 10% Acetonitrile in 0.01 *M* sodium acetate, pH 4 | 2.0 | UV | 254 | Amb. | μ-Bondapak C-18 | Waters | 10 |
| *18* | 20% Methanol in water containing 1% proponic acid, pH 5 | 1.7 | UV | 280 | Amb. | μ-Bondapak C-18 | Waters | 10 |
| *19* | 25% Methanol in 0.01 *M* sodium acetate, pH 4.0 | | UV | 280 | Amb. | Spherisorb ODS | Spectra Physics | 10 |
| *20* | Chloroform/heptane/methanol, 39/56/5 (v/v) | 3 | UV | 254 | Amb. | Partisil 10 ODS | Whatman | 10 |

| | | | | | | | | |
|---|---|---|---|---|---|---|---|---|
| | | | *Lidocaine* | | | | | |
| *23* | 10% Acetonitrile in 0.1 *M* phosphate buffer, pH 6.0 | 2 | UV | 205 | 40°C | ODS Sil-X-I | P-E | 13 |
| *24* | 10% Acetonitrile in 0.1 *M* phosphate, pH 4.0 | 0.5 | UV | 205 | 40°C | ODS-Sil-X-I | P-E | 13 |
| *26* | Methanol/water/acetic acid, pH 5.5 600/400/10 (v/v) | | UV | 280 | Amb. | μ-Bondapak C-18 | Waters | 10 |
| | | | *Propranolol* | | | | | |
| *29* | Methanol/water/phosphoric acid, 37/63/.1 (v/v) | 1 | FL | 243 Ex, 375 Em | 65°C | ODS-Sil-X-I | P-E | 13 |
| *30* | 70% Acetonitrile in 0.02 *M* acetate buffer | 2 | FL | 270 Ex, 340 cut off filter | Amb. | μ-Bondapak CN | Waters | 10 |
| *31* | 25% Acetonitrile in phosphate buffer, pH 2.7 | 1.5 | FL | 285 Ex, 350 Em | 50°C | Partisil 10 ODS | Whatman | 10 |
| *32* | 27% Acetonitrile/water containing 0.06% phosphoric acid | 2 | FL | 205 Ex, Em fil KV 340 | Amb. | μ-Bondapak Alkyl phenyl | Waters | 10 |
| | | | *Quinidine* | | | | | |
| *34* | Acetonitrile in 2.5% acetic acid, 12/88 (v/v) | 1.8 | FL | Ex 340, Em fil, KV 418 | Amb. | μ-Bondapak C-18 | Waters | 10 |
| *35* | Methanol/acetic acid, water, pH 2.6, 25/4/71 (v/v) | 2 | UV | 254 | Amb. | μ-Bondapak C-18 | Waters | 10 |
| *36* | Dichloromethane/hexane/methanol/perchloric acid, 60/35/3.5/0.1 (v/v) | 3 | FL | 325, 420 fil | Amb. | LiChrosorb SI-60 | Merck | 5 |
| *37* | 4% (v/v) Of a 1% (v/v) aqueous ammonium carbonate, 31% (v/v) MeOH and 65% (v/v) acetonitrile | 3.5 | FL 450 | 350 | Amb. | μ-Bondapak C-18 | Waters | 10 |
| | | | *Disopyramide* | | | | | |
| *43* | 53% Methanol in 0.005 *M* heptane sulfonic acid, pH 3.5 | 1 | UV | 254 | Amb. | Micropak SI 10 | Varian | 10 |
| *44* | 35% Acetonitrile in 0.05 *M* phosphate, pH 3 | 2 | UV | 258 | Amb. | μ-Bondapak C-18 | Waters | 10 |
| *45* | 15% Methanol in 0.04 *M* | 1.9 | UV | 254 | Amb. | μ-Bondapak CN | Waters | 10 |

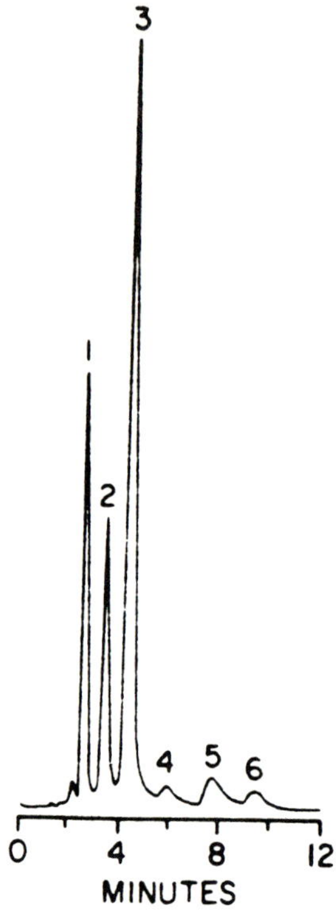


FIG. 9. Chromatogram of plasma from a patient dosed with quinidine sulfate. Peaks represent 3-hydroxyquinidine (a quinidine metabolite) (1) as yet unidentified metabolites of quinidine or dihydroquinidine (2, 4, 5), quinidine, 3.8 $\mu$g/L (3), and dihydroquinidine (6).

quinidine, dihydroquinidine, and several other metabolic products of dihydroquinidine. An example of this separation for 0.5 mL of serum extracted with 2 mL benzene/isoamyl alcohol, 1/1 (v/v), is shown in Fig. 9.

Although fluorescence detection combined with liquid chromatography has been widely used, there have been a number of procedures that use UV detection (*38–41*). Powers and Sadee (*38*) used protein precipitation for sample pretreatment, combined with reversed-phase chromatography. Quinidine was detected at 330 nm to improve selectivity and sensitivity over that obtainable at 254 nm. Peat and Jennison (*39*) analyzed quinidine using liquid-solid chromatography. One mL serum samples were extracted with diethyl ether. Detection was at 280 nm, with cinchonidine used as an internal standard.

## D. Disopyramide

Disopyramide (Norpace®) is a relatively new antiarrhythmic drug reported to be less toxic than quinidine. The therapeutic concentration is in the range of 2–4 mg/L. This drug has been analyzed previously by fluorescence, but the assay is nonspecific.

GC has also been utilized; however, multiextraction techniques and derivatization are usually required. The measurement of the active metabolites of this drug has also proved difficult by gas chromatography (*42*).

Meffin (*43*) used ion-pair reversed-phase chromatography with 53% methanol in water containing 0.005*M* heptanesulfonic acid. Samples were extracted with ether and back extracted to yield cleaner extracts. Detection was at 254 nm. The results showed that the drugs were not adequately retained and that lidocaine interfered.

Illet (*44*) utilized a reversed-phase column with 35% acetonitrile/phosphate buffer and detection at 258 nm. One-mL plasma samples were extracted with 5 mL dichloromethane. Chlorodisopyramide was used as an internal standard. No data was reported for separation of the active metabolite.

Recently, Lime (*45*) used a cyano column with 35% acetonitrile/water for the analysis of disopyramide and the *N*-dealkylated metabolite. One-mL plasma samples, after dilution with 25 mL water, were extracted with 1.2 mL of chloroform. The compounds were detected at 254 nm and *p*-chlorodisopyramide was used as an internal standard. A patient serum processed according to these procedures is shown in Fig. 10.

# IV. Summary

I have summarized many of the LC analytical techniques reported for theophylline and a number of the most important antiarrhythmic agents. Primary factors that need to be considered when developing an assay are:

*1. Chromatography.* The reversed-phase chromatographic mode is the preferred method owing to its broad applicability in clinical analysis. However, there are situations where another chromatographic mode, either a polar bonded phase or an adsorption technique, may be more suitable.

*2. Method of Detection.* The detector should provide adequate sensitivity and selectivity for the analysis, be easy to use and

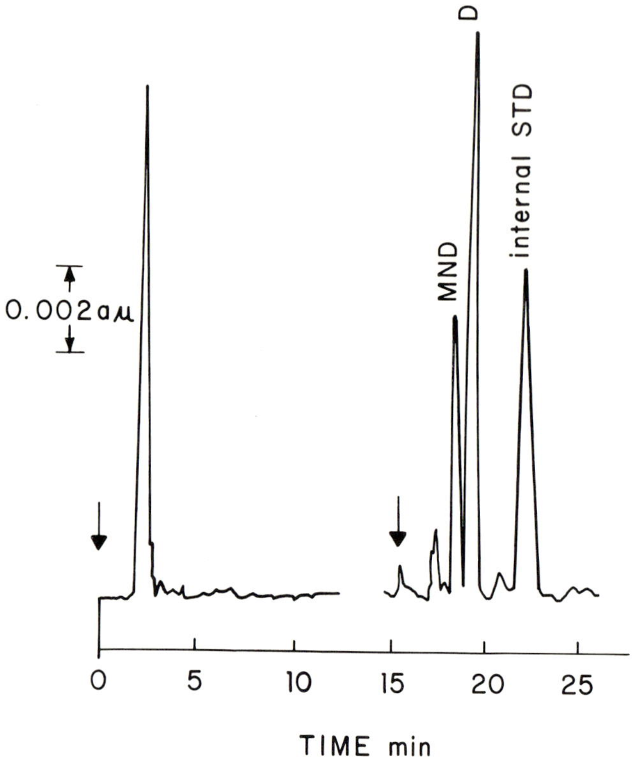


FIG. 10. Typical chromatograms of a pooled serum sample from patients not receiving disopyramide (D) (left), and from the analysis of serum from a patient taking 150 mg of disopyramide phosphate every 6 h (right).

reliable in operation. It is for these reasons that the variable wavelength UV spectrophotometer is a very versatile detector for clinical laboratories.

*3. Sample Preparation.* Appropriate sample pretreatment is important. The most sensitive detector and the best chromatographic column are useless if the sample is treated inappropriately prior to chromatography. Sample pretreatment should be simple and convenient yet give good recoveries and be reasonably selective.

I have discussed many of the beneficial aspects of liquid chromatography as well as its limitations and drawbacks. Some of the limitations that we experience today may be overcome as improvements in technology occurs. With the development of more efficient columns, separations that are difficult or impossible to achieve today may be performed with relative ease in the future.

Another area of potential improvement is that of sample preparation. Many of the time consuming manipulations could be automated by microprocessor controlled devices. Some of these devices are available at this time, and improvements in hardware will make this a more versatile and useful tool.

As I have pointed out, there will always be the potential of interference in chromatography as well as in spectrophotometric procedures. There needs to be much improved interfacing between chromatography and on-line spectrophotometry to help eliminate the nonspecificity of each technique. Combining these two techniques will greatly improve confidence in the reported analytical result.

## Acknowledgment

I would like to thank Mrs. Mary Lynn Koller for her typing of the manuscript.

## References


1. Schack, J. A., and Waxler, S. H., *J. Pharmacol. Exp. Ther.* **97,** 283 (1949).
2. Johnson, G. F., Dechtiaruk, W. A., and Soloman, H. M., *Clin. Chem.* **21,** 144 (1975).
3. Pranskevich, C. A., Swihart, J. I., and Thoma, J. J., *J. Anal. Tox.* **2,** 3 (1978).
4. Least Jr., C. J., Johnson, G. F., and Solomon, H. M., *Clin. Chem.* **22,** 765 (1976).
5. Sitar, D. S., Piafsky, K. M., Rangno, R. E., and Oglivie, R. I., *Clin. Chem.* **21,** 1774 (1975).
6. Evenson, M. A., and Warren, B. L., *Clin. Chem.* **22,** 851 (1976).
7. Weddle, O. H., and Mason, W. D., *J. Pharm. Sci.* **6,** 865 (1976).
8. Adams, R. F., Vandemark, F. L., and Schmidt, G. J., *Clin. Chem.* **22,** 1903 (1976).
9. Orcutt, J. J., Kozak, Jr., P. P., Gillman, S. A., and Cummins, L. H., *Clin. Chem.* **23,** 599 (1977).
10. Kelly, R. C., Prentice, D. E., and Hearne, *Clin. Chem.* **24,** 838 (1978).
11. Soldin, S. J., and Hill, J. G., *Clin. Biochem.* **10,** 74 (1977).
12. Robinson Jr., C. A., Mitchell, B., Vasiliades, J., and Siegel, A. L., *Clin. Chem.* **24,** 1847 (1978).
13. Robinson, Jr., C. A., and Dobbs, J., *Clin. Chem.* **24,** 2208 (1978).
14. Popovich, D. J., Butts, E. T., and Lancaster, C. J., *J. Liq. Chrom.* **1,** 469 (1978).

15. Manion, C. V., Shoeman, D. W., and Azarnoff, D. L., *J. Chrom.* **101,** 169 (1974).
16. Weinberger, M., and Chidsey, C., *Clin. Chem.* **21,** 834 (1975).
17. Franconi, L. C., Hawk, G. L., Sandmann, B. J., and Haney, W. G., *Anal. Chem.* **48,** 372 (1976).
18. Hill, R. E., *J. Chrom.* **135,** 419 (1977).
19. Peat, M. A., and Jennison, J. A., *J. Anal. Tox.* **1,** 204 (1977).
20. Maijub, A. G., Stafford, D. T., and Chamberlain, R. T., *J. Chrom. Sci.* **14,** 52 (1976).
21. Aggarwal, V., and Bath, R., in *Methodology for Analytical Toxicology,* Sunshine, I., ed., CRC Press, Cleveland, Ohio, 1975.
22. Atkinson Jr., A. J., Parker, M., and Strong, J., *Clin. Chem.* **18,** 643 (1972).
23. Adams, R. F., Vandemark, F. L., and Schmidt, G., *Clin. Chem. Acta.* **69,** 515 (1976).
24. Schmidt, G., Vandemark, F. L., and Adams, R. F., *Chromatography Newsletter* **4,** 32 (1976).
25. Elson, J., Strong, J. M., Lee, W. K., and Atkinson, Jr., A. J., *Clin. Pharmacol. Ther.* **17,** 134 (1975).
26. Rocco, R. M., Abbott, D. C., Giese, R. W., and Karger, B. L., *Clin. Chem.* **23,** 705 (1977).
27. Black, J. W., Duncan, W. A. M., and Shanks, R. G., *Brit. J. Pharmacol.* **25,** 596 (1965).
28. DiScalle, E., Baker, K. M., Bareggi, S. R., Watkins, W. D., Chidsey, C. A., Frigerio, A., and Morselli, P. L., *J. Chrom.* **84,** 347 (1973).
29. Schmidt, G. J., and Vandemark, F. L., *Chromatography Newsletter* **5,** 42 (1977).
30. Nygard, G., Shelver, W. H., and Wahba Khalil, S. K., *J. Pharm. Sci.* **68,** 379 (1979).
31. Jatlow, P., Bush, W., and Hochster, H., *Clin. Chem.* **25,** 777 (1979).
32. Nation, R. L., Peng., G. W., and Chiou, W. L., *J. Chrom. Biomed. App.* **145,** 429 (1978).
33. Brodie, B. B., and Udenfriend, S., *J. Pharmacol. Exp. Ther.* **78,** 154 (1943).
34. Drayer, D. E., Restivo, K., and Reidenberg, M. M., *J. Lab. Clin. Med.* **90,** 816 (1977).
35. Crouthamel, W. G., Kowarski, B., and Narang, P. K., *Clin. Chem.* **23,** 2030 (1977).
36. Sved, S., McGilveray, I. J., and Beaudoin, N., *J. Chrom. Biomed. App.* **145,** 437 (1978).
37. Kline, B. J., Turner, U. A., and Barr, W. H., *Anal. Chem.* **51,** 499 (1979).
38. Powers, J. L., and Sadee, W., *Clin. Chem.* **24,** 299 (1978).
39. Peat, M. A., and Jennison, T. A., *Clin. Chem.* **24,** 2166 (1978).
40. Weidner, H., Ladenson, J. H., Larson, L. Kessler, G., and McDonald, J. M., *Clin. Chem. Acta.* **91,** 7, (1979).
41. Bonora, M. R., Gunetert, T. W., Upton, R. A., and Riegelman, S., *Clin. Chem. Acta.* **91,** 277 (1979).

42. Hayler, A. M., and Flanagan, R. J., *J. Chrom.* **153,** 461 (1978).
43. Meffin, P. J., Harapat, S. R., and Harrison D. C., *J. Chrom.* **132,** 503 (1977).
44. Ilett, K. F., Hackett, L. P., Dusci, L. J., and Tjokrosetio, R., *J. Chrom.* **154,** 325 (1978).
45. Lima, J. J., *Clin. Chem.* **25,** 405 (1979).

## Editors' Note

We have recently completed the development of a method that is able to analyze all of the antiarrhythmic agents mentioned in this chapter (procainamide, *N*-acetylprocainamide (NAPA), lidocaine, quinidine, disopyramide, *N*-desisopropyl, and propranolol) utilizing a newly introduced $C_8$ column (Ultrasphere-octyl 5 $\mu$M, Altex Scientific, Berkeley, CA 94710) (Kabra, Chen, and Marton, *Therapeutic Drug Monitoring,* in press). All seven drugs can be analyzed simultaneously by using gradient LC, or in two panels, isocratically. Procainamide and NAPA can be analyzed together using a low acetonitrile concentration, while the remaining drugs can be analyzed simultaneously at a higher acetonitrile concentration using the same column. The method is sensitive enough to monitor low therapeutic concentrations of propranolol and quinidine at 216 nm. If additional sensitivity is desired (i.e., for pediatric samples), it is readily achieved by utilizing fluorescence detection. Because of the infrequent need to analyze some of these drugs, the availability of a single versatile analytical method has distinct advantages.

# Chapter 7

## Antibiotics

John P. Anhalt

*Department of Laboratory Medicine*
*Mayo Clinic and Mayo Foundation*
*Rochester, Minnesota*

## I. Introduction

### A. Case Histories

A 28-year-old man was transferred to our hospital and underwent surgery for resection of an aortic graft infected with *Klebsiella pneumoniae*. Antimicrobial therapy consisted of amikacin, cefazolin, chloramphenicol, sulfamethoxazole, and trimethoprim. A request for amikacin and sulfamethoxazole assays was received by the laboratory along with information that the patient had received tobramycin until 24 h before the serum was obtained.

A 32-year-old woman with a complicated medical history developed bacteremia with *Klebsiella pneumoniae* while receiving tobramycin and cefazolin. Amikacin was given on the suspicion that the organism was resistant to tobramycin. She was hypotensive and anuric. Serum drug levels for each aminoglycoside were requested.

### B. Need for Specificity

Although these cases involved complicated medical and surgical decisions, the involvement of the laboratory arose in essentially a typical manner, namely through a test request. The role of the laboratory was to provide the requested services commensurate with the medical urgency and to interpret the results if necessary. The role of the laboratory in therapeutic drug monitoring is often limited to these technical and interpretive functions. The decisions regarding which drugs to measure and at what times usually are made by the physician directly responsible for the patient. However, advice from the laboratorian may be requested by the clinician, particularly when the laboratorian has demonstrated competence, and ordering practices can be influenced by the services offered. The questions that must be addressed in this expanded role cannot be answered independently. Obviously, which antimicrobics should be measured is related to why measurements should be made, and available methodology and laboratory size influence which assays can be done. Less obviously, the reason for an assay can affect the choice of methodology. Pharmacological studies require accurate, but not necessarily rapid or specific assays, whereas clinical assays can often be less accurate, but must be specific and sufficiently rapid to allow dosage adjustments based on the results to be made. Specificity in particular is of great importance. Often, patients receive multiple antimicrobics that may be chemically similar or which may have overlapping spectra of activity. The two cases described above illustrate situations in which typical bioassay procedures would be inadequate because of poor specificity. The need for specificity is underscored by the fact that a laboratory cannot rely on having accurate information concerning antimicrobial mixtures. Reeves and Holt (*1*), for example, found undisclosed antimicrobics in 19% of the sera submitted to their laboratory for assay. In addition, microbiologically active and inactive metabolites of antimicrobics may be present in sera. For example, the metabolites of sulfonamides and chloramphenicol are inactive; the metabolites of cephalothin, cephapirin, and rifampin have only a portion of the activity of the parent drugs; and degradation of carbenicillin to benzylpenicillin gives a product with a different spectrum of activity (*2, 3*). Accuracy is less important in clinical assays than in pharmacological studies. Reeves and Wise (*2*) concluded that for clinical assays of drugs such as aminoglycosides, a 95% confidence limit of ±25–30% (a CV of about 10–15%) was adequate. For drugs with a low potential for toxicity, a 95% confidence limit of ±50% was considered adequate. These limits are within the capabilities of most assay procedures. Speed is often emphasized for clinical assays. What is actually needed in most cases is a result within the time limit of a dosing interval.

## II. Efficient Utilization of Resources

Liquid chromatographic methods with adequate specificity, accuracy, and rapidity have been developed for most antimicrobics (Table 1), and additional methods are being developed at a rapid pace. However, many antimicrobics are assayed either too infrequently or the tolerable limits for accuracy are so great that to maintain liquid chromatographic procedures for these drugs would result in poor utilization of resources. The practical use of liquid chromatography may be limited to assays of those drugs that constitute a significant percentage of the laboratory workload or which have had a high potential for toxicity.

### A. Reasons to Monitor

The purpose of monitoring drug levels is to ensure that dosage is sufficient for adequate therapy, while avoiding the excessive levels that may be associated with increased toxicity.

The determination of the level of antimicrobic in serum or other fluids can be justified only when that level can be correlated with

Table 1
Liquid Chromatographic Assays of Antimicrobics in Serum, CSF, or Urine[a]

| | | |
|---|---|---|
| Aminoglycoside group | β-Lactam group | |
| Gentamicin | Cephalothin | Cefatrizine |
| Tobramycin | Cephalexin | Cefuroxime |
| Amikacin | Cephradine | Cefoxitin |
| Netilmicin | Cephaloridine | Amoxycillin |
| | Cefazolin | Ampicillin |
| | Cefamandole | |
| | Miscellaneous | |
| Amphotericin B | Isoniazid | Rifampin |
| Chloramphenicol | Mefloquine | Sulfonamides |
| Dapsone | Metronidazole | Tetracyclines |
| Econazole | Misonidazole | Tinidazole |
| Erythromycin | Nalidixic acid | Trimethoprim |
| Flucytosine | Niridazole | Vancomycin |
| Griseofulvin | Nitrofurantoin | |

[a]References are contained in a review by Gerson and Anhalt (*4*), except for cefatrizine (*5*) and erythromycin (*6*).

benchmarks of efficacy or excess (*7*). The relationship of efficacy to serum levels of antimicrobics is complex and difficult to study. Most antimicrobics have a large therapeutic index, and the tendency has been to administer relatively large doses to give levels several-fold the assumed minimal effective level (*2*). Antimicrobics that have a narrow therapeutic index, such as aminoglycosides, have had their effective and toxic levels defined more precisely (*2, 8*). All drugs can produce adverse effects, and the incidence of these effects generally increases with the concentration of the drug in the patient. Therefore, even for drugs with a high therapeutic index, analyses can be justified when there is evidence that the serum level may be markedly different from usual or cannot be predicted even roughly owing to the complexity of the clinical situation. Related to the subject of excessive levels are questions of whether serum levels can be used to predict and to avoid toxicity, and whether other, perhaps better, indicators of impending toxicity exist (*7*). Alternative indicators of acute toxicity often do not exist for antimicrobics. For example, with total renal failure, serum creatinine increases at the rate of only 1.1 mg/dL (100 μmol/L) per day (*2*). Similarly, clinical indications are inadequate to predict impending cardiovascular collapse (grey syndrome) from chloramphenicol toxicity (*9*). In contrast, Lau et al. (*10*) found that serum creatinine rose sooner than trough levels in chronic toxicity from amikacin and gentamicin. This observation, however, does not eliminate the need to measure nephrotoxic drugs, because serum creatinine can increase owing to a variety of causes.

Lastly, antimicrobic concentrations in serum or other fluids should be measured when they are not adequately predicted by dosage. Compliance in taking a prescribed dose is rarely a significant problem with antimicrobic therapy of hospitalized patients. Medication errors do occur, however, and can result in extremely unusual levels (*11*). Various methods have been developed to calculate dosages based on desired serum levels and measured or estimated values for absorption, distribution, and excretion. The adequacy of these predictions varies with the clinical situation and the acceptability of error. These calculations are particularly inaccurate for critically ill patients in whom absorption from an intramuscular injection may vary, in patients with impaired or changing renal function, in patients with extensive burns, and when rapid attainment of adequate levels of a toxic drug is needed to treat a life-threatening infection. In addition, the calculations do not account for the effects of fever on decreasing peak levels, the effects of metabolic variations, the effects of dialysis, or the effects of drugs that change the rate of elimination. Antimicrobic levels in fluids other than serum must be determined by direct measurement. Chronic or prolonged (>10 days) administration of a drug is another indication for an occasional serum assay to guard against unsuspected accumulation.

Implied in the measurement of serum levels is the assumption that adjustment of dosage can be used to change the level. This assumption holds true for most antimicrobics, but may not be true for amphotericin B, which is a colloid. Bindschadler and Bennett (*12*) found that immediately after an intravenous infusion, only 10% of the dose could be accounted for in serum, and the loss was not a consequence of excretion. There was also poor correlation between the amount administered and concentration at various times after infusion. They proposed a mechanism in which drug was rapidly removed from the circulation during infusion and then slowly liberated to maintain a constant low level for several hours. The highly predictable serum levels and the absence of a relationship between dose and serum level decreases the significance for routine determinations of this drug, even though it has significant toxicity.

### B. Mayo Clinic Experience

One measure of the drugs for which a laboratory should provide assays is given by the ordering practices of the physicians served. Data from the Mayo Clinic for 1978 are shown in Table 2. As expected, the aminoglycosides outnumber all other antimicrobics and represent 83% of the assays. The β-lactam antimicrobics represent 12% of the assays. This high percentage probably reflects the high frequency of use, rather than concern that they are more toxic than less frequently used drugs, such as vancomycin and chloramphenicol.

## III. Current Scope of Liquid Chromatographic Assays

Table 1 lists the antimicrobics for which assays of serum, CSF, or urine have been reported. In most of the procedures, sample preparation involves either protein precipitation followed by analysis of the protein-free fluid, or extraction of the antimicrobic into an organic solvent. Reversed-phase chromatography has been used in almost all of the recently described procedures, and with the notable exception of the aminoglycosides, detection has been by ultraviolet absorption. The major shortcomings of many of the procedures have been a failure to incorporate an internal standard and an emphasis on individual drugs rather than on a class of drugs, such as the β-lactam antimicrobics.

Table 2
Antimicrobic Assays at the Mayo Clinic in 1978[a]

| Antibiotic | Number | |
|---|---|---|
| *Aminoglycosides* | | 1363 |
| Gentamicin | 839 | |
| Tobramycin | 421 | |
| Amikacin | 53 | |
| Kanamycin | 2 | |
| Streptomycin | 18 | |
| Neomycin | 30 | |
| *Cephalosporins* | | 60 |
| Cephalothin | 42 | |
| Cefazolin | 8 | |
| Cephalexin | 10 | |
| *Penicillins* | | 133 |
| Penicillin G | 27 | |
| Ampicillin | 29 | |
| Amoxicillin | 2 | |
| Oxacillin | 18 | |
| Methicillin | 4 | |
| Dicloxacillin | 14 | |
| Nafcillin | 12 | |
| Carbenicillin | 26 | |
| Ticarcillin | 1 | |
| *Other* | | 86 |
| Vancomycin | 46 | |
| Chloramphenicol | 24 | |
| Trimethoprim | 8 | |
| Clindamycin | 7 | |
| Tetracycline | 1 | |

[a]Data for sulfonamides are not available.

## A. β-Lactam Antimicrobics

Procedures for various β-lactam antimicrobics are summarized in Table 3. As a group, the β-lactam drugs represent a major portion of the assays performed in our laboratory; however, the individual members of this class represent only a relatively minor part of the workload. A method that was applicable to the entire class would greatly increase the usefulness of liquid chromatography for analysis of these drugs in a clinical laboratory.

Table 3
Analysis of $\beta$-Lactam Antimicrobics in Serum or Urine[a]

| Antibiotic | Extraction | Mobile phase | Column[b] | Limit[c] |
|---|---|---|---|---|
| Cephalothin | Ion-pair extraction from serum into ethyl acetate; urine injected directly | Sodium dihydrogen phosphate-sodium nitrate buffer, pH 4.8 | AS-Pellionex-SAX™ | 1 |
| | Protein precipitated with trichloroacetic acid | Methanol in aqueous ammonium acetate | PhenylCorasil™ | 10 |
| | Protein precipitated with dimethylformamide | Methanol in aqueous acetic acid | $\mu$-Bondapak $C_{18}$™ | 1 |
| | Urine injected directly | Sodium acetate buffer, pH 5.0 | Zipax SAX™ | 2 |
| Cephalexin | Urine injected directly | Methanol in aqueous ammonium carbonate | $C_{18}$ on LiChrosorb™ Si 100 | 5 |
| | Urine injected directly | Methanol in 0.5% aqueous acetic acid | $\mu$-Bondapak $C_{18}$ | 1 |
| Cephradine | Serum and urine injected directly | Methanol in aqueous ammonium carbonate | $C_{18}$ on LiChrosorb Si 100 | 5 |
| Cephradine | Serum and urine injected directly | Methanol in aqueous ammonium carbonate | $C_{18}$ on LiChrosorb Si 100 | 5 |
| Cephaloridine | Protein precipitated with trichloroacetic acid | Methanol in aqueous ammonium acetate | PhenylCorasil | 2 |
| Cefazolin | Protein precipitated with trichloroacetic acid | Methanol in aqueous ammonium acetate | PhenylCorasil | 10 |
| | Protein precipitated with trichloroacetic acid | Methanol in 1% aqueous acetic acid | PhenylCorasil | 1.6 |
| Cefamandole | Plasma acidified with acetic acid; urine acidified with sodium acetate buffer, pH 5.0 | Acetonitrile in 1% aqueous acetic acid | Vydac™ reverse phase | 5 |
| Cefatrizine | Protein precipitated with trichloroacetic acid | Acetonitrile in sodium phosphate buffer, pH 7 | Octadecylsilyl reversed-phase | 0.1 |
| Cefuroxime | Protein precipitated with dimethylformamide | Methanol in 1% aqueous acetic acid | $\mu$-Bondapak $C_{18}$ | 1 |
| Cefoxitin | Urine injected directly | Sodium acetate buffer, pH 5.0 | Zipax SAX | 2 |
| Ampicillin | Protein precipitated with perchloric acid | Potassium dihydrogen phosphate buffer, pH 4.6 | LiChrosorb RP-8 | 0.5 |
| Amoxycillin | Protein precipitated with perchloric acid | Mixture of methanol and potassium dihydrogen phosphate buffer, pH 4.6 | LiChrosorb RP-8 | 0.5 |

[a]References are contained in a review by Gerson and Anhalt (*4*), except for cefatrizine (*5*).

[b]Trademarks: AS-Pellionex-SAX™, Whatman, Clifton, NJ; PhenylCorasil™ and $\mu$-Bondapak $C_{18}$™, Waters Associates, Milford, MA; LiChrosorb™, E. Merck, Darmstadt, F. R. Germany; Vydac™ reverse phase, The Separations Group, Hesperia, CA; Zipax SAX™, Dupont, Wilmington, DE.

[c]Minimal concentration measured in serum, mg/L.

## B. Aminocyclitol Antimicrobics

Aminocyclitols are a class of antimicrobics that include spectinomycin (Fig. 1) and the aminoglycosides. The common aminoglycosides used parenterally are gentamicin, tobramycin, amikacin, and streptomycin (Figs. *2–4*). Among these, gentamicin is unique in that it is not a pure chemical substance, but is a complex mixture of chemically similar components. The principal component of this mixture is the C-complex, which in turn can be separated into approximately equal amounts of gentamicin $C_1$ and $C_2$, a lesser amount of gentamicin $C_{1a}$, and two minor components, gentamicin $C_{2a}$ and $C_{2b}$ (*13*). Gentamicin $C_{2a}$ is a stereoisomer of gentamicin $C_2$ at the 6′-position. All of the C-complex components have similar antimicrobial activity. Neomycin and kanamycin are also mixtures of related compounds. Neomycin consists of approximately 87% neomycin B and 13% neomycin C;

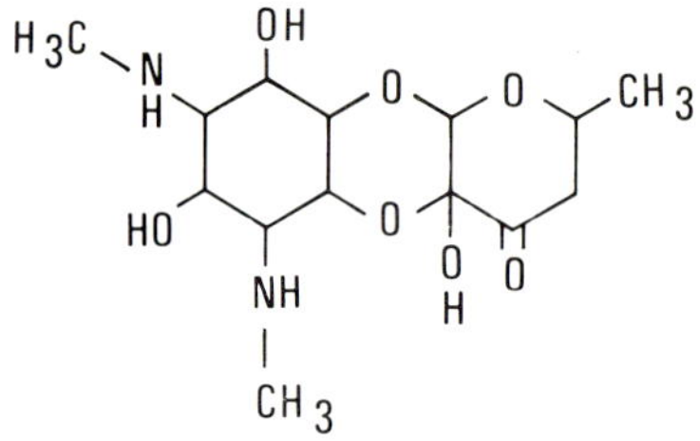


FIG. 1. Structure of the aminocyclitol, spectinomycin.

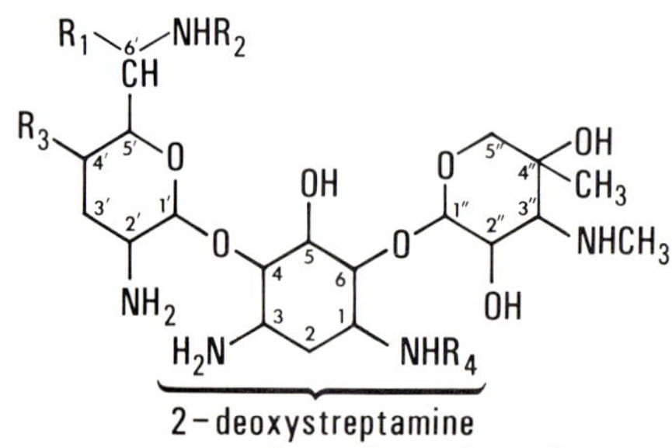


| | $R_1$ | $R_2$ | $R_3$ | $R_4$ |
|---|---|---|---|---|
| Gentamicin $C_{1a}$ | H | H | H | H |
| Gentamicin $C_2$ | $CH_3$ | H | H | H |
| Gentamicin $C_{2a}$ | $CH_3$ | H | H | H |
| Gentamicin $C_{2b}$ | H | $CH_3$ | H | H |
| Gentamicin $C_1$ | $CH_3$ | $CH_3$ | H | H |
| Sisomicin | H | H | $\Delta^{4'}$ | H |
| Netilmicin | H | H | $\Delta^{4'}$ | $C_2H_5$ |

FIG. 2. Structures of the gentamicin-like aminoglycosides.

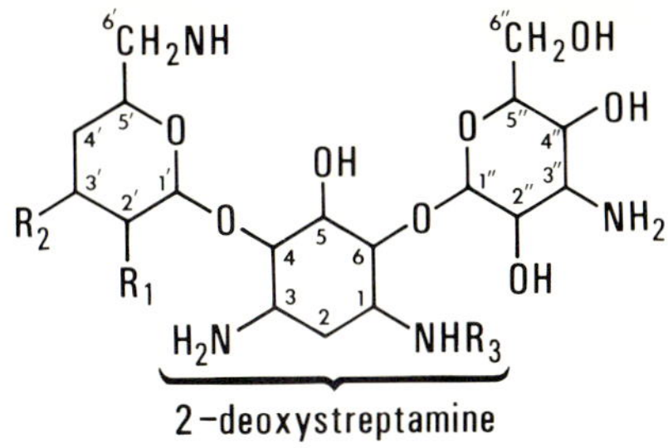


| | $R_1$ | $R_2$ | $R_3$ |
|---|---|---|---|
| Kanamycin A | OH | OH | H |
| Kanamycin B | $NH_2$ | OH | H |
| Tobramycin | $NH_2$ | H | H |
| Amikacin | OH | OH | $COCH(OH)CH_2CH_2NH_2$ |

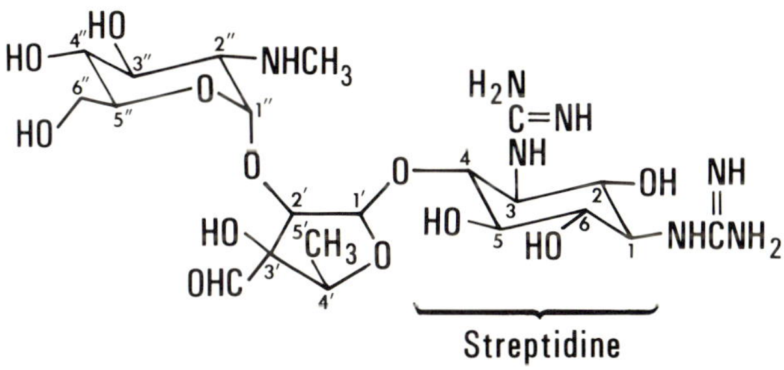


however, it is usually used topically and the need rarely arises to measure serum levels. Kanamycin consists predominately of kanamycin A with less than 5% of kanamycin B. Kanamycin is rarely used today as a parenteral drug. All of the aminoglycosides can potentially cause ototoxicity and nephrotoxicity, although the precise relationship of serum levels to development of toxicity is poorly defined (*8, 14*). Aminoglycosides share many pharmacological properties (*8*). They are poorly absorbed after oral administration; however, toxic levels can accumulate after oral or rectal administration to patients with impaired renal function. Streptomycin is usually given intramuscularly, while the other parenteral agents may be given intramuscularly or intravenously. Absorption is complete after intramuscular injection, and peak levels can occur from 0.5 to 3 h after a dose. The level 1 h after an injection, however, will usually be at least 70% of the peak level. Aminoglycosides distribute in the extracellular fluid volume (about 30% of lean body weight) and are excreted almost entirely through the kidney by glomerular filtration. They are not metabolized, and the elimination half-life with normal renal function from serum is 2–3 h. With the exception of streptomycin, about 30–35% of which is protein bound, other aminoglycosides have been reported not to be protein bound (*8*). This conclusion, however, has been disputed. Recent studies with gentamicin showed that binding may occur to the extent of 20% in normal serum, and that heparin in plasma can greatly increase the extent of apparent binding (*15*). The desirable levels for aminoglycosides are shown in Table 4.

Spectinomycin is unlike the aminoglycosides in that it lacks an amino sugar and is of low toxicity (*16*). It is now used to treat infections caused by penicillin-resistant *Neisseria gonorrhoeae,* and a single 2-g intramuscular dose will give a peak serum concentration of about 100 μg/mL. The serum elimination half-life is 1 h and elimination is predominately via the kidney, although only 74% of a dose can be recovered in urine as active drug.

The aminocyclitols are basic, water-soluble drugs that cannot be extracted from serum into nonaqueous organic solvents. The liquid chromatographic assay of these drugs presented a challenge in that they cannot be detected at clinically significant concentrations using ultraviolet absorption. For the aminoglycosides, this problem was readily solved by the preparation of derivatives that were detected by either ultraviolet absorption or fluorescence. The assay of spectinomycin was more complex from a technical standpoint, because the drug not only lacked a suitable chromophore, but also lacked a primary amino group.

Table 4
Desirable Aminoglycoside Levels[a]

| | Desirable level, μg/mL | | |
|---|---|---|---|
| Antibiotic | Peak | Trough | Toxic range, μg/mL |
| Gentamicin | 5–8 | 1–2 | >10–12 |
| Tobramycin | | | |
| Sisomicin | | | |
| Netilmicin | | | |
| Kanamycin | 20–25 | 5–10 | >30–35 |
| Amikacin | | | |
| Streptomycin | 5–20 | <5 | >40–50 |

[a]Data according to Barza and Scheiffe (*8*).

Derivatization can be performed either before chromatography (pre-column derivatization) or after analytical separation (post-column derivatization). Post-column derivatization requires that the chemical reaction proceed rapidly if the method is to be used with a continuous-flow reactor and excessive band broadening is to be avoided. If offers the advantage, however, that the chemical reaction need not give a single product derived from the analyte. Thus, a reaction that gives a mixture of degradation products or a reaction in which the detected entity is not derived from the analyte can be used as long as the detector response is proportional to the amount of analyte injected. Pre-column derivatization allows the use of slower chemical reactions, but generally requires that a single product be derived from the analyte. Derivatization should go to completion because of the difficulty in precisely controlling conditions, and the derivative should be stable. Neither of these latter conditions are required with post-column derivatization.

Mays and associates (*17*) studied several possible post-column derivatization reactions for the liquid chromatographic analysis of kanamycin. All of the reactions were well known methods for derivatization of primary amines. Reaction with ninhydrin was unsatisfactory because of the lengthy reaction time required. Reaction with trinitrobenzene sulfonic acid at 80° C was faster, but the method was plagued by technical problems and was not as sensitive as methods based on reaction with fluorescamine or *o*-phthalaldehyde. The latter reagents also reacted at ambient temperature and performed about equally, but fluorescamine was considerably more expensive. All of these studies were done with aqueous solutions of the antimicrobic and used ion-exchange chromatography.

Assays of aminoglycosides in clinical specimens have all used reversed-phase chromatography, either *o*-phthalaldehyde or 5-dimethylamino-1-naphthalenesulfonyl chloride (dansyl chloride) for derivatization, and a fluorescence detector. A summary of these procedures is given in Table 5.

The method developed in my laboratory for aminoglycoside assays (*18, 19*) uses post-column derivatization with *o*-phthalaldehyde. A diagram of the chromatographic system is shown in Fig. 5. Chromatography is performed on a $C_{18}$-bonded-phase column with a mobile phase containing sodium pentane sulfonate (0.02 mol/L) as an ion-pairing reagent, sodium sulfate (0.2 or 0.1 mol/L), and 0.1% (vol/vol) acetic acid in a water–methanol mixture. The sodium sulfate acts to increase ionic strength, which was predicted to be necessary for separation by ion-pair chromatography of analytes with several ionic groups (*18*). In the absence of this salt, the chromatographic peaks

Table 5
Summary of Liquid Chromatographic Methods for Aminoglycosides

| Aminoglycoside | Internal standard | Extraction | Separation | Derivatization | Ref. |
|---|---|---|---|---|---|
| Gentamicin | *N*-acetylgentamicin $C_1$ or tobramycin | CM-Sephadex ion exchange | Reversed-phase, ion-pair | Post-column, *o*-phthalaldehyde | *18, 19* |
| | None | Solvent extraction | Reversed-phase | Pre-column, dansyl chloride | *21, 23* |
| | None | Silica gel adsorption | Reversed-phase | Pre-column, *o*-phthalaldehyde | *25* |
| | None | Solvent extraction | Reversed-phase | Pre-column, *o*-phthalaldehyde | *24* |
| Tobramycin | Amikacin | CM-Sephadex ion exchange | Reversed-phase, ion-pair | Post-column, *o*-phthalaldehyde | *19* |
| | None | Silica gel adsorption | Reversed-phase | Pre-column, *o*-phthalaldehyde | *26* |
| | None | Solvent extraction | Reversed-phase | Pre-column, *o*-phthalaldehyde | *24* |
| Amikacin | Tobramycin | CM-Sephadex ion exchange | Reversed-phase | Post-column, *o*-phthalaldehyde | *19* |
| | None | Silica gel adsorption | Reversed-phase | Pre-column, *o*-phthalaldehyde | *27* |
| Netilmicin | None | Solvent extraction | Reversed-phase | Pre-column, dansyl chloride | *22* |
| | None | Solvent extraction | Reversed-phase | Pre-column, *o*-phthalaldehyde | *24* |

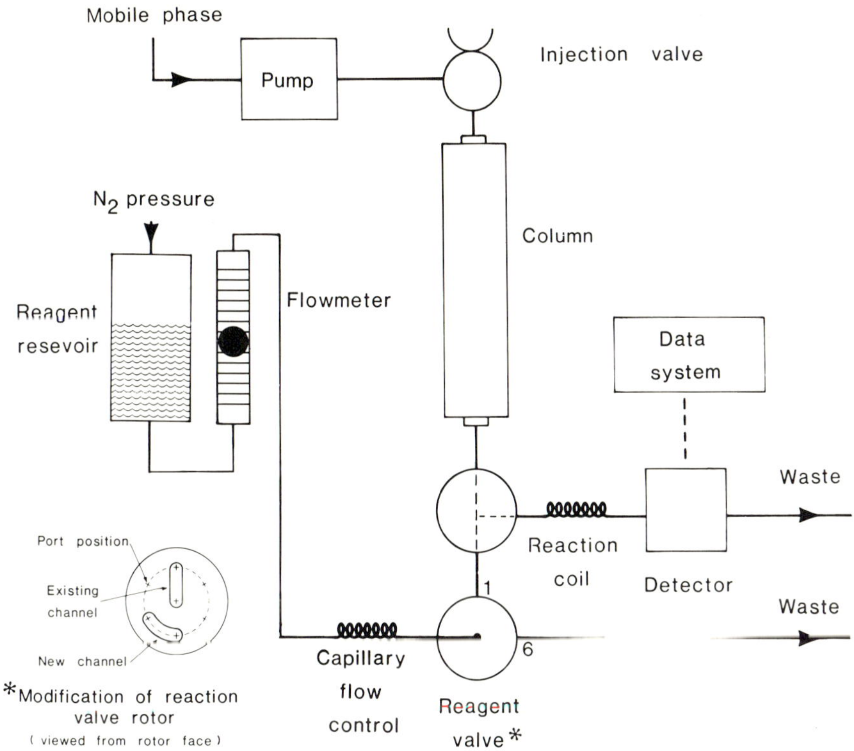


FIG. 5. Diagram of chromatographic system used for post-column derivatization.

were skewed, high concentrations of methanol were required to elute the aminoglycosides, and the components of gentamicin were not resolved. A weak cation-exchange resin, such as CM-Sephadex (Pharmacia Fine Chemicals, Inc., Piscataway, N.J.), is used to extract the aminoglycosides from serum. The resin is washed with excess sodium sulfate solution, and the bound aminoglycosides are then eluted from the resin with a dilute solution of sodium hydroxide. The extracts must be acidified and ion-pairing reagent must be added before analysis. If these steps are omitted, retention times may vary, and chromatographic peaks may be distorted or even split factitiously. An internal standard is added to the sample before addition to the cation-exchange resin. Originally, the internal standard was prepared in sodium sulfate solution; however, our experience with large numbers of clinical specimens showed that a diluent containing 0.2 mol/L sodium phosphate at pH 6.9 gave better reproducibility. More than 2000 clinical specimens have been analyzed to date in my

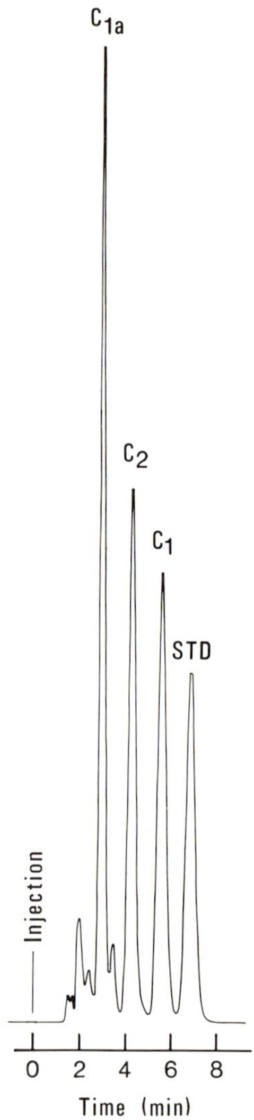


FIG. 6. Representative chromatogram of gentamicin (10 mg/L) and 1-*N*-acetylgentamicin (STD) obtained by reversed-phase, ion-pair chromatography and post-column derivatization.

laboratory with this modified procedure. A typical chromatogram of a serum extract is shown in Fig. 6.

Pre-column derivatization for clinical analyses has been done with either *o*-phthalaldehyde or dansyl chloride as reagents (Table 5). An analytical method for neomycin, gentamicin, and kanamycin using

1-fluoro-2,4-dinitrobenzene as a derivatization reagent and ultraviolet absorption detection has also been reported, but this method has been applied only to aqueous standards and has not been used for analysis of serum or other biological fluids (*20*).

The method of Peng et al. (*21, 22*) for analysis of gentamicin and netilmicin in clinical specimens used pre-column derivatization with dansyl chloride. Sample preparation involved dilution and alkalinization of serum followed by addition of acetonitrile to precipitate proteins. The protein-free solution was then extracted with methylene chloride. A solution of dansyl chloride was added to the aqueous phase, and the mixture was heated at 75° C for 5 min. The derivatized aminoglycosides were extracted from this reaction mixture into ethyl acetate, which was injected directly into the liquid chromatograph. Analysis used a reversed-phase column and an acetonitrile–water mixture as mobile phase. No internal standard was used for either analysis, and gentamicin components $C_{1a}$ and $C_2$ were not resolved from each other. A later note on this method (*23*) cautioned that a column that had been exposed to an acidic mobile phase was rendered unsuitable for this particular analysis. This observation of a "memory effect" had not been published previously for aminoglycoside analyses, but is worth considering when troubleshooting any procedure.

The method of Bäck et al. (*24*) for analysis of gentamicin, tobramycin, and netilmicin used essentially the same method of sample preparation as used by Peng et al. (*21*), except that *o*-phthalaldehyde was used instead of dansyl chloride to derivatize the aminoglycosides. This substitution eliminated the necessity to heat the reaction mixture and also allowed chromatographic resolution of the three major gentamicin components. In contrast to the elution order shown in Fig. 6, the elution order for *o*-phthalaldehyde derivatives was gentamicin $C_1$ before $C_{1a}$ before $C_2$. No internal standards were used, and the mobile phase was basically a water–methanol mixture with a buffer or triethylamine added to improve chromatographic characteristics. Analytical recovery was insensitive to sample volume for gentamicin and netilmicin, but recovery of tobramycin decreased if sample size was increased. The authors postulated that this problem resulted from co-precipitation of tobramycin with serum proteins.

A third approach for sample preparation and pre-column derivatization of aminoglycosides was developed by Maitra et al. (*25*). This method was used originally for gentamicin and was applied later to tobramycin (*26*) and amikacin (*27*). The aminoglycosides were adsorbed from serum onto silica gel, which was then washed with a buffer to remove unadsorbed components. While still adsorbed to the silica gel, the aminoglycosides were derivatized by addition of *o*-phthalaldehyde. The derivatized compounds were then eluted with either ethanol or isopropyl alcohol and chromatographed using a reversed-phase column. The three gentamicin components were resolved using a methanol–water mobile phase containing a small amount of buffer. As in the method of Bäck et al. (*24*), gentamicin $C_1$ eluted before $C_{1a}$, which eluted before $C_2$. Essentially the same procedure was used for tobramycin and amikacin with modification of the mobile phase. Although these drugs are single components, each gave at least two chromatographic peaks after extraction from serum and derivatization. When the extracts were heated, however, one of the peaks could be made to disappear while the other increased in intensity. No internal standard was used in these analyses, and Bäck et al. (*24*) later reported that recovery of tobramycin and netilmicin was incomplete. Bäck et al. (*24*) also observed only a single chromatographic peak from tobramycin when derivatization was performed at ambient temperature in solution in contrast to the results when derivatization was done while the drug was adsorbed to silica gel. It has not been shown whether this difference owes to formation of different products or whether the chromatographic conditions allowed resolution of the products in one case but not in the other.

Analysis of the aminocyclitol, spectinomycin (Fig. 1), presented an added challenge because of the absence of a primary amino group. Myers and Rindler (*28*) used a two-step, post-column reaction to solve the detection problem. The first step was oxidation with sodium hypochlorite at 100° C. The reaction was rapid and liberated primary amines. These were then detected by fluorescence after reaction with *o*-phthalaldehyde. The procedure was used with aqueous solutions of drug and was not applied to analysis of serum or other biological fluids. The chromatographic mobile phase was similar to that described above for ion-pair chromatography of aminoglycosides, except sodium heptanesulfonate was used instead of sodium pentanesulfonate as the ion-pairing reagent.

## C. Vancomycin

Vancomycin is a bactericidal antimicrobic active against a great variety of gram-positive bacteria and some gram-negative cocci (*29*). The serum half-life is 6 h in patients with normal renal function and approximately 10% is bound to protein (*30, 31*). The usual dosage schedule is 1 g intravenously every 12 h, which results in peak levels of 30–40 mg/L and minimum levels of 5–10 mg/L in adults (*30, 31*). Vancomycin is excreted almost entirely by the kidney, and 90–100% of administered drug activity can be recovered in urine. The most serious dose-related toxicity affects the auditory nerve, which occurs only rarely when levels are kept below 30 mg/L (*30*). Higher levels have been associated with toxicity in patients with impaired renal function in whom the levels would be expected to remain high for a prolonged period (*32*). In patients with normal renal function, levels as high as 90 mg/L were tolerated without toxicity (*31*).

The vancomycin hydrochloride used in therapy contains a mixture of compounds. The major component, vancomycin (Fig. 7), constitutes at least 85% of the mixture. The other components appear to be derived from vancomycin by partial hydrolysis (*33*).

Uhl and Anhalt (*34*) developed a clinical assay for vancomycin that used ristocetin as an internal standard. Serum and a solution of internal standard in an acidic buffer were mixed and passed through a column containing a weak cation-exchange resin to which the drugs bind. The column was washed with sodium sulfate solution to remove interfering substances, and the drugs were then eluted with an alkaline borate buffer at pH 9.45. The more highly alkaline buffer used for extraction of aminoglycosides (*18, 19*) could not be used because vancomycin and ristocetin decomposed rapidly at the higher pH. A typical chromatogram of a serum extract is shown in Fig. 8. The analysis used ultraviolet absorption at 210 nm for detection and a reversed-phase column with a mixture of acetonitrile and phosphate buffer at pH 6.0 for the mobile phase.

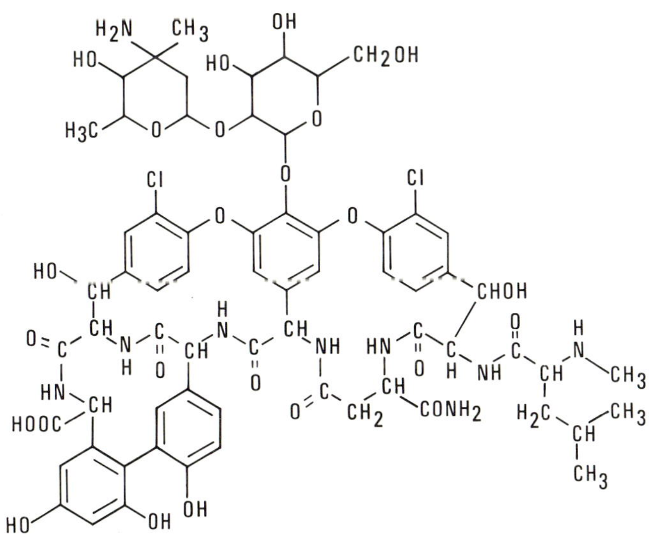

FIG. 7. Structure of vancomycin.

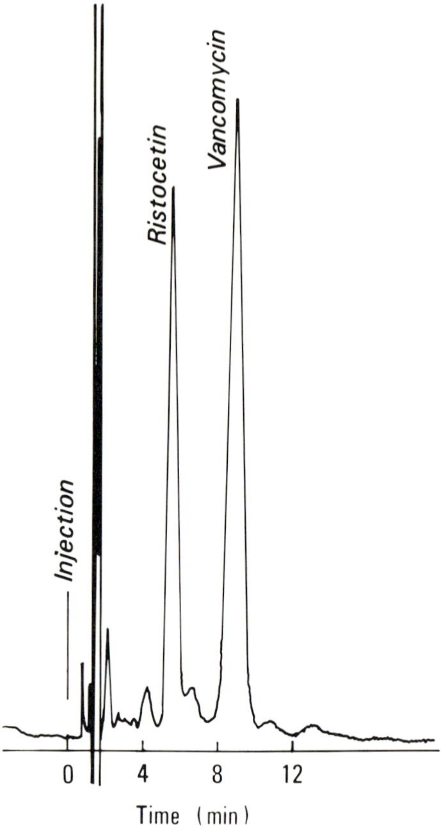


FIG. 8. Representative chromatogram of vancomycin (16 mg/L) and ristocetin (16 mg/L).

## D. Chloramphenicol

Chloramphenicol (Fig. 9) is primarily a bacteriostatic agent and is active against common aerobic and anaerobic bacteria. It is not active against *Serratia* spp. or *Pseudomonas aeruginosa.* Chloramphenicol is absorbed well following oral administration, and an oral dose of 1.0 g gives peak levels of 10–20 mg/L after 1–2 h. Similar levels are achieved by intravenous administration; intramuscular administration is not recommended (*35*). Chloramphenicol penetrates well into extravascular fluids and the concentration in cerebrospinal fluid or bile may be as much as one-half of the blood concentration. Approximately 60% is protein bound in serum. It is metabolized in liver principally to an inactive glucuronide, in which form it is excreted by the kidney. The elimination half-life from serum is 1.5–3.5 h in adults (*36*). This half-

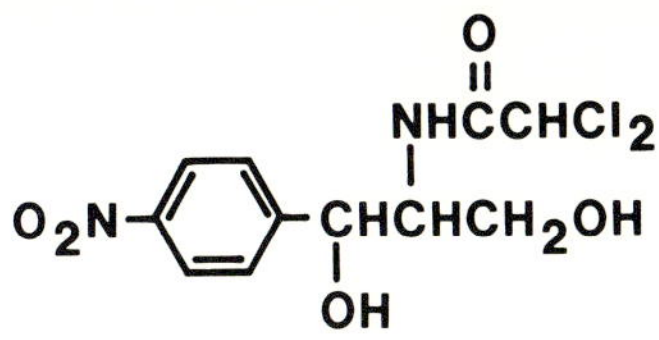

FIG. 9. Structure of chloramphenicol.

life may be markedly prolonged when hepatic and renal dysfunction coexist (*36*) and is difficult to predict in newborn infants because of variation in hepatic metabolism. The major toxic problems with chloramphenicol are blood dyscrasias and cardiovascular collapse (grey syndrome). Blood dyscrasias are of two types. A fatal aplastic anemia occurs with a frequency of 1 in 24,000 to 1 in 40,000. This anemia is not related directly to dose and may occur up to one year after therapy is stopped. The other form of anemia is related to dose and is reversed upon stopping the medication. Doses of more than 4 g/d or sustained levels of more than 25 mg/L have been associated with the latter form of anemia (*37*). Cardiovascular collapse is also related to blood level and occurs primarily in newborn infants, but may also occur in adults (*38*). This toxicity results from accumulation of the unconjugated chloramphenicol. In infants, this form of toxicity has been related to total serum chloramphenicol levels greater than 50 mg/L; however, it is difficult to attain therapeutic levels of 10–20 mg/L without risking toxicity if dosages are not adjusted by reference to measured serum levels (*39*).

Several liquid chromatographic assays have been developed for chloramphenicol (*40*–*47*). The methods are summarized in Table 6. Each of the methods has adequate sensitivity for clinically relevant levels of chloramphenicol and uses a reversed-phase column. In each method, therefore, polar metabolites would be expected to elute earlier than chloramphenicol. The method of Wal et al. (*40*) involves a complex procedure for serum extraction. In the absence of an internal standard, one would expect this method to be less precise than the other examples described in Table 6. The other procedures may be differentiated on the basis of whether an internal standard and solvent extraction (*41, 44, 45, 47*) or a simple protein precipitation with an organic solvent (*42, 43, 46*) are used. From the published accounts of these procedures, a choice cannot be made based on an obvious superiority of one procedure over another. Koup et al. (*44*) found that oxacillin eluted closely to chloramphenicol, but did not interfere with the assay because it was removed during extraction. Unfortunately, the procedures that do not use extraction did not study possible interference from oxacillin. Phenobarbital should also be studied as a

Table 6
Analysis of Chloramphenicol in Body Fluids

| Extraction | Internal standard | Mobile phase (composition by vol) | Column[a] | Wavelength, nm[b] | Ref. |
|---|---|---|---|---|---|
| Dilute serum in phosphate buffer; extract with ethyl acetate; evaporate ethyl acetate; extract with acetonitrile saturated with isooctane; separate isooctane and discard; evaporate acetonitrile; dissolve residue in ethyl acetate and analyze. | None | Water–methanol (70/30) | Spherisorb ODS™ | 254 | *40* |
| Dilute specimen with internal standard in methanol; add tris (hydroxymethyl) aminomethane buffer; extract with diethyl ether; evaporate ether; dissolve residue in methanol and analyze. | Mephenesin | Water–methanol (70/30) | Partisil-10 ODS™ | 278 | *41* |
| Dilute specimen with methanol; separate precipitate and analyze supernatant. | None | Water–methanol–acetic acid (62/37/1) | μBondapak $C_{18}$™ | 280 | *42* |
| Dilute plasma with acetonitrile; separate precipitate and analyze supernatant. | None | Ethanol–0.5% phosphoric acid (1/4) | HC ODS Sil-X-I™ | 280 | *43* |
| Dilute serum with sodium acetate buffer; add internal standard in ethyl acetate and mix; separate and evaporate to dryness ethyl acetate layer; dissolve residue in methanol and analyze. | 5-Ethyl-5-*p*-tolylbarbituric acid | Acetonitrile–acetate buffer, pH6.0 (25/75) | μBondapak $C_{18}$ | 270 | *44* |
| Mix serum, ethyl acetate with internal standard, and solid sodium chloride; separate ethyl acetate and evaporate to dryness; dissolve residue in methanol and analyze. | Thiamphenicol | Methanol–water (40/60) | μBondapak $C_{18}$ | 280 | *45* |
| Mix serum with acetonitrile which contains internal standard; separate precipitate and analyze supernatant. | *p*-Nitroacetanilide | Methanol–phosphate buffer, pH 7.0 (35/65) | μBondapak $C_{18}$ | 278 | *46* |
| Mix serum with chloroform-2-propanol (1/1) which contains internal standards; evaporate organic phase (lower) to dryness; dissolve residue in methanol and analyze. | *N*-acetylchloramphenicol and sulfamethoxazole | Acetonitrile–acetate buffer, pH 4.0 (20/80) | μBondapak $C_{18}$ | 280 | *47* |

[a]Trademarks: Spherisorb ODS, Phase Separations Ltd., Queensferry, Flintshire, England; Partisil il-10 ODS, Whatman, Clifton, NJ; μBondapak $C_{18}$, Waters Associates, Milford, MA; HC ODS Sil-X-I, Perkin-Elmer Corp., Norwalk, CT.
[b]Optimal wavelength for chloramphenicol is 270 nm (*41*).

possible interference (*47*). The simplicity of precipitation methods is appealing, particularly for analysis of cerebrospinal fluids; however, interferences are less likely to be encountered with an extraction method.

## IV. Conclusions

Assays using liquid chromatography that are clinically applicable have been developed for almost all antimicrobics. Earlier limitations that resulted from detector insensitivity have been bypassed by the use of derivatives. The $\beta$-lactam antimicrobics represent a large number of drugs, which as a class are measured frequently, but individually are measured only infrequently. Simple procedures applicable to the class as a whole are needed. Lastly, alternative methods that are cost-effective for small sample workloads (e.g., enzyme immunoassays) may compete with liquid chromatography as these nonchromatographic methods become available.

## References


*1.* Reeves, D. S., and Holt, H. A., *J. Clin. Pathol.* **28,** 435 (1979).

*2.* Reeves, D. S., and Wise, R., "Antibiotic Assays in Clinical Microbiology," in *Laboratory Methods in Antimicrobial Chemotherapy,* Reeves, D. S., Phillips, I., Williams, J. D., and Wise, R., eds., Churchill Livingston, London, 1978, pp. 137–143.

*3.* Drayer, D. E., *Am. J. Med.* **62,** 486 (1977).

*4.* Gerson, B., and Anhalt, J. P., *High-Pressure Liquid Chromatography and Therapeutic Drug Monitoring,* American Society of Clinical Pathologists, Chicago, Ill., 1980.

*5.* Crombez, E., Van Der Weken, G., Van Den Bossche, W., and De Moerloose, P., *J. Chromatogr.* **177,** 323 (1979).

*6.* Tsuji, K., *J. Chromatogr.* **158,** 337 (1978).

*7.* Werner, M., Sutherland, III, E. W., and Abramson, F. P., *Clin. Chem.* **21,** 1368 (1975).

*8.* Barza, M., and Scheiffe, R. T., *Am. J. Hosp. Pharm.* **34,** 723 (1977).

*9.* McCracken, Jr., G. H., *Am. J. Dis. Child.* **128,** 407 (1974).

*10.* Lau, W. K., Young, L. S., Black, R. E., Winston, D. J., Linne, S. R., Weinstein, R. J., and Hewitt, W. L., *Am. J. Med.* **62,** 959 (1977).

*11.* Ho, P. W. L., Pien, F. D., and Kominami, N., *Ann. Intern. Med.* **91,** 227 (1979).

*12.* Bindschadler, D. D., and Bennett, J. E., *J. Infect. Dis.* **120,** 427 (1969).

*13.* Byrne, K. M., Kershner, A. S., Maehr, H., Marquez, J. A., and Schaffner, C. P., *J. Chromatogr.* **131,** 191 (1977).

*14.* Brewer, N. S., *Mayo Clin. Proc.* **52,** 675 (1977).
*15.* Myers, D. R., DeFehr, J., Bennett, W. M., Porter, G. A., and Olsen, G. D., *Clin. Pharmacol. Ther.* **23,** 356 (1978).
*16.* Wagner, J. G., Novak, E., Leslie, L. G., and Metzler, C. M., *International J. Clin. Pharmacol.* **1,** 261 (1968).
*17.* Mays, D. L., Van Apeldoorn, R. J., and Lauback, R. G., *J. Chromatogr.* **120,** 93 (1976).
*18.* Anhalt, J. P., *Antimicrob. Agents Chemother.* **11,** 651 (1977).
*19.* Anhalt, J. P., and Brown, S. D., *Clin. Chem.* **24,** 1940 (1978).
*20.* Tsuji, K., Goetz, J. F., Van Meter, W., and Gusciora, K. A., *J. Chromatogr.* **175,** 141 (1979).
*21.* Peng, G. W., Gadalla, M. A. F., Peng, A., Smith, V., and Chiou, W. L., *Clin. Chem.* **23,** 1838 (1977).
*22.* Peng, G. W., Jackson, G. G., and Chiou, W. L., *Antimicrob. Agents Chemother.* **12,** 707 (1977).
*23.* Chiou, W. L., Nation, R. L., Peng, G. W., and Huang, S. M., *Clin. Chem.* **24,** 1846 (1978).
*24.* Back, S.-E., Nilsson-Ehle, I., and Nilsson-Ehle, P., *Clin. Chem.* **25,** 1222 (1979).
*25.* Maitra, S. K., Yoshikawa, T. T., Hansen, J. L., Nilsson-Ehle, I., Palin, W. J., Schotz, M. C., and Guze, L. B., *Clin. Chem.* **23,** 2275 (1977).
*26.* Maitra, S. K., Yoshikawa, T. T., Hansen, J. L., Schotz, M. C., and Guze, L. B., *Am. J. Clin. Pathol.* **71,** 428 (1979).
*27.* Maitra, S. K., Yoshikawa, T. T., Steyn, C. M., Guze, L. B., and Schotz, M. C., *Antimicrob. Agents Chemother.* **14,** 880 (1978).
*28.* Myers, H. N., and Rindler, J. V. *J. Chromatogr.* **176,** 103 (1979).
*29.* McCormick, M. H., Stark, W. M., Pittenger, G. E., Pittenger, R. C., and McGuire, J. M., "Vancomycin, a New Antibiotic. I. Chemical and Biologic Properties," in *Antibiotics Annual 1955–1956,* Medical Encyclopedia, New York, 1956, pp. 606–611.
*30.* Geraci, J. E., *Mayo Clin. Proc.* **52,** 631 (1977).
*31.* Cook, F. V., and Farrar, Jr., W. E., *Ann. Intern. Med.* **88,** 813 (1978).
*32.* Geraci, J. E., Heilman, F. R., Nichols, D. R., and Wellman, W. E., *Proc. Staff Meet. Mayo Clinic* **33,** 172 (1958).
*33.* Best, G. K., Best, N. H., and Durham, N. N., *Antimicrobial Agents Chemother.-1968,* 115 (1969).
*34.* Uhl, J. R., and Anhalt, J. P., *Therapeut. Drug Monitoring* **1,** 75 (1979).
*35.* Wilson, W. R., *Mayo Clin. Proc.* **52,** 635 (1977).
*36.* Bennett, W. M., Singer, I., Golper, T., Feig, P., and Coggins, C. J., *Ann. Intern. Med.* **86,** 754 (1977).
*37.* Scott, J. L., Finegold, S. M., Belkin, G. A., and Lawrence, J. S., *N. Engl. J. Med.* **292,** 1137 (1965).
*38.* Cannon, G. H., and Lietman, P. S., *Johns Hopkins Med. J.* **143,** 60 (1978).
*39.* Black, S. B., Levine, P., and Shinefield, H. R., *J. Pediatr.* **92,** 235 (1978).
*40.* Wal, J. M., Peleran, J. C., and Bories, G., *J. Chromatogr.* **145,** 502 (1978).
*41.* Thies, R. L., and Fischer, L. J., *Clin. Chem.* **24,** 778 (1978).

42. Nilsson-Ehle, I., Kahlmeter, G., and Nilsson-Ehle, P., *J. Antimicrob. Chemother.* **4,** 169 (1978).
43. Peng, G. W., Gadalla, M. A. F., and Chiou, W. L., *J. Pharm. Sci.* **67,** 1036 (1978).
44. Koup, J. R., Brodsky, B., Lau, A., and Beam, Jr., T. R., *Antimicrob. Agents Chemother.* **14,** 439 (1978).
45. Crechiolo, J., and Hill, R. E., *J. Chromatogr.* **162,** 480 (1979).
46. Petersdorf, S. H., Raisys, V. A., and Opheim, K. E., *Clin. Chem.* **25,** 1300 (1979).
47. Sample, R. H. B., Glick, M. R., Kleiman, M. B., Smith, J. W., and Oei, T. O., *Antimicrob. Agents Chemother.* **15,** 491 (1979).

# Chapter 8

# Tricyclic Antidepressants

Gary J. Schmidt

*Analytical Chemistry Department*
*Perkin-Elmer Corporation*
*Norwalk, Connecticut*

## I. Introduction

The use of tricyclic antidepressant drugs is becoming increasingly prevalent for the treatment of depressed patients. It has been suggested that, analogous to many other drug substances, the tricyclic drugs exhibit clinical effectiveness within a defined therapeutic concentration range (*1–10*). Very recently, both Dito (*11*) and Orsulak and Schildkraut (*12*) have summarized the usefulness of measuring serum concentrations of these drugs. These authors suggest that knowledge of the plasma concentrations of these drugs aid the physician in determining patient compliance and initiating the best possible drug treatment.

Most patients receiving tricyclic drug therapy show plasma drug concentrations between 20 and 200 μg/L (*5*). Since the major tricyclic drugs are partly metabolized to the corresponding pharmacologically active *N*-desmethyl metabolites, the measurement of both parent drug compound and metabolite is required. Chromatographic procedures fulfill this requirement.

The tricyclic drugs have been determined using a variety of analytical techniques. These have included ultraviolet spectroscopy

(*13, 14*), fluorescence spectrophotometry (*15–18*), and thin-layer chromatography (*19–22*). Procedures based upon these analytical techniques are generally time-consuming, lack adequate sensitivity for measuring low therapeutic concentrations, and are prone to interference. They have, however, proven invaluable in strengthening the foundation upon which many current chromatographic procedures have been built.

In recent years, gas chromatographic procedures have been described for determining these drugs (*23–35*). Many of these procedures have made use of flame ionization detectors, which have been useful for determining high therapeutic or toxic drug concentrations. Unfortunately, these procedures do not possess sufficient sensitivity for measuring the tricyclics over the entire therapeutic concentration range.

More sensitive gas chromatographic detection systems have recently been described for determining these drugs. Both nitrogen selective (*31–33*) and electron capture (*34, 35*) detectors have been used. These procedures often permit detection of the tricyclic drugs at concentrations as low as 1 μg/L. However, these procedures usually entail lengthy pretreatment, may require derivatization of the drugs, and in the case of electron capture detection of amitriptyline, oxidation of the drug to anthraquinone (*34*).

Over the past five years, liquid chromatographic procedures have become an increasingly valuable adjunct to procedures utilizing other analytical techniques for determining the tricyclic antidepressants. These procedures have primarily taken two different chromatographic approaches: that of ion-pair chromatography and that of liquid–solid chromatography on unmodified silica surfaces. These chromatographic systems will be discussed in detail in Section III. A third and promising chromatographic system utilizes high pH mobile-phase solutions. This technique will be discussed in Section IV.

## II. The Tricyclics

Figure 1 lists the chemical structures of the major tricyclic drugs, amitriptyline and imipramine. Also included are the pharmacologically active *N*-desmethyl metabolites of these drugs, nortriptyline and desipramine, respectively. Protriptyline is often used as an internal standard in liquid chromatographic procedures.

The tricyclic drugs are basic compounds, a fact that poses certain difficulties for the chromatographer. Since the tricyclics possess basic p*K* values, they are ionized in acidic or neutral pH mobile-phase solutions, preventing good chromatographic separations.

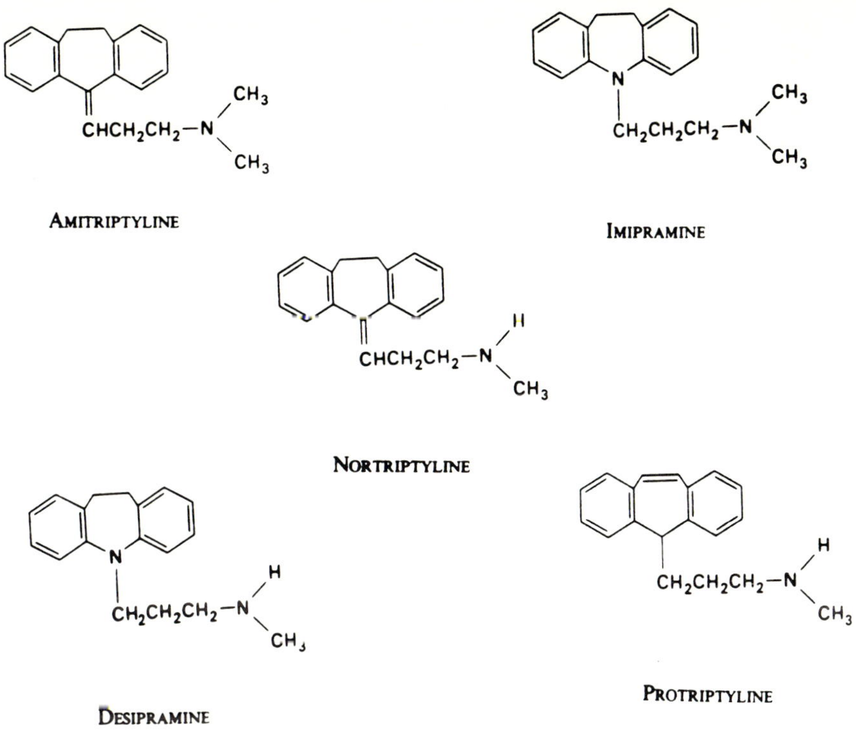


FIG. 1. Chemical structures of the major tricyclic antidepressant drugs.

The chromatography of compounds with ionic functional groups is best accomplished using a mobile phase pH that approximates the p$K$ of the compounds. For acidic compounds, this is readily achieved by using an acidic mobile phase for effective ionic suppression. However, the use of basic mobile phase pH for suppressing the ionization of basic compounds is not so readily realized. This results from the dissolution of silica-based liquid chromatographic column packing materials when high pH mobile phases are used. As a result, most liquid chromatographic procedures that have been developed utilize the specific chromatographic systems mentioned.

## III. Determination of Tricyclics in Physiological Samples

### A. Sample Pretreatment

Since the tricyclic drugs are present at very low concentrations in serum, a suitable sample pretreatment procedure must be used. These procedures serve two purposes, isolation of the drugs from the serum matrix and concentration of the drugs such that they are suitably adjusted for analysis. Plasma or serum volumes of 2 mL are usually required.

There are two methods in current use for isolating and concentrating the tricyclics from the serum matrix. These are liquid–liquid or solvent extraction, and liquid–solid or adsorption–desorption extraction on solid surfaces. In general, liquid–liquid extraction systems have been the most widely used and require the adjustment of the serum pH followed by extraction of the drugs into a suitable organic solvent that is immiscible with water.

Important criteria for developing a suitable liquid–liquid extraction procedure include the choice of extraction pH and of the extraction organic phase. In most cases, drugs are extracted as the uncharged species. Since the tricyclic drugs are basic compounds, they are extracted at basic pH. Extraction of the neutral drug species will minimize the hydrophilic nature of the charged compound, thereby favoring the partition of the drug into the organic solvent phase.

An appropriate extraction pH may be estimated by using a pH that is close to the p*K* value of the ionic species of the drug being extracted. When a class of structurally similar drugs, such as the tricyclic antidepressants, are to be extracted simultaneously, a single extraction at basic pH is usually suitable.

The primary requirement of the organic solvent used for extraction is to effectively solvate the drugs from the serum matrix. That is, the drugs should be readily soluble in the organic solvent so that effective partitioning between the aqueous and organic phases can occur.

Useful extraction solvents may be chosen on the basis of their polarity. As a general rule, polar drug compounds will partition most effectively into polar solvents; nonpolar drugs into nonpolar solvents. Extraction solvents that possess moderate hydrogen bonding are useful for the extraction of compounds containing hydrogen-accepting groups. For moderately hydrophilic compounds, solvents that form strong hydrogen bonds, such as ethyl acetate, may be useful. For extremely water-soluble compounds, ion-pair extraction techniques may be employed.

An additional consideration that must be taken into account is the amount of manipulation required of the sample during extraction. Handling of the sample should be minimized to avoid losses of the drugs. Ideally, a single extraction, perhaps using a ten volume excess of organic solvent, is preferable. However, in complicated sample matrices, it may be necessary to use more extensive sample pretreatment techniques. In extraction methods where interfering endogenous compounds are co-extracted, classical acid–base back extraction techniques may be employed. In these instances, the basic nature of the tricyclic drugs may be used to selectively isolate the compounds of interest from interfering substances. For example, the serum can be adjusted to a basic pH and the drugs extracted into an organic solvent. If additional sample cleanup is required, the drugs may be re-extracted into a small volume of acid, leaving potentially interfering substances behind.

A procedure developed by our laboratory for determining the tricyclic drugs in serum utilizes liquid–liquid extraction (*36*). The procedure includes adding 0.5 mL of saturated sodium carbonate solution to 2 mL of serum and extracting the drugs into 5 mL of hexane/isoamyl alcohol (98/2, v/v). After centrifugation to separate the layers, the extraction solvent is evaporated to dryness. The dried residue is then redissolved in 10 μL of the mobile phase and 5 μL is injected into the chromatograph.

This relatively simple extraction procedure provided recoveries of about 65% for the four major tricyclic drugs from a 2-mL serum sample. The procedure incorporated the use of an internal standard, protriptyline, to compensate for procedural and injection size variations. Using adsorption chromatography, we found the procedure to be linear over a serum concentration range from 10 to 800 μg/L. The chromatography will be described in more detail in Part B of this Section.

There are other liquid chromatographic procedures that make use of liquid–liquid extraction in a procedure similar to that described above. For almost all procedures, hexane, or hexane containing a small percentage of isoamyl alcohol, is used for extraction. In those procedures where interfering compounds might be co-extracted, back extraction techniques into acid are often employed.

An alternative to liquid–liquid extraction is to adsorb the drugs selectively from the serum onto a solid adsorptive surface. Sometimes this technique involves column chromatography and the adsorbent material is contained in a small flow-through column. For example, these materials include XAD-2 resin or diatomaceous earth.

In general, the serum sample is buffered to an appropriate pH and then applied to the top of the extraction column. If the correct conditions are used, the drugs interact with the adsorbent and are retained. Excess serum and many endogenous compounds which do not interact with the adsorbent pass through the column. The column may be washed to remove excess serum, followed by desorption of the drugs using an organic solvent. An eluting solvent is chosen to completely desorb the drugs from the column packing material.

An example of this procedure has been given by Bondo et al. (*37*). Two mL of serum is buffered to pH 10 and applied to the adsorption pretreatment column (Clin Elut®; Analytichem Int'L, Lawndale, CA). The drugs are eluted from the column using hexane, followed by back-extraction into 100 μL of 0.1*N* HCl. The acid phase is dried and the extraction residue is redissolved in 100 μL of the mobile phase prior to injection.

## B. Chromatography

*1. Ion-Pair.* A commonly used separation mode for determining the tricyclics is ion-pair chromatography. The basis of this chromatography mode is the formation of an ion-pair complex of the ionic tricyclic molecule with an appropriate counter-ion or ion-pair reagent. The formation of this ion-pair complex negates the ionic attraction of the tricyclic drug to the column packing, resulting in good chromatographic efficiency.

Recently, Proelss et al. (*38*) described the use of ion-pair chromatography for determining the major tricyclic drugs in serum. Two mL of serum was adjusted to pH 14 and the drugs were extracted into 10-mL hexane/isoamyl alcohol (99/1, v/v). The drugs were back-extracted from the organic phase into 200 μL of 0.1 mol/L HCl and 85 μL of the aqueous phase was injected into the chromatograph.

These workers usee a 10-μm $C_{18}$ reversed-phase column for chromatographing the serum extracts. The mobile phase consisted of methanol/acetonitrile/0.1 mol/L phosphate buffer, pH 7.6 (41/15/44, v/v) containing 5 mmol/L pentanesulfonic acid. The mobile phase flow rate was 1.5 mL/min.

Figure 2 illustrates a chromatogram obtained from the analysis of a serum standard containing 50 μg/L of the major tricyclic drugs. Also included in this chromatogram is β-napthylamine (peak 1), which was used as an internal standard. Linearity of the method was adequate over a serum concentration range from 25 to 1500 μg/L. The detection limit for determining these drugs in serum was claimed to be approximately 3 μg/L, using an ultraviolet detector set at 254 nm.

Mellstrom and co-workers (*39, 40*) used ion-pair chromatography for determining chlorimipramine and desmethylchlorimipramine in plasma. These workers prepared the separation columns by coating 10-μm silica gel with 0.1 mol/L hydrochloric acid containing 0.01 mol/L tetrapropylammonium hydrogen sulfate. They extracted the drugs from 1 mL of serum into diethyl ether. Following this initial ether

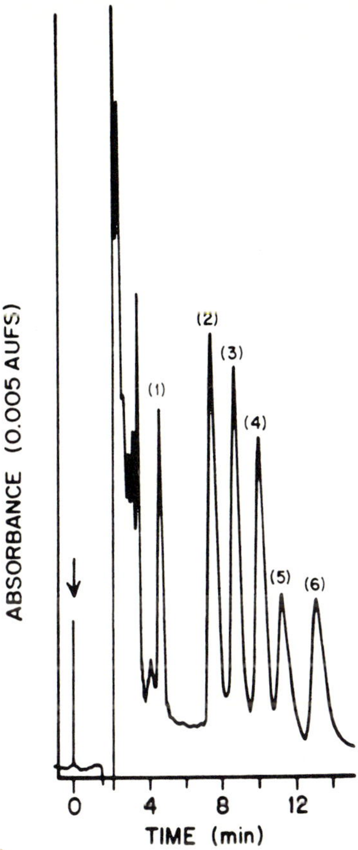


FIG. 2. Chromatogram illustrating the analysis of a serum standard by ion-pair chromatography on a $C_{18}$ reversed-phase column. Mobile phase was methanol/acetonitrile/phosphate buffer (41/15/44) containing 5 mmol/L pentanesulfonic acid. Flow rate was 1.5 mL/min. 1, β-napthylamine; 2, doxepin; 3, desipramine; 4, nortriptyline; 5, imipramine; 6, amitriptyline (reproduced from ref. *38*).

extraction, the drugs were back extracted into 1-mL 0.25*N* sulfuric acid, the pH adjusted to basic, and the drugs extracted into 75 μL of the mobile phase. Forty μL of the mobile phase solution was then injected into the chromatograph.

Figure 3 illustrates a chromatogram showing the separation of chlorimipramine, desmethylchlorimipramine, and four major tricyclic drugs. These workers also used similar chromatography techniques for determining amitriptyline and metabolites in serum samples (*41*).

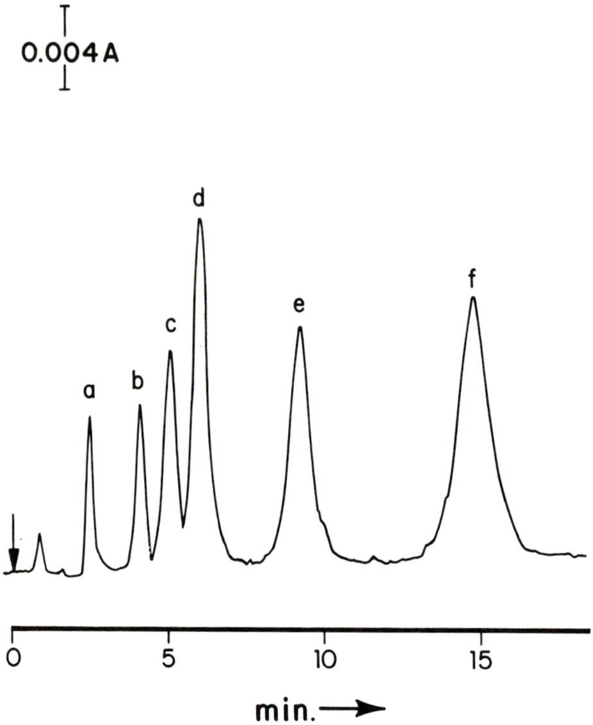


FIG. 3. Ion-pair chromatography of tricyclic drugs using tetrapropylammonium hydrogen sulfate as the ion-pair reagent. A, desmethylchlorimipramine; b, nortriptyline; c, chlorimipramine; d, desipramine; e, amitriptyline; f, imipramine (reproduced from ref. *39*).

Persson and Lagerstrom (*42*) applied ion-pair chromatography using methanesulfonic acid to the analysis of a wide range of biogenic compounds including some of the tricyclic drugs in plasma and urine. Knox and Jurand (*43*) separated twenty tricyclics as perchlorate ion-pairs on a 10-$\mu$m silica packing using organic eluents.

*2. Adsorption.* An alternative to the use of ion-pair chromatography for determining the tricyclic drugs is adsorption chromatography. Since the tricyclics are ionized at neutral or acidic mobile phase pH, a small amount of a base added to the mobile phase can suppress this ionization. This technique is well-suited to the use of adsorption chromatography columns used in conjunction with organic mobile phase solutions. The use of organic eluents prevents the rapid dissolution of the silica packing, providing acceptable column lifetime.

We developed an adsorption chromatography procedure for determining the four major tricyclic drugs in 2 mL of serum. The sample pretreatment included liquid–liquid extraction of the drugs as described earlier. Chromatography was done on a 5-μm silica column using a mobile phase consisting of acetonitrile/isopropanol/ammonium hydroxide, 89.8/10/0.2 (v/v). The mobile-phase flow rate was 1.5 mL/min and the column temperature was maintained at 65° C.

We found that the use of a variable wavelength ultraviolet detector set to 211 nm provided sufficient sensitivity to quantitate the drugs at concentrations as low as 10 μg/L. The tricyclics exhibited a well-defined absorption maxima at approximately 245 nm, however the absorption increased at lower detection wavelengths. Detection at 211 nm provided an enhancement in sensitivity of about fourfold over that obtained at 245 nm.

Figure 4 illustrates a chromatogram from the injection of a drug standard mixture containing 50 ng of each of the four major tricyclic drugs. Chromatography was complete in approximately 15 min, including the elution of the internal standard, protriptyline. Figure 5

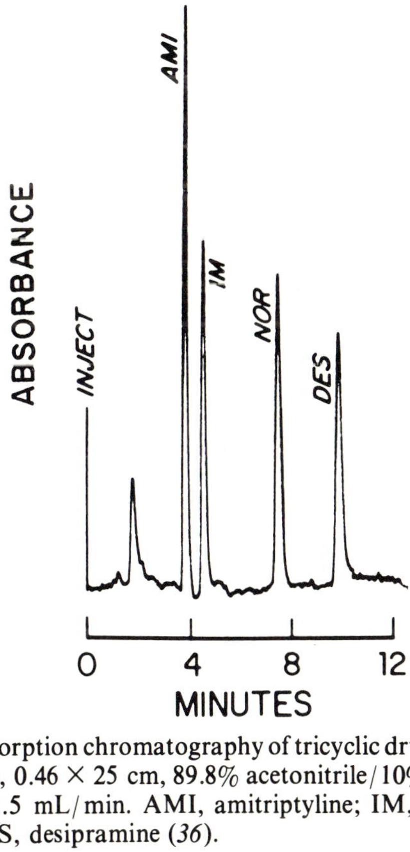


FIG. 4. Adsorption chromatography of tricyclic drug standards. Five-μm silica column, 0.46 × 25 cm, 89.8% acetonitrile/10% isopropanol/0.2% $NH_4OH$, 65°C, 1.5 mL/min. AMI, amitriptyline; IM, imipramine; NOR, nortriptyline; DES, desipramine (*36*).

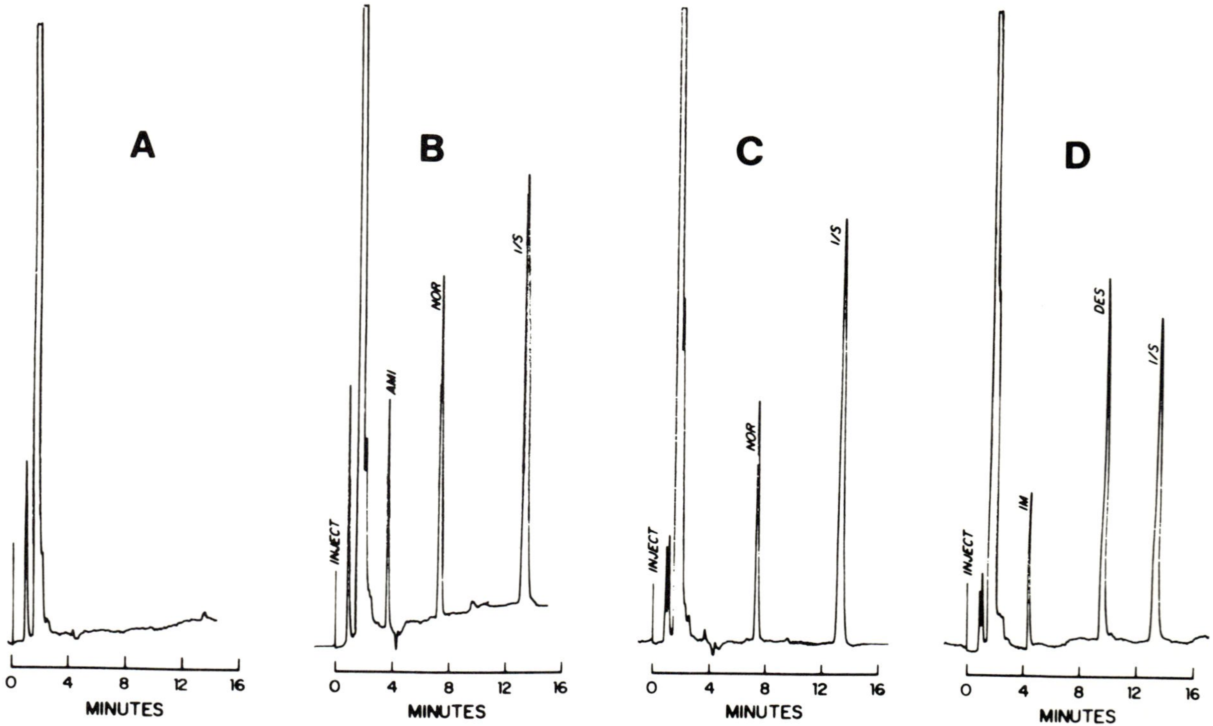


FIG. 5. Chromatograms of serum extracts. (A) Drug-free serum. (B) Patient receiving amitriptyline; concentration in plasma calculated to be 119 $\mu$g of amitriptyline and 156 $\mu$g of nortriptyline per liter. (C) Patient receiving nortriptyline; concentration in plasma calculated to be 127 $\mu$g/L. (D) Patient receiving imipramine; concentration in plasma calculated to be 75 $\mu$g of imipramine and 206 $\mu$g of desipramine per liter. Conditions and legend as in Fig. 4 (*36*).

illustrates chromatograms of serum samples from patients on tricyclic drug therapy.

This chromatographic procedure is linear over a serum concentration range from 25 to 800 μg/L, although most patient samples we have analyzed contain drugs in the range of 50–300 μg/L. Within-run and day-to-day precision were approximately 6% for drug concentrations of 100 μg/L. Diphenhydramine and doxepin interfere with the analysis of amitriptyline, however, combined use of these drugs is rare except in cases of drug abuse.

Recently, Sutheimer (*44*) used similar chromatographic conditions to separate simultaneously a large number of the tricyclic drugs and related drug substances. All chromatographic conditions were the same as we described, except for the substitution of diethylamine for ammonium hydroxide as the base added to the mobile phase. The use of diethylamine permitted greater flexibility in the adjustment of the chromatographic separation and provided complete resolution between doxepin and amitriptyline. Figure 6 illustrates this separation.

De Zeeuw and Westenberg (*45*) described the use of adsorption chromatography for determining overdose concentrations of the tricyclics. They used a 5-μm silica column with a ternary mobile phase mixture consisting of hexane/dichloromethane/methanol, 8/1/1 (v/v). The hexane contained 0.001% methylamine. Using a UV detector set at 250 nm, the detection limit was 100 μg/L for tertiary tricyclics and 250 μg/L for the secondary tricyclic drugs. The total analysis time was approximately 30 min.

Van Den Berg et al. (*46*) also used a 5-μm silica column for separating the tricyclic drugs. The mobile phase was 99.8% ethyl acetate of which 20% was saturated with water and 0.2% methylamine. Figure 7 shows the separation they obtained. For this analysis, monochlorobenzene (peak 1) was added for calculation of the unretained peak volume. These workers found that the retention of the drugs could be decreased considerably by increasing the amount of methylamine added to the mobile phase.

Numerous other adsorption chromatography procedures have been reported for determining the tricyclics. Watson and Stewart (*47, 48*) developed a procedure for determining amitriptyline and nortriptyline in 2 mL of serum. Westenberg et al. (*49*) used a silica column for determining clomipramine and desmethylclomipramine in 1–4 mL of serum, and observed detection limits of 2 μg/L for clomipramine and 10 μg/L for the *N*-desmethyl metabolite. Detaevernier et al. (*50*) separated eight tricyclic drugs on a 10-μm silica column with propylamine as the base added to the mobile phase. Very

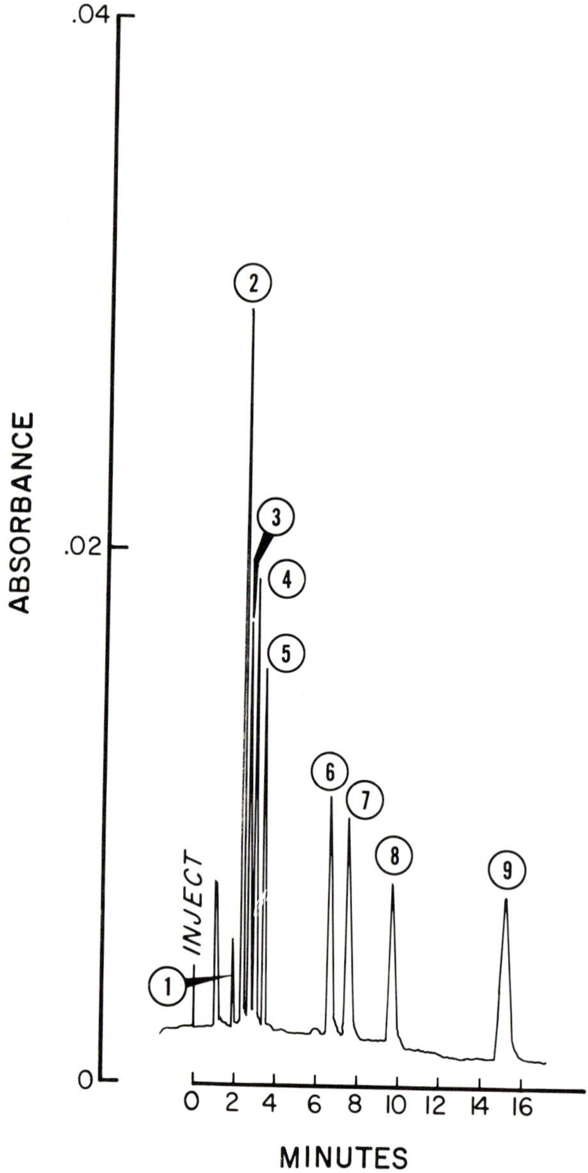


FIG. 6. Adsorption chromatography of tricyclic drug standards. Conditions similar to those described in Fig. 4 except diethylamine substituted for ammonium hydroxide as the base. 1, phencyclidine; 2, chlorpromazine; 3, doxepin; 4, amitriptyline; 5, imipramine; 6, nordoxepin; 7, nortriptyline; 8, desipramine; 9, protriptyline (reproduced from ref. *44*).

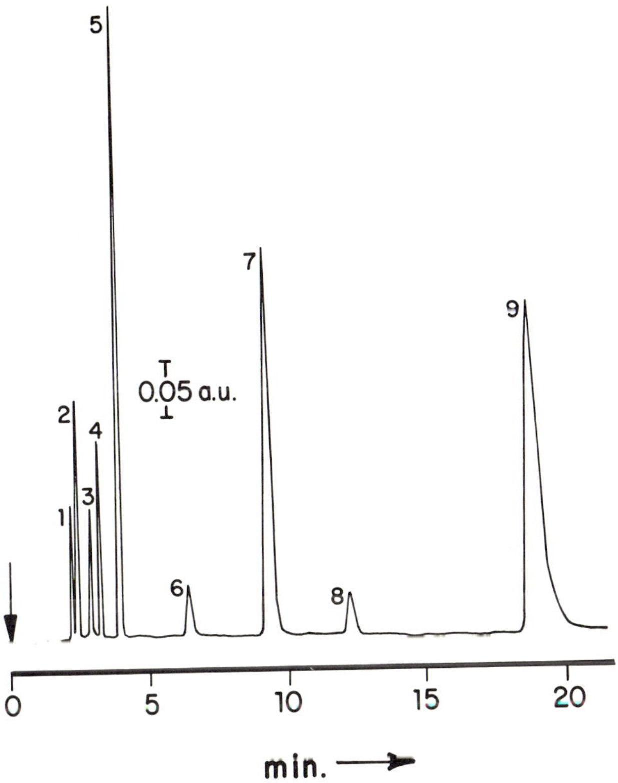


FIG. 7. Adsorption chromatography of tricyclic drugs using methylamine added to the mobile phase. Five-μm silica column, 0.46 × 30 cm, 99.8% ethyl acetate (20% saturated with water) + 0.2% methylamine, 1.7 mL/min. 1, monochlorobenzene; 2, trimipramine, 3, clomipramine; 4, amitriptyline; 5, imipramine; 6, promazine; 7, nortriptyline; 89, desipramine; 9, protriptyline (reproduced from ref. *46*).

recently, Reece and Zacest (*51*) used a fluorescence detector for measuring imipramine and desipramine and were able to detect concentrations as low as 500 pg/mL.

## IV. Use of High pH Mobile Phases

A third and potentially useful chromatographic system for separating the tricyclic drugs utilizes high pH mobile-phase solutions. As mentioned earlier, the use of a mobile phase pH approximating the p*K* of the tricyclics aids good chromatographic efficiency. However, since the tricyclics have basic p*K* values, the use of a basic pH aqueous-based mobile phase results in premature column failure owing to the dissolution of the silica packing.

We have developed a technique in our laboratory that greatly increases the lifetime of silica-based LC column packings when high pH mobile phases are used (*52*). This is achieved by presaturating the mobile phase with silica before the analytical column by placing a silica saturation column in the mobile phase flow before the sample injection valve. Mobile-phase solutions passing through this silica saturation column become saturated with silica, thereby greatly reducing further dissolution of silica from the analytical column. By mounting the silica saturation column in the column oven with the analytical column, changes in the silica saturation concentration with changes in separation temperature are automatically compensated. A schematic illustrating the use of this technique is shown in Fig. 8.

We have applied this technique to the separation of the major tricyclic antidepressant drugs using both bonded-phase and adsorption chromatography columns. The poor peak symmetry associated with chromatographing the tricyclic drugs having charged

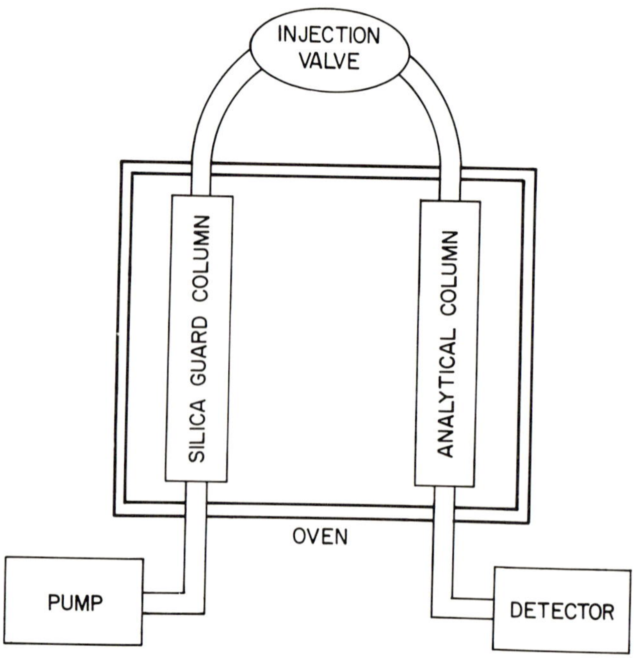


FIG. 8. Schematic layout of the silica saturating guard column in the LC system (*52*).

ionic species is controlled by adjustment of the mobile-phase pH. This reduces the attraction of ionized functional groups on the tricyclic molecule to active sites on the column packing.

Theoretically, the use of high-pH mobile phases should prove useful whenever basic compounds are chromatographed. However, our initial results with the use of bonded-phase packings with high-pH mobile phases for separating the tricyclics have been unsuccessful. Elution of the tricyclic drugs from $C_{18}$, $C_8$, and phenyl bonded-phase packings requires a strong solvent and results in incomplete separation and tailing of the chromatographic peaks.

The most successful reversed-phase separation was done using a phenyl-bonded phase column. The separation of five tricyclics is shown in Fig. 9. The mobile phase consisted of 90% acetonitrile and

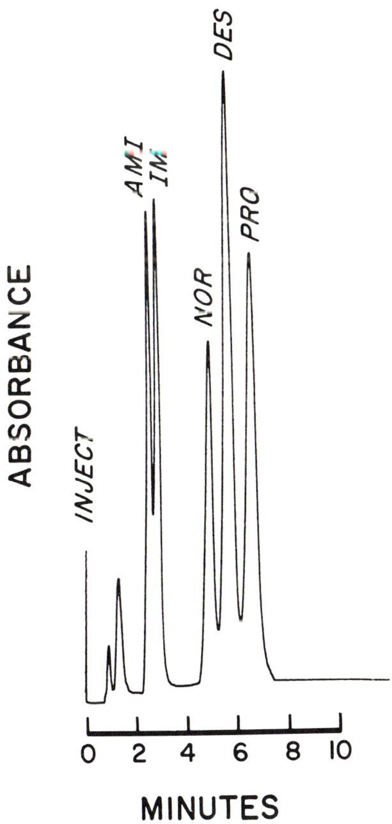


FIG. 9. Separation of some tricyclic drug standards. Phenyl-bonded phase column, 0.26 × 25 cm, 90% acetonitrile: 10% 0.01 $M$ $(NH_4)_2CO_3$, 70° C, 1 mL/min. Figure legend as in Fig. 4 (*53*).

10% 10 m*M* ammonium carbonate solution. Although peak shape was considerably improved over that obtained when using the $C_8$ or $C_{18}$ bonded phases, the resolution was not as good as those systems using ion-pair or adsorption chromatograpy.

The use of adsorption chromatography columns with high pH aqueous-based mobile phases has resulted in good chromatographic separations. Figure 10 illustrates the separation of the important tricyclic drugs on a 5-μm silica column. The mobile phase was 35% acetonitrile in 65% 8 m*M* sodium borate buffer, pH 9.06. The separation was quite acceptable although the peaks do tail slightly. Note that even when using a relatively low solvent strength of 35% acetonitrile, analysis time was not lengthy.

Altering the mobile-phase pH has a dramatic effect on the separation. At pH 9.06, peak tailing resulted from partial ionization of the drugs. By increasing the pH to 10.06 by addition of ammonium

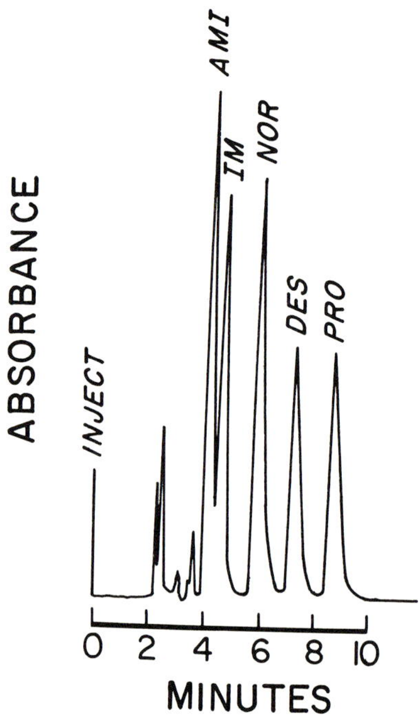


FIG. 10. Separation of some tricyclic drug standards. Five-μm silica column, 0.46 × 25 cm, 35% acetonitrile: 65% 8 m*M* sodium borate (pH 9.06), 23°C, 1 mL/min. Legend as in Fig. 4 (*53*).

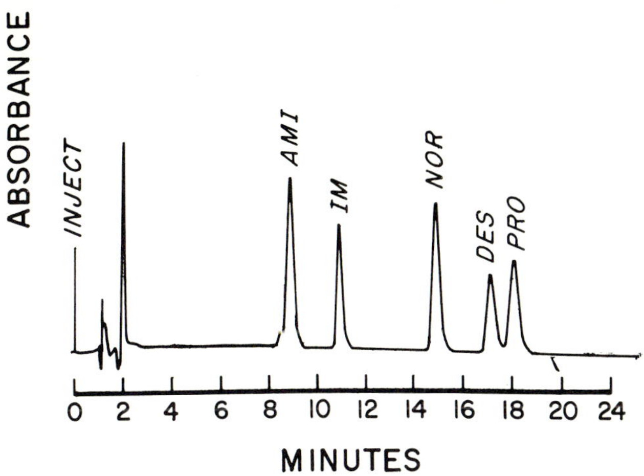


FIG. 11. Separation of some tricyclic drug standards. Five-μm silica column, 0.46 × 25 cm, 50% acetonitrile/50% water (pH 10.06 with ammonium hydroxide), 40°C, 2 mL/min. Figure legend as in Fig. 4 (*53*).

hydroxide, we obtained excellent peak shape and complete resolution of the drugs. Figure 11 illustrates the separation at a mobile phase pH of 10.06 and 50% acetonitrile. The flow rate was 2 mL/min and the separation temperature was 40°C.

Whenever ionization of a compound of interest occurs in acidic or neutral solutions, the use of high-pH mobile phases should be considered. The use of this technique provided good chromatographic performance for the tricyclic drugs. A more detailed discussion for separating these drugs using high pH mobile phases has been published (*53*).

## V. Determination of Hydroxy Metabolites

Although the major metabolic pathway of the tricyclic antidepressants is to the corresponding demethylated analogs, a number of other metabolites have been identified. The most significant of these are the hydroxylated metabolites that have been identified in both the serum and urine of patients on tricyclic drug therapy. Monitoring these metabolites may prove useful in differentiating between fast metabolic degradation and poor bioavailability.

Watson and Stewart (*54*) described the use of adsorption chromatography for determining the tricyclics and many conjugated and unconjugated metabolites in the urine of patients receiving tricyclic drug therapy. Unconjugated metabolites were extracted from a 10-mL urine specimen, adjusted to pH 12, and dispersed into 25 mL of dichloromethane. After extraction of the unconjugated metabolites, the aqueous phase was acidified and hydrolyzed with limpet acetone powder for 48 h at 37° C. After hydrolysis, the remaining metabolites were extracted using the same procedure as for the initial extraction of the unconjugated metabolites. These workers used a 5-μm silica column and a mobile phase consisting of dichloromethane/2-propanol/ammonium hydroxide in the proportions 100/10/0.2 (v/v). The flow rate was 1 mL/min and the separation was performed at room temperature. An ultraviolet absorption detector set at 240 nm was used.

Figure 12 shows a typical separation of unconjugated metabolites from the urine of a patient receiving amitriptyline. Using these

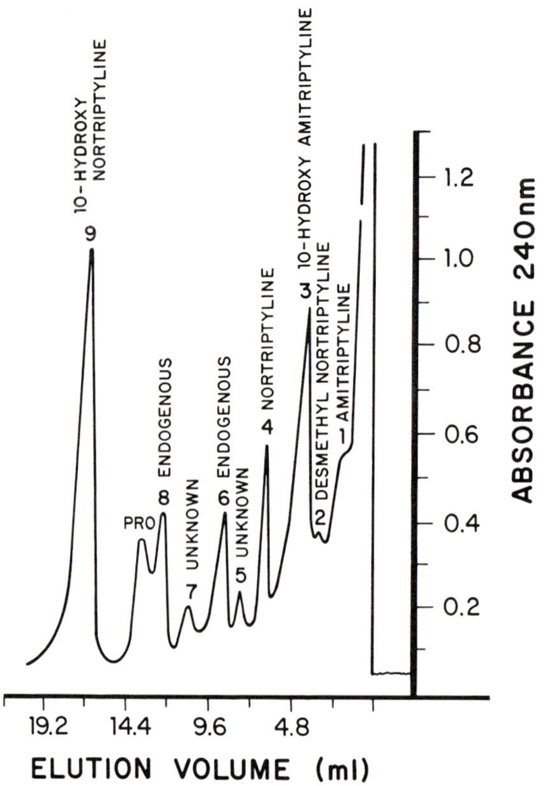


FIG. 12. Separation of unconjugated metabolites from the urine of a patient receiving amitriptyline. Five-μm silica column; dichloromethane/2-propanol/ammonium hydroxide, 100/10/0.2 (v/v), 1 mL/min; UV detection at 240 nm (reproduced from ref. *54*).

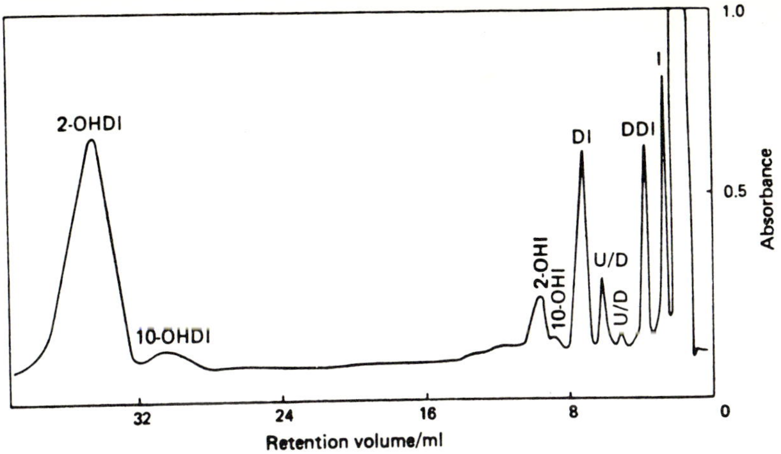


FIG. 13. Separation of urinary imipramine and metabolites on a 5-$\mu$m silica column. Same conditions as in Fig. 12, except for UV detection at 251 nm (reproduced from ref. *55*).

chromatographic conditions, amitriptyline elutes very quickly, since a strong solvent was used to elute the more polar metabolites. Peak 3 and 9 correspond to the 10-hydroxy metabolites of amitriptyline and nortriptyline, respectively.

More recently, these workers applied the use of adsorption chromatography to the analysis of both urinary and serum imipramine and metabolites (*55*). The chromatographic conditions were similar to those described above for determining amitriptyline metabolites. Figure 13 shows the separation of imipramine and its metabolites from urine.

Kraak and Bijster (*56*) used a $C_8$ column for separating amitriptyline metabolites from serum. The metabolites were extracted from 1 mL of serum into 5 mL of hexane. Figure 14 illustrates a typical separation of serum drug metabolites from a patient receiving 150 mg of amitriptyline per day orally.

Mellstrom and Braithwaite (*41*) also separated amitriptyline metabolites in plasma. However, these workers used ion-pair chromatography on 5-$\mu$m silica columns. Figure 15 shows two chromatograms using this form of chromatography. The chromatogram to the left is of a blank serum sample and the chromatogram to the right is of a serum from a patient receiving amitriptyline. The mobile phase consisted of methanol/dichloromethane/diisopropyl ether/0.1 perchloric acid in the proportions 5.2/10/30/0.9 (v/v) delivered at a flow rate of 1.9 mL/min. The drugs were extracted from a 1-mL serum sample by liquid–liquid extraction into hexane.

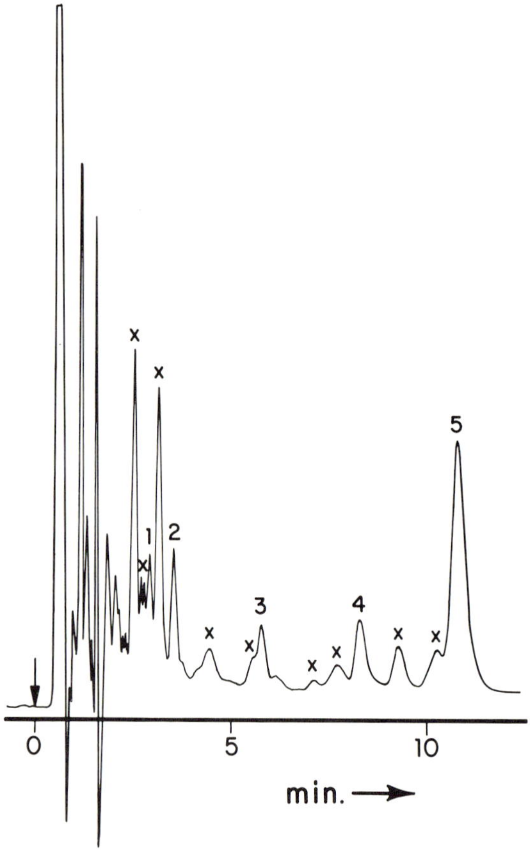


FIG. 14. Separation of amitriptyline metabolites from serum on a $C_8$ reversed-phase column. The mobile phase was water/methanol/dichloromethane, 13/8/3 (v/v) containing 1% propylamine. 1, trans-10-hydroxynortriptyline; 2, *trans*-10-hydroxyamitriptyline; 3, desmethylnortriptyline; 4, nortriptyline; 5, amitriptyline (reproduced from ref. *56*).

## VI. Conclusions

The use of liquid chromatography has greatly extended the field of therapeutic drug monitoring. Major developments in the area of instrument flexibility, column performance, and availability, and the increased understanding of the mechanisms of the chromatographic separation are continuing at an ever quickening rate. These advances continue to open up new areas of investigation.

The analysis of the tricyclic antidepressant drugs has presented certain difficulties. These include poor chromatographic efficiency

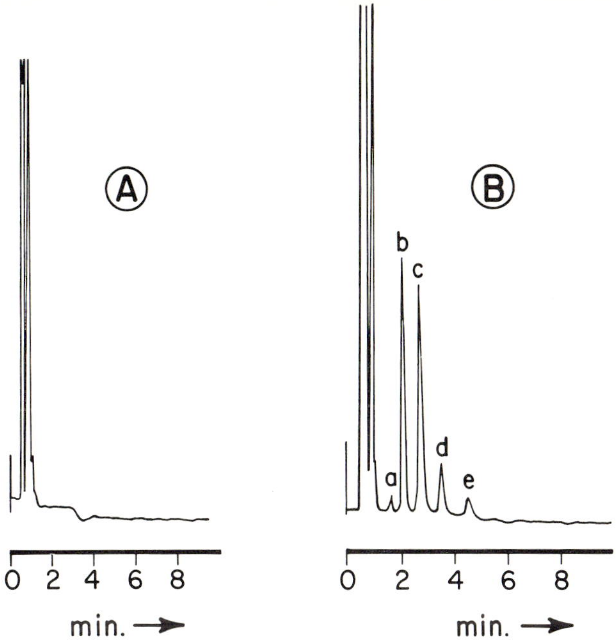


FIG. 15. Chromatograms of serum sample using ion-pair chromatography on a 5-μm silica column. Chromatogram A is a blank serum sample and chromatogram B is a serum sample from a patient receiving amitriptyline. The mobile phase was methanol/dichloromethane/diisopropyl ether/0.1 *M* perchloric acid, 5.2/10/30/0.9 (v/v), delivered at a flow rate of 1.9 mL/min. Peaks: a, desmethylnortriptyline; b, nortriptyline; c, internal standard; d, amitriptyline; e, trans-10-hydroxynortriptyline (reproduced from ref. *41*).

owing to the ionization of the drugs in neutral or acidic mobile phases and the sensitivity required to detect the drugs at the low therapeutic concentrations at which they are present. The problems associated with the chromatography of these drugs have been resolved by the use of ion-pair or adsorption chromatography techniques. These procedures result in good peak quality and simultaneous separation of many of the major antidepressants. Detection of these drugs is considerably facilitated by the use of low detection wavelengths.

The use of high-pH mobile phases offers a considerable increase in chromatographic flexibility. Preliminary results using this technique for determining the tricyclics have been promising. The significance of using this technique is not restricted to the chromatography of the tricyclics. In fact, whenever basic drug compounds are to be chromatographed, the technique may prove useful.

## Acknowledgments

I would like to thank Walter Slavin, my colleague at Perkin-Elmer, for his valuable suggestions for improving this manuscript. I would also like to acknowledge Ms. Janice Goodchild for her careful typing of the manuscript.

## References

1. Biggs, J. T., and Ziegler, V. E., *Clin. Pharmacologist Ther.* **22,** 269 (1977).
2. Braithwaite, R. A., Goulding, R., Theano, G., Bailey, J., and Coppen, A., *Lancet* **1,** 1297 (1972).
3. Burrows, G. D., Davis, B., and Scoggins, B. A., *Lancet* **2,** 619 (1972).
4. Kragh-Sorensen P., Hansen C. E., Beastrut P. C., and Hvidberg, E. F., *Psychopharmacologia* **45,** 305 (1976).
5. Braithwaite, R. A., Montgomery, S., and Dawling, S., *Clin. Pharmacol. Ther.* **23,** 303 (1978).
6. Khalid, R., Amin, M. M., and Ban, T. A., *Psychopharmacol. Bull.* **14,** 43 (1978).
7. Applebaum, P. S., Vasile, R. G., Orsulak, P. J., and Schildkraut, J. J., *Am. J. Psychiatry* **736,** 339 (1979).
8. Gram, L. F., and Christiansen, J., *Clin. Pharmacol. Ther.* **17,** 555 (1975).
9. Glassman, A. H., and Perel, J. M., *Clin. Pharmacol. Ther.* **16,** 198 (1974).
10. Glassman, A. H., Shostak, M., Kantor, S. J., and Perel, J. M., *Psychopharmacol. Bull.* **11,** 27 (1975).
11. Dito, W. R., *Diagnostic Medicine* **5,** 48 (1979).
12. Orsulak, P. J., and Schildkraut, J. J., *Therapeutic Drug Monitoring* **1,** 199 (1979).
13. Wallace, J. E., and Dahl, E. V., *J. Forensic Sci.* **12,** 484 (1967).
14. Henwood, C. R., *J. Forensic Sci.* **15,** 147 (1975).
15. Westerlund, D., and Borg, K. O., *Acta Pharm. Suecica* **7,** 267 (1970).
16. Moody, J. P., Whyte, S. F., and Naylor, G. J., *Clin. Chim. Acta* **43,** 355 (1973).
17. Persson, B. A., *Acta Pharm. Suecica* **7,** 337 (1970).
18. Moody, J. P., Tait, A. C., and Todrick, A., *Br. J. Psychiatry* **113,** 183 (1967).
19. Facino, R., and Corona, G. L., *J. Pharm. Sci.* **58,** 764 (1969).
20. Faber, D. B., Mulder, C., and Man in't Veld, W. A., *J. Chromatogr.* **100,** 55 (1974).
21. Oliver, J. S., and Smith, H., *J. Forensic Sci.* **3,** 181 (1974).
22. Nagy, A., and Trieber, L., *J. Pharmacol.* **25,** 599 (1973).
23. Nyberg, G., and Martensson, E., *J. Chromatogr.* **143,** 491 (1977).
24. Dubois, J. P., King, W., Theobald, W., and Wirz, W., *Clin. Chem.* **22,** 892 (1976).
25. Weder, H. J., and Bickel, M. H., *J. Chromatogr.* **37,** 181 (1968).

26. Braithwaite, R. A., and Widdop, B., *Clin. Chim. Acta* **35,** 461 (1971).
27. Jorgensen, A., *Acta Pharmacol. Toxicol.* **36,** 79 (1975).
28. Hucker, H. B., and Stauffer, S. C., *J. Pharm. Sci.* **63,** 296 (1974).
29. Hammar, E. G., Alexandersson, B., Holmstedt, B., and Sjogvist, F., *Clin. Pharmacol. Ther.* **12,** 496 (1971).
30. Biggs, J. T., Holland, W. H., Chang, S. S., Hipps, P. P., and Sherman, W. R., *J. Pharm. Sci.* **65,** 261 (1976).
31. Bailey, D. N., and Jatlow, P. I., *Clin. Chem.* **22,** 1697 (1976).
32. Dorrity, F., Jr., Linnoila, M., and Habig, R. L., *Clin. Chem.* **23,** 1326 (1977).
33. Bailey, D. N., and Jatlow, P. I., *Clin. Chem.* **22,** 777 (1976).
34. Wallace, J. E., Hamilton, H. E., Goggin, L. K., and Blum, K., *Anal. Chem.* **47,** 1516 (1975).
35. Borga, O., and Garle, M., *J. Chromatogr.* **68,** 77 (1972).
36. Vandemark, F. L., Adams, R. F., and Schmidt, G. J., *Clin. Chem.* **24,** 87 (1978).
37. Bondo, P. B., Thoma, J. J., and Beltz, G. A., *Clin. Chem.* **25,** 1118 (1979).
38. Proelss, H. F., Lohmann, H. J., and Miles, D. G., *Clin. Chem.* **24,** 1948 (1978).
39. Mellstrom, B., and Tybring, G., *J. Chromatogr. Biomed. Appl.* **143,** 597 (1977).
40. Mellstrom, B., and Eksborg, S., *J. Chromatogr.* **116,** 475 (1976).
41. Mellstrom, B., and Baithwaite, R., *J. Chromatogr.* **157,** 379 (1978).
42. Persson, B. A., and Lagerstrom, P. O., *J. Chromatogr.* **122,** 305 (1976).
43. Knox, J. H., and Jurand, J., *J. Chromatogr.* **103,** 311 (1975).
44. Sutheimer, C., *Chromatogr. Newslett.* **7,** 38 (1979).
45. De Zeeuw, R. A., and Westenberg, H. G. M., *J. Anal. Tox.* **2,** 229 (1978).
46. Van Den Berg, J. H. M., De Ruwe, H. J. J. M., Deelder, R. S., and Plomp, Th. A., *J. Chromatogr.* **138,** 431 (1977).
47. Watson, I. D., and Stewart, M. J., *J. Chromatogr.* **132,** 155 (1977).
48. Watson, I. D., and Stewart, M. J., *J. Chromatogr.* **110,** 389 (1975).
49. Westenberg, H. G. M., Drenth, B. F. H., De Zeeuw, R. A., De Cuyper, H., Van Praag, H. M., and Korf, J., *J. Chromatogr.* **142,** 725 (1977).
50. Detaevernier, M. R., Dryon, L., and Massart, D. L., *J. Chromatogr.* **128,** 204 (1976).
51. Reece, P. A., Zacest, R., and Barrow, C. G., *J. Chromatogr., Biomed. Appl.* **163,** 310 (1979).
52. Atwood, J. G., Schmidt, G. J., and Slavin, W., *J. Chromatogr.* **171,** 109 (1979).
53. Schmidt, G. J., and Vandemark, F. L., *Chromatogr. Newslett.* **7,** 25 (1979).
54. Watson, I. D., and Stewart, M. J., *J. Chromatogr.* **134,** 182 (1977).
55. Watson, I. D., *Proc. Anal. Div. Chem. Soc.* **16,** 293 (1979).
56. Kraak, J. C., and Bijster, P., *J. Chromatogr., Biomed. Appl.* **143,** 499 (1977).

*Appendix 1*
Summary of Procedures

| Column | Mobile | Flow (mL/min) | Column | Detector | Sample | Extractant | Ref. |
|---|---|---|---|---|---|---|---|
| Silica B/5 (Perkin-Elmer) | Acetonitrile: 89.8 | 2 | 65°C | UVB/211 | 2-mL plasma | Hexane/isoamyl alcohol | *36* |
| $\mu$Bondapak $C_{18}$ (Waters) | Methanol: 41<br>Acetonitrile: 15<br>Water: 44<br>containing 5 mmol/L pentanesulfonic acid | 1.5 | amb. | UV/254 | 2-mL serum | Hexane/isoamyl alcohol | *38* |
| Partisil 5 (Whatman) | 0.1*M* perchloric acid/0.9<br>Methanol/5.2<br>Dichloromethane: 10<br>Diisopropyl ether: 30 | 1.9 | amb. | UV/254 | 1-mL plasma | Hexane | *41* |
| LiChrosorb SI 60 (E. Merck) | Hexane: 8<br>Dichloromethane: 1<br>Methanol: 1<br>containing 10 ppm methylamine | 1.2 | amb. | UV/250 | 1-mL plasma | Ether | *45* |
| Micropak-5 $\mu$m (Varian) | Dichloromethane: 100<br>2-propanol/2,<br>$NH_4OH$/0.25 | 1 | amb. | UV/240 | 2-mL plasma | Dichloro-methane | *47* |
| LiChrosorb SI 60 (E. Merck) | Hexane/8<br>Dichloromethane: 1<br>Methanol/1 | 1.2 | amb. | UV/250 | 1–4 mL plasma | Hexane | *49* |
| $\mu$-Bondapak phenyl (Waters) | Acetonitrile: 71<br>0.015% $H_3PO_4$:29 | 2 | amb. | Fluores-cence<br>$E_x = 252$<br>$E_m = 360$ | 2-mL plasma | Hexane/isoamyl alcohol | *51* |

# Chapter 9

# Antineoplastic Drugs

Wolfgang Sadée and Yousry Mahmoud El Sayed

*Departments of Pharmacy and Pharmaceutical Chemistry*
*School of Pharmacy, University of California*
*San Francisco, California*

## I. Drug-Level Monitoring in Cancer Chemotherapy

### A. Investigational Clinical Trials

The limited scope of therapeutic drug-level monitoring in cancer chemotherapy results from the often complex biochemical mechanisms that contribute to antineoplastic activity and obscure the relationships among drug serum levels and therapeutic benefits. Moreover, new agents for cancer chemotherapy are being introduced at a more rapid rate than for the treatment of other diseases, although the successful application of therapeutic drug-level monitoring may require several years of intensive study of the significance of serum drug levels. However, drug level monitoring can be of considerable value during phase I clinical trials of new antineoplastic agents in order to assess drug metabolism, bioavailability, and intersubject variability; these are important parameters in the interpretation of clinical studies, but have no immediate benefit to the patient. High performance liquid chromatography (HPLC) probably represents the most versatile and easily adaptable analytical technique for drug metabolite screening (*1*). HPLC may therefore now be the method of choice during phase I clinical trials of antineoplastic drugs. For example, within a single week we developed an HPLC assay—using a $C_{18}$ reverse-phase column, UV detection, and direct serum injection after protein precipitation—for the new radiosensitizer, misonidazole (*2*). During the actual phase I patient studies, we detected two new metabolites of misonidazole and measured the pharmacokinetic disposition of these agents. This study served as a guide to optimizing the dosing schedule and time of radiation therapy (*3*).

## B. Routine Therapeutic Applications: Methotrexate

Methotrexate (MTX) serum-level monitoring after the administration of high dosages represents one of the few clinical drug assays, and the only anticancer drug assay, for which the therapeutic utility is generally acknowledged. Therefore, a brief review of the purposes for MTX level monitoring is included with this chapter (see also ref. *4*). Methotrexate is usually given intraventricularly in low doses, and intravenously in low or high doses (up to 9 $g/m^2$ over 4–6 h) in a variety of cancers (*5, 6*). High-dose MTX therapy (*7*) was introduced to overcome tumor resistance that may be secondary to a deficiency of an active membrane-associated MTX transport system (*9*), and to establish "free" intracellular MTX levels in excess of the MTX bound to the target enzyme, dihydrofolate reductase. Such "free" intracellular MTX levels have been associated with the cell killing effects of MTX (*6*). We have recently reviewed the rationale of high-dose MTX in more detail (*4*).

Administration of MTX in high dosages broadens the antitumor spectrum of this agent; for example, osteogenic sarcoma is insensitive to low doses, but does respond to high doses of MTX (*5, 6*). However, the high dosages of MTX used clinically also increase the risk of serious myelotoxicity to the point that it may be fatal. Therefore, high-dose MTX ($> 0.5$ $g/m^2$) is always given together with citrovorum factor (CF, leucovorin), a tetrahydrofolate derivative that bypasses the depletion of tetrahydrofolate through dihydrofolate reductase inhibition (*7, 8*). Clear evidence has accumulated that the individual risk of severe MTX toxicity, despite CF rescue treatment, is associated with prolonged MTX retention in the body (*4, 6*). Methotrexate is primarily eliminated, unchanged, in the urine. Therefore, patients with impaired renal function are excluded from high-dose MTX protocols. However, deterioration of kidney function secondary to MTX administration can occasionally occur, and this is often not detectable by changes in serum creatinine until 3–4 days post-therapy. Consequently, determination of MTX serum levels offers the safest procedure to identify patients at risk of toxicity (*4*–6). Serum levels of

MTX persisting above $10^{-6}$ *M* for 48 h, and above $10^{-7}$ *M* for 72 h, are indicative of impending severe toxicity that can be successfully averted by continuing the CF rescue treatment at an increased dosage level. Since the toxicity risk can be assessed, and potential adverse effects can usually be prevented, the routine measurement of MTX serum levels is imperative with the clinical use of high-dose MTX therapy.

## II. Analytical Procedures

### A. Review of Liquid Chromatographic Analysis of Antineoplastic Agents

A summary of recent LC procedures that are suitable for the analysis of biological samples is contained in Table 1. The prevalence of reverse-phase chromatographic procedures over ion exchange or silica gel chromatography of anticancer drugs, as well as of other drug classes (*1*), demonstrates the versatility of this technique. Most assay procedures involve UV or fluorescence detection, while the potentially suitable electrochemical detectors have been used rather infrequently as yet. Table 1 also includes the extraction procedure and special analytical conditions that are required to achieve suitable assay methods. The extraction of serum or plasma samples may consist of simple protein precipitation using trichloroacetic acid (TCA), methanol, ethanol, or acetonitrile. The direct LC analysis of such samples is possible because of the high separation capabilities of the analytical columns, coupled with UV or fluorescence detection that is optimized for the drug using variable wavelength detectors. Another important extraction procedure employs organic solvents or solvent mixtures that are immiscible with water. Because of the polar nature of many anticancer drugs, polar solvent mixtures may be required for satisfactory extraction yields, e.g., chloroform/ methanol (adriamycin) or ether/*n*-propranol (5-fluorouracil). Additional use of high salt concentrations, referred to as "salting out," further increases the extraction yield (ammonium sulfate for 5-fluorouracil). Even more polar agents can be extracted as ion pairs (e.g., methotrexate as its perchlorate ion pair).

Special chromatographic conditions include reverse-phase paired-ion chromatography for highly polar drugs (bleomycin $A_2$, 6-mercaptopurine, methotrexate). Furthermore, chelating salts ($Mg^{2+}$ for adriamycin) or sulfhydryl protecting agents (dithioerythritol for 6-mercaptopurine) are used to improve extraction yield and chromatographic behavior of these drugs.

Table 1
Review of HPLC Procedures That Can Detect or Separate Antineoplastic Drugs in Biological Fluids at Therapeutic Concentrations

| Drug | Extraction | Column[a] | Detection | Special conditions | References |
|---|---|---|---|---|---|
| Bleomycin $A_2$ | Protein precipitation (TCA) | RP | UV (254 nm) | Eluent contains 0.0085 *M* heptanesulfonic acid (paired-ion chromatogr.) | *25* |
| Cyclophosphamide | None | $C_{18}$-RP | UV (200 nm) | Suitable for aqueous (nonbiological) solutions | *26* |
| Doxorubicin | $CHCl_3/CH_3OH$(4:1) at pH 9.8 | SG (7 $\mu$) | Fluorescence (480→560 nm) | Daunorubicin serves as internal standard; $Mg^{2+}$ chelate improves separation; sensitivity 1 ng/mL plasma | *20* |
| | Organic solvents | SG (5 $\mu$) | UV (490 nm) | Sensitivity, 10 ng/mL plasma | *24* |
| | | RP | UV | Study of chromatographic behavior on several columns | *22* |
| | Organic solvents | RP | | Collected fractions are analyzed by RIA | *23* |
| | Organic solvents | SG | Fluorescence (475→580 nm) | Eluent contains $NH_3$ and $H_2O$; sensitivity: 1 ng/mL plasma | *21* |
| 5-Fluorouracil | Ether/*n*-propanol + $(NH_4)_2SO_4$ | AE | UV (254 nm) | 5-Chlorouracil is the internal standard; sensitivity, 100 ng/mL plasma | *16* |
| | Ethyl acetate | $C_{18}$-RP | UV (254 + 280 nm) | Sensitivity, 20 ng/mL plasma; ftorafur is also measured | *18* |
| | AE preseparation and ethyl acetate extraction | RP | UV | $^3$H-FUra is added to measure extraction recovery; sensitivity, 1 to 10 ng/mL plasma | *17* |
| Melphalan | Protein precipitation with $CH_3OH$ | $C_{18}$-RP | UV (254 nm) | Dansylproline is the internal standard; sensitivity, 50 ng/mL serum | *27* |
| | | RP | UV | Mono- and dihydroxy metabolites are separated; sensitiovity 10 ng/mL plasma | *28* |

| | | | | | |
|---|---|---|---|---|---|
| 6-Mercaptopurine | Ethyl acetate at pH 5.1 | $C_{18}$-RP | UV (235 nm) | Dithioerythritol is added to prevent loss of 6-MP; sensitivity, 5 ng/mL plasma, azathioprin can also be analyzed | *29* |
| | | $C_{18}$-RP | UV (313 nm) | The mobile phase contains 0.005 *M* *m*-heptanesulfonic acid for paired ion chromatography; sensitivity, 0.1 μg/mL plasma | *30* |
| Methotrexate | Solvent extractions | RP (alkyl-phenyl) | UV (303 nm) | The sensitivity for MTX and 7-OH-MTX is 10–20 n*M* in serum | *11* |
| | On column concentration | AE | UV | Sensitivity is 20 n*M* | *14* |
| | $HClO_4$ ion-pair extraction | AE | UV | Sensitivity for MTX and 7-OH-MTX is 100 n*M* in serum. | *16* |
| | | $C_{18}$-RP | UV | The eluent contains tetrabutyl-ammonium hydroxide for paired ion chromatography; applicable to urine samples | *13* |
| | | $C_{18}$-RP | Fluorescence | MTX is oxidized by $MnO_4^-$ to the fluorescent 2,4-diaminopteridine-6-carboxylic acid; sensitivity, 20 n*M* in serum | *15* |
| Metronidazole, Misonidazole | Protein precipitation with ethanol | $C_{18}$-RP | UV (324 nm) | Sensitivity is 0.5 μg/mL with a 10 μL serum sample; also suitable for misonidazole and metronidazole metabolites | *2* |
| Procarbazine | Ether | $C_{18}$-RP | UV (254 nm) | Unchanged drug is not detectable in plasma by HPLC; sensitivity limit for drug and metabolite, 200 ng/mL | *31* |

[a]RP, reverse phase; SG, silica gel; AE, anion exchange.

Published sensitivity limits are in the order of 10 nmol of the drug in serum for both fluorescence and UV LC assays. Fluorescence detection can be considerably more sensitive and new UV detectors may enhance the sensitivity limit above that of currently available detectors.

## B. Liquid Chromatographic Analysis of Selected Drugs

*1. Methotrexate.* The routine analysis of methotrexate (MTX) in the clinical laboratory is complicated by the following factors:

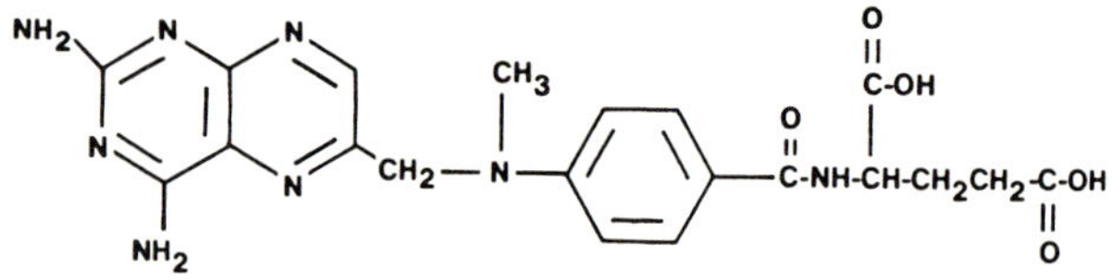

*a.* The serum concentration range extends over five orders of magnitude, from 10 nm to 1 mM, which makes serial dilutions necessary if competitive protein binding assays are used (*1*). LC analysis, however, has a broad dynamic range, a distinct advantage in the assay of MTX serum levels.

*b.* The MTX metabolite, 7-hydroxy-MTX (7-OH-MTX) can accumulate in the body to levels far exceeding those of the parent MTX (*11*), which necessitates a clear separation between the two compounds. This can be readily achieved by LC (*11–13*). Competitive protein binding assays (radioimmuno-assays employing dihydrofolate reductase) are also specific for MTX in the presence of 7-OH-MTX, which has a 100-fold lower affinity for the enzyme dihydrofolate reductase (*1*). However, the potentially important metabolite, 4-amino-4-deoxypteroic acid (APA), possesses a high cross-affinity to most antibodies and 2–4% cross-affinity to dihydrofolate reductase (for reviews see refs. *1, 4*); therefore, APA may interfere with binding assays, but not with LC assays because of its complete separation from MTX (*11*).

*c.* The combined use of high-dose MTX with CF rescue frequently results in high levels of CF (5-formyl-tetrahydrofolate) in the presence of low levels of MTX. Both competitive protein binding and LC assays are capable of distinguishing between MTX and CF (*1*). Endogenous folates are usually present at rather low levels and do not interfere with the MTX assays.

*d.* The specificity requirements for MTX assays in patient sera must reflect potential interferences from concomitant drug therapy and varying pathophysiological conditions.

The LC analysis of MTX can be readily performed with anion exchange chromatography (*12, 14*). In combination with an on-column concentration step, using a disposable pre-column, anion-exchange LC can be sensitive to 20 nM MTX in serum (*14*). Reverse phase $C_{18}$ LC is hampered by asymmetric peaks; this can be overcome with the addition of tetrabutyl ammonium hydroxide to the eluent for paired-ion LC (*13*). Nelson et al. (*15*) oxidized MTX to the fluorescent 2,4-diaminopteridine-6-carboxylic acid, which can be readily chromatographed on a $C_{18}$ reverse-phase column; fluorescence detection affords increased sensitivity over UV detection. We employed an alkylphenyl reverse phase column that has suitable properties for the analysis of MTX and 7-OH-MTX (Fig. 1) (*11*). Furthermore, we used a solvent extraction with a combination of acetonitrile, *n*-butanol, and ether that deproteinizes the serum and concentrates aqueous samples from 2 mL to approximately 100 μL. The entire remaining aqueous phase, containing most (~80%) of the highly polar MTX and 7-OH-MTX, can be subjected to LC analysis. The procedure results in a sensitivity of 10–20 nM for both agents (11).

The therapeutic or toxic significance of 7-OH-MTX remains unknown. At present, it is sufficient to monitor MTX levels alone, for which purpose competitive protein binding assays and LC procedures are suitable.

*2. 5-Fluorouracil.* The LC assay of 5-fluorouracil (FUra) by anion-exchange chromatography has limited sensitivity (100 ng/mL plasma) (*16*). However, reverse phase LC assays employing UV detection between 254 and 280 nm are quite sensitive for FUra. Preseparation of FUra on a low pressure anion-exchange column followed by ethyl acetate extraction and $C_{18}$ reverse-phase LC afford a sensitivity of 1–10 ng FUra/mL plasma (*17*). We have employed a simple ethyl acetate extraction of plasma and subsequent $C_{18}$ reverse-phase LC with a sensitivity of 20 ng FUra/mL (*18*). This assay was designed to also detect the FUra prodrug, ftorafur (1-(tetrahydro-2-furanyl)-5-fluorouracil, FT), and several other FT metabolites.

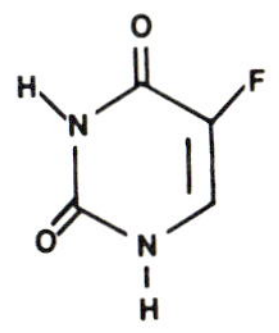

FUra plasma levels may be a poor indicator of the exposure of the body to the drug, since the relationship between plasma FUra and intracellular active FUra metabolites is nonlinear. For example, after administration of therapeutically active FT doses the FUra plasma

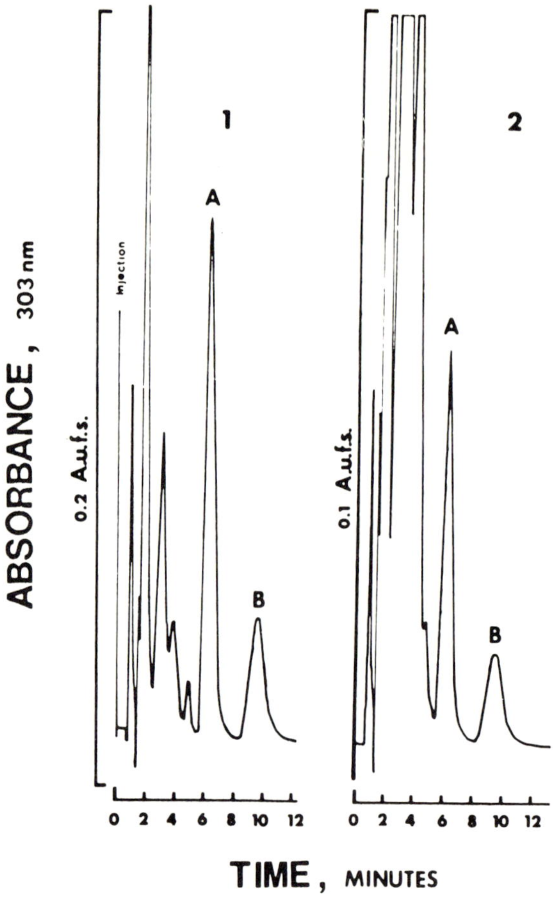


FIG. 1. Liquid chromatography record of MTX (peak A) and 7-OH-MTX (peak B) extracted from human serum for 30 h (1) and 54 h (2) after a dose of 9 g of MTX/m$^2$. The MTX concentrations were $8.6 \times 10^{-6}$ *M* and $3.1 \times 10^{-6}$ *M* respectively, recorded at 0.2 aufs (1) and 0.1 aufs (2). (Reproduced with permission from *Cancer Treatment Rept.* ref. *11*.)

concentrations are exceedingly low (< 100 ng/mL), indicating that the FUra that is generated from FT within the cells does not reach the systemic circulation and, therefore, does not redistribute throughout the body (*19*). Under these conditions, the enzymatic mechanism of FT activation to FUra assumes an important role in determining tissue selective toxicity of FUra (*19*). This example of FUra kinetics points to a serious complication in interpreting plasma level data that may be common to many antineoplastic agents. A large fraction of the anticancer drugs form active intermediate metabolites that do not circulate through the blood and may therefore escape detection.

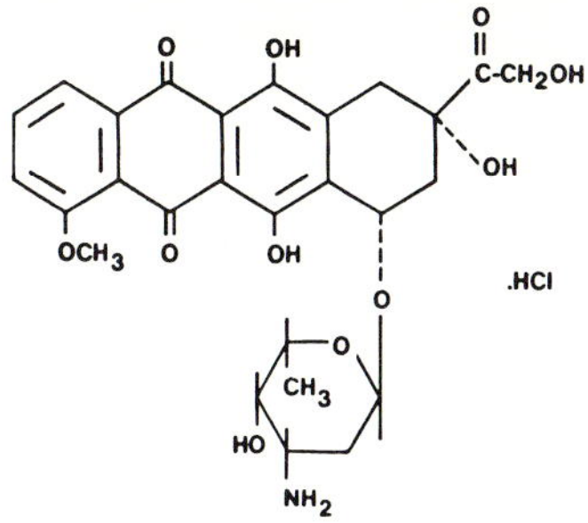


*3. Doxorubicin.* Doxorubicin (adriamycin, Doxo) is an anthracycline antibiotic that intercalates with DNA. The development of LC assays of DOX in biological samples is complicated by the low levels of DOXO that prevail only a few hours after the dose (*20, 21*), and by its physiochemical characteristics. Doxo readily adsorbs to glass surfaces unless they are silanized; moreover, tailing LC peaks may be observed because of complexation of Doxo with divalent cations. Inclusion of $Mg^{2+}$ with the eluent improves the chromatographic behavior of Doxo on silica gel columns (*20*). Reverse phase HPLC (*22, 23*) results in an inferior resolution of Doxo from its metabolites when compared to silica gel LC (*20, 21, 24*). Detection by UV affords a sensitivity of 10 ng Doxo/mL plasma (*24*), while fluorescence detection is sensitive to 1 ng Doxo/mL (*20, 21*). Langone and van Vunakis (*23*) employ reverse phase LC; appropriate LC eluent fractions are collected and analyzed by radioimmunoassay. Such a combination of analytical techniques can be highly sensitive and specific.

## III. Trends in Liquid Chromatographic Analysis of Antineoplastic Agents

The use of LC in anticancer drug level monitoring is certain to increase in the future. Many of the newly developed techniques of drug isolation, improved separation by high efficiency columns and versatile detection have not yet been applied to the anticancer drugs. Suitable assay methods are needed in animal pharmacology and early clinical trails, since many new agents are already waiting to be tested for their antitumor efficacy. We can expect LC methods to contribute significantly to the development of these agents. In addition, therapeutic MTX serum level monitoring by LC may be introduced to more clinical laboratories because of its versatility in detecting metabolites and its broad analytical range.

## References


1. Sadée, W., and Beelen, G. C. M., *Drug Level Monitoring,* Wiley, New York, 1980.
2. Marques, R. A., Stafford, B., Flynn, N., and Sadée, W., *J. Chromatogr.* **146,** 163 (1978).
3. Wasserman, T. H., Philips, T. L., Johnson, R. L., Gomer, C. J., Lawrence, G. A., Sadée, W., and Marques, R. A., *Int. J. Rad. Oncol. Biol. Phys.* **5,** 775 (1979).
4. **Sadée, W., *Therap. Drug Monitor.,* in press.**
5. Bleyer, W. A., *Cancer Treat. Rev.* **4,** 87 (1977).
6. Tattersall, M. H. N., Parker, L. M., Pitman, S. W., and Frei, E. III, *Cancer Chemother. Rep. (Part 3)* **6,** 25 (1975).
7. Stoller, R. G., Hande, K. R., Jacobs, S. A., Rosenberg, S. A., and Chabner, B. A., *N.E.J. Med.* **297,** 630 (1977).
8. Frei, E. III, Jaffe, N., Tattersall, M. H. N., Pitman, S., and Parker, L., *N.E.J. Med.* **292,** 846 (1975).
9. Dedrick, R. L., Zaharko, D. S., and Lutz, R. J., *J. Pharm. Sci.* **62,** 882 (1973).
10. Goldman, D., *Cancer Treat. Rep. (Part 3)* **6,** 51 (1975).
11. Canfell, C. and Sadée, W., *Cancer Treat. Rep.,* in press (1980).
12. Watson, E., Cohen, J. L., and Chan, K. K., *Cancer Treat. Rep.* **62,** 381 (1978).
13. Wisnicki, J. L., Tong, W. P., and Ludlum, D. B., *Cancer Treat. Rep.* **62,** 529 (1978).
14. Lankelma, J., and Poppe, H., *J. Chromatogr.* **149,** 587 (1978).
15. Nelson, J. R., Harris, B. A., Decker, W. J., and Farquhar, D., *Cancer Res.* **37,** 3970 (1977).
16. Cohen, J. L., and Brown, R. E., *J. Chromatogr.* **151,** 237 (1978).
17. Jones, R. A., Buckpitt, A. R., Londer, H. H., Myers, C. E., Chabner, B. A., and Boyd, M. R., *Bull. Cancer* **66,** 75 (1979).
18. Au, J. L., Wu, A. T., Friedman, M. A., and Sadée, W., *Cancer Treat. Rep.* **63,** 343 (1979).
19. Au, J. L., and Sadée, W., *Cancer Res.* **39,** 4289 (1979).
20. Baurain, R., Deprez, D., De-Campeneere, and Trouet, A., *Analyt. Biochem.* **94,** 112 (1979).
21. Peters, J. H., and Murray, Jr., J. F., *J. Liquid Chromatogr.* **2,** 45 (1979).
22. Staffan, E., *J. Chromatogr.* **149,** 225 (1978).
23. Langone, J. J., and van Vunakis, H., *Biochem. Med.* **12,** 283 (1975).
24. Hulhoven, R., and Desager, J. P., *J. Chromatogr.* **125,** 369 (1976).
25. Shiu, G. K., Goehil, T. J., and Pitlick, W. H., *J. Pharm. Sci.* **68,** 232 (1979).
26. Terry, T. K., Belime, R. J., and Brooke, D., *J. Pharm. Sci.* **68,** 172 (1979).
27. Chang, S. Y., Alberts, D. S., Melvick, L. R., Walson, P. D., and Salman, S. E., *J. Pharm. Sci.* **5,** 679 (1978A).
28. Furner, R. L., Mellet, L. B., Brown, R. K., and Duncan, G., *Drug Metab. Dispos.* **4,** 577 (1976).

*29.* Ding, T. L., and Benet, L. Z., *J. Chromatogr. Biomed. Applic.* **163,** 281 (1979).
*30.* Day, J. L., Tterlikkis, L., Niemann, R., Mobley, A., and Spikes, C., *J. Pharm. Sci.* **67,** 1027 (1978).
*31.* Weinkam, R. J., and Shiba, D. A., *Life. Sci.* **22,** 937 (1978).

# Chapter 10

## Hypnotics and Sedatives

Pokar M. Kabra, Howard Y. Koo,
and Laurence J. Marton

*Department of Laboratory Medicine*
*School of Medicine*
*University of California*
*San Francisco, California*

### I. Introduction

In recent years, most large hospitals have observed a marked increase in the admission of patients suffering from drug overdose. Overdose of narcotic drugs, such as the opiates, represent less of a problem on a day-to-day basis than do overdoses of prescribed drugs, such as sedatives and hypnotics. Clinical signs and symptoms for a narcotic drug overdose are very distinct, and in the majority of cases can be easily recognized by the attending physicians without the help of a toxicology laboratory. Loomis (*1*) reported that the majority of fatal poisonings owed to one, or a combination, of four agents: barbiturates, carbon monoxide, ethyl alcohol, and salicylates. Berry (*2*) estimated that 5-5′-disubstituted barbiturates were the second commonest cause of fatal poisoning in England, and that the frequency of their use was increasing. Other nonbarbiturate hypnotics involved in coma-producing incidents include glutethimide (Doriden®), methyprylon (Noludar®), and meprobamate (*3, 4*). In the last five years, diazepam (Valium®) has become one of the leading misused drugs (*5*).

Between August 1974 and December 1976, the Drug Assay Laboratory at Stanford University Hospital reported that out of four thousand drugs found in positive toxicology screens, the ten most commonly found drugs were ethanol, barbiturates, salicylates, diazepam, phenothiazines, acetone, codeine, chlordiazepoxide, phencylidine, and phenytoin (*6*). Other hypnotics, such as methaqualone, glutethimide, methyprylon, and ethchlorvynol, constituted 3.0% of the total drugs found in blood and urine. In 1977, the National Institute of Drug Abuse (*7*) estimated that barbiturates were associated with nearly five thousand deaths a year in the United States, and that users of barbiturates make about 25,000 visits to hospital emergency rooms each year. It is also estimated in this report that in 1976 Americans received about twenty-seven million sleeping pill prescriptions, including barbiturates and many commonly prescribed nonbarbiturates.

## II. Toxicological Effects of Sedative–Hypnotic Poisoning

Barbiturates are derivatives of malonylurea or barbituric acid. Barbituric acid, itself, is not hypnotic, but its 5,5′-disubstituted derivatives possess this property. Barbiturates exert their action on the central nervous system, and the extent of depression of the CNS is dependent on the individual barbiturate. For this reason, it is imperative that the individual barbiturates be identified, as well as quantitated, in blood or serum. The choice of treatment will, in most cases, be dependent on the laboratory findings. The therapeutic and lethal levels for short-acting and long-acting barbiturates are given in Table 1. McBay (*8*) reported that when the blood concentration of ethanol is 1 g/L (0.10%), as little as 5 mg/L of barbiturate in blood is sufficient to cause death.

The nonbarbiturate hypnotics are composed of a large number of compounds with diverse chemical and pharmacological properties. However, like the barbiturates, they possess the ability to produce central nervous system depression. Table 2 summarizes the pharmacological properties and the lethal levels of nonbarbiturate hypnotics. Glutethimide is lipid soluble and only slowly absorbed from the stomach and small intestine (*9*). Consequently, in acute overdose, coma may be unnecessarily prolonged unless the stomach and upper gastrointestinal tract are promptly cleaned of unabsorbed drug. Unlike glutethimide, methaqualone is readily absorbed from the gastrointestinal tract. It is concentrated primarily in the liver and adipose tissue (*10*).

Table 1
Pharmacologic Properties and Toxic Levels of Barbiturates

| Barbiturates | Type of barbiturates | Average concentration in blood produced by 600 mg oral doses, mg/L | Lethal levels: Barbiturates alone, mg/L | Lethal levels: Barbiturates plus ethanol, mg/L |
|---|---|---|---|---|
| Pentobarbital | Short acting | 3.3 | 15–44 | 10–24 |
| Secobarbital | Short acting | 4.8 | 15–24 | 10–24 |
| Amobarbital | Intermediate acting | 9.6 | 30–54 | 10–29 |
| Butabarbital | Intermediate acting | 14.0 | 30–39 | 15–19 |
| Phenobarbital | Long acting | 23.0 | 105–134 | 10–54 |
| Butalbital | Intermediate | Not reported | 13 | Not reported |

Table 2
Pharmacologic Properties and Lethal Levels of Some Sedatives and Hypnotics

| Drugs | Average blood concentration, mg/L | Oral doses | Lethal blood levels, mg/L |
|---|---|---|---|
| Primidone | 8–12 | 1.5 g | Not reported |
| Glutethimide | 7 | 1 g | 30 |
| Ethchlorvynol | 2 | 200 mg | 60 |
| Phenytoin | 10 | 600 mg | 70 |
| Methaqualone | 2 | 250 mg | 5 |
| Methyprylon | 10 | 650 mg | 90 |

## III. Review of Analytical Methods

Several analytical techniques have been employed for the analysis of hypnotics. These include thin layer chromatography (*11–17*), ultraviolet spectroscopy (*18–24*), gas–liquid chromatography (*23*–35), gas chromatography/mass spectrometry (*36–38*), radioimmunoassay (*39*), enzyme multiplied immuno techniques (EMIT) (*40*), and liquid chromatography (*46–49*).

Thin layer chromatography (TLC) is a widely used technique for identifying many of these drugs, but suffers from its inability to allow for accurate and precise quantitation. These drugs can be isolated from biological fluids by a variety of extraction methods. The concentrates of these extractions are then applied to several TLC plates and developed using different solvent systems. The drugs are identified by employing chromogenic sprays that develop specific colors with individual drugs. Mule (*17*) suggested the following chromogenic spray sequence for the detection of barbiturates: (A) 10% ammonium hydroxide followed by observation under ultraviolet light (no reaction is observed with nonfluorescent indicator plates—with indicator plates, the barbiturates appear blue on an orange background); (B) 1% aqueous $KMnO_4$ (all unsaturated substituted barbiturates, such as secobarbital, appear as light yellow spots); (C) 1.0% silver acetate spray (barbiturates appear as white spots); and (D) 0.1% diphenylcarbazone (barbiturates appear blue on a yellow background). Berry and Grove (*52*) used a mercuric chloride-diphenylcarbazone mixture as a single spray. White spots on a lilac background indicated the presence of barbiturates or related compounds.

Cochin and Daly (*12*) developed a TLC method for the analysis of sixteen barbiturate and four nonbarbiturate hypnotics (methyprylon, glutethimide, ethinamate, and ethchlorvynol) in urine, blood, and tissues. The extracted residues were spotted on silica gel plates. A mercurous nitrate spray allowed for the detection of 1–5 μg of the barbiturates and glutethimide (white spot on light gray background), but was less sensitive for the other three nonbarbiturate hypnotics (10–25 μg).

Paper chromatography was one of the earliest forms of chromatography used in separating different barbiturates from other materials. After barbiturates and other acidic compounds are isolated from the biological matrix, paper chromatography has been used in separating and identifying the barbiturates. Both ascending and descending paper-chromatographic techniques have been utilized for the analysis of barbiturates in biofluids (*41*).

Barbiturates may be analyzed spectrophotometrically after extracting the biofluid into organic solvent, followed by an alkali extraction of the solvent. The 5,5′-disubstituted barbiturates show a maximum at approximately 255 nm at pH 13. When the pH is changed to 10 with a buffer, the maximum shifts to approximately 240 nm. The difference in absorption between the strongly alkaline and the buffered solution at 260 nm is proportional to the concentration of barbiturate in the sample. This method is reasonably specific if properly performed and interfering substances are absent; however, there are several drawbacks. The main disadvantage is that drugs that do not have sufficiently strong absorption across their ultraviolet spectrum cannot be analyzed by absorption spectrometry. This eliminates drugs such as methyprylon and ethchlorvynol. Compounds with an $E^{1\%}_{1cm}$ of <200 cannot be effectively analyzed in blood (*18*). Even for drugs with relatively intense UV absorption, a sufficiently high blood level must be reached for UV spectrophotometric analysis. Finally, UV spectrophotometry will not resolve mixtures of drugs with similar absorption spectra.

Jatlow (*21*) critically evaluated interfering substances likely to occur in ultraviolet spectrophotometric analysis of barbiturates. To minimize interference of acidic drugs, he recommended extraction at neutral pH and quantitation by utilizing the net difference between the absorptivities at 260 and 240 nm.

There are several reports in the literature concerning the determination of glutethimide by UV spectrophotometry (*22, 23*). The basic principle of all these methods is the measurement of alkaline hydrolysis product of glutethimide at 230 nm. Methaqualone is generally analyzed by extraction of serum in hexane and back extraction into HCl (*24*). Methaqualone is then identified and quantitated from its characteristic spectrum at 235 nm.

Gas–liquid chromatography (GLC) has proven invaluable to the toxicologist for the resolution and identification of drugs and metabolites in biological fluids. Most GLC methods for the analysis of barbiturates and other acidic or neutral drugs of interest utilize OV-17, OV-1, Dexil 300, and SE-30 liquid phase columns (*42*). The use of two columns, one with a nonpolar liquid phase and another with a polar phase, for distinguishing barbiturates has been recommended by several investigators. Problems with tailing and adsorption to the column have been alleviated by employing derivatization procedures. Alkylation or silyation are the two most frequently used derivatization procedures for the analysis of barbiturates (*42, 43*). More recently, Street (*44*) recommended the use of nonderivatized barbiturates for quantitative analysis by GLC, and "on column" formation of methyl and trimethylsilyl derivatives for qualitative identification. Berry (*2*) critically evaluated a number of the commonly used GLC columns in an effort to find a single column able to give reliable, accurate, and specific identification of barbiturates at both therapeutic and toxic concentrations. Out of twelve columns tested, Berry determined that cyclohexanedimethanol succinate (CDMS) was the best column for his purpose, and that OV-225 proved to be the second most useful column.

The availability of a wide variety of columns allows one to choose a second, chemically different, column to confirm the identity of unknown peaks eluted by the first column. This practice prevents false positive identification of compounds. A comprehensive dual column (3% OV 1 and 3% OV 17) GLC analysis for sedative drugs was developed by Thoma et al. (*34*). The procedure provided for the identification and quantitation of barbiturates, glutethimide, meprobamate, methyprylon, benzodiazepines, methaqualone, and propoxyphene as the free drug in serum. The drugs were first identified on OV-17 column, using a temperature program from 165 to 280°C. Confirmation of the identified peak was then achieved by derivatizing the same sample with trimethylaniline hydroxide (TMAH) on an OV-1 column, and using a second temperature program from 135 to 250°C. With the relative retention times from two columns that had different characteristics of separation, significant improvement in identification of compounds of toxicological importance was achieved.

The combination of gas chromatography and mass spectrometry has given the analyst a most powerful and versatile tool. The recent increasing use of this technique in the toxicology laboratory has largely been the result of the introduction of smaller, lower priced, medium-resolution mass spectrometers linked to a gas chromatograph.

Mass spectrometry functions by producing ions whose mass to charge ratios are quantitated. Reference mass spectral data have been accumulated by spectroscopists, and can be conveniently used either manually or in a computerized form for the rapid identification of barbiturates and hypnotics (*36–38*). Fales et al. (*32*) presented a method for the mass spectrometric identification of barbiturates using chemical ionization techniques. Electron impact techniques produced unstable molecular ions of the barbiturates, thus creating detection problems. Fragmentation of barbiturates usually occurs by loss of CO and one of its two side chains. However, many barbiturates possess similar side chains, thus yielding identical molecular ions. Fales et al. (*38*) demonstrated that chemical ionization mass spectrometry with methane provided intense quasimolecular ($QM^+$) ions at $M/e(M+1)^+$ in all eight representative barbiturates. Isomeric barbiturates, such as amobarbital and pentobarbital, could not be distinguished, but could be identified from their electron impact mass spectra.

The homogenous enzyme immunoassay technique has been adapted to the quantitative determination of barbiturates in biological fluids (*45*). A secobarbital derivative is labeled with the enzyme lysozyme. When this enzyme-labeled drug is complexed with antibarbiturate antibody, the enzyme is rendered inactive. In the presence of free drug, or related compounds, the lysozyme drug conjugate and free drug compete for antibody binding sites. Thus, the presence of free drug causes some of the lysozyme drug conjugate to remain uncomplexed and enzymatically active. The amount of enzyme activity is directly proportional to the concentration of free drug. The EMIT assay does not differentiate between various barbiturates. Various other immunological techniques, such as radioimmunoassay, free radical assay, and hemagglutination inhibition, have been developed for the analysis of various barbiturates present in serum, urine, and tissue (*41*).

Although the use of liquid chromatography has been extensively reported upon for the analysis of various therapeutic drugs (see other chapters in this book), there are only a very few reports concerning the utilization of LC for sedative and hypnotic drug screening. Dixon and Stoll (*46*) developed a LC method for the detection of six barbiturates (barbital, phenobarbital, butabarbital, amobarbital, pentobarbital, and secobarbital) using an octadecylsilane column. The barbiturates were extracted from biofluids by a single solvent extraction, similar to that of common GC methods, and detected at 216 nm with good sensitivity. Recovery varied from 68% for barbital to 100% for secobarbital. The isomeric barbiturates (amobarbital and pentobarbital) could not be separated. Figure 1 illustrates the separation of some representative barbiturates on $C_{18}/\mu$ Porasil and $C_{18}$ Corasil columns. Although the separation of five of the barbiturates is adequate on the $C_{18}$ Corasil column, they are not completely resolved unless the $C_{18}/\mu$ Porasil (smaller size packing) column is used. Tjaden et al. (*47*) reported a LC method for the analysis of barbiturates in blood and saliva using a methylsilica column. The extraction of these drugs from biofluids was rather complex and time consuming. Blank serum extraction showed the presence of substances with chromatographic retention times similar to those of the barbiturates. In order to remove these interfering serum constituents, the extraction procedure was amended to include a back extraction step. The barbiturates were detected at 220 nm. The methylsilica column showed good selectivity for the barbiturates; however, these columns are not available commercially. Early eluting metabolites are interfered with by serum constituents.

In our laboratory, during the development of a reversed-phase LC method for the analysis of anticonvulsants, we observed that the same reversed-phase chromatographic conditions were also applicable to the analysis of some of the common weakly acidic and neutral hypnotics (*48*). Based on these observations, we developed a LC method for the simultaneous analysis of twelve commonly used hypnotics (*49*). Sample preparation was minimal and solvent

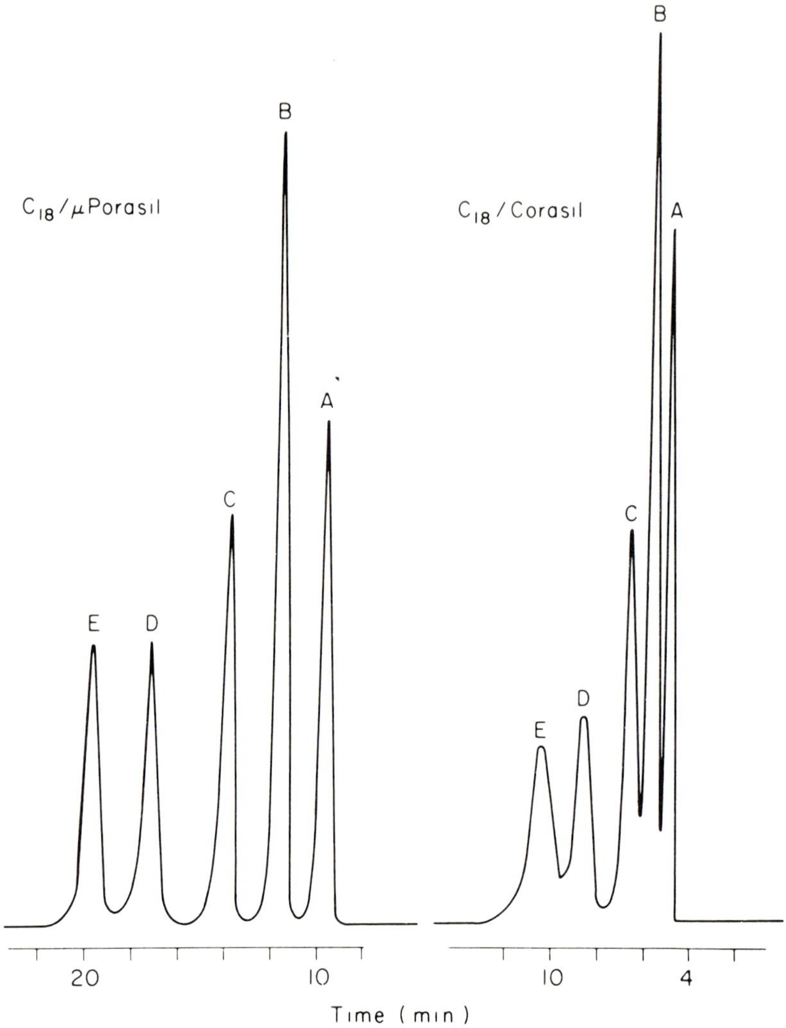


FIG. 1. Separation of barbiturates on reversed-phase partition mode showing the effect of particle size. A, barbital; B, phenobarbital; C, butabarbital; D, amobarbital; E, secobarbital. Reprinted from reference *46* by courtesy of Academic Press.

extraction was not mandatory. UV "end absorption" at 195 nm enabled us to detect methyprylon and ethchlorvynol; both of these compounds have negligible absorption at 254 nm. Elevated column temperature (50° C) allowed for increased resolution, selectivity, and overall efficiency of the chromatographic technique.

LC offers several advantages over other techniques in that sample manipulation prior to chromatography is minimal. LC is capable of analyzing several different classes of hypnotics simultaneously with good specificity, precision, and accuracy. A major advantage of LC for toxicological purposes is the ability to collect the column effluent for further drug identification and characterization.

## IV. Collection and Preparation of Samples

Serum or plasma samples are usually preferred for the liquid chromatographic analysis of barbiturates and hypnotics. However, whole blood can be used if suitable extraction methods are employed. Extraction methods for isolationg and concentrating hypnotics from a sample matrix may vary from a simple one step solvent extraction, to complicated back extractions, column extractions, and charcoal adsorption techniques. The type of extraction and the amount of sample cleanup is dictated by the efficiency and selectivity of the chromatographic technique used for the analysis. The more specific and efficient the chromatographic system, the less sample extraction and clean-up are necessary to obtain the desired results. Sometimes extraction techniques also serve to concentrate the analyte in the sample, thus improving the sensitivity of the assay.

A method using a single solvent extraction procedure was reported by Dixon and Stoll (*46*). They extracted 1 mL of serum with 2 mL of chloroform for 5 min. After centrifugation, the chloroform layer was separated and evaporated. The residue was redissolved in 0.1 mL 60% aqueous methanol. Ten μL of the sample was injected into a $C_{18}$ μ-Porasil column. Protein precipitation is probably the simplest and most rapid method for LC sample preparation. This technique precipitates serum proteins in the sample using a water-miscible organic solvent, such as acetonitrile, methanol, or ethanol. Although the proteins precipitate, the drugs remain freely soluble in the supernatant. Acetonitrile is probably the best solvent for this purpose because it yields a clear supernatant in a short period of time. The procedure involves pipetting 25–500 μL of serum into a tube to which an equal, or double, volume of acetonitrile is added; the sample is vortexed and then centrifuged. We recommend that the sample be centrifuged at high speed, preferably greater than 10,000*g*, to pack the protein into a tight pellet at the bottom of the sample tube, and to assure a clear supernatant. The major advantage of this technique is its simplicity and speed.

Other often used extraction methods include XAD-2 non-ionic and adsorbent resin, charcoal, and cation exchange loaded paper. Prantitis et al. (*50*) reported the use of XAD-2 resin for the extraction of several drugs from biofluids. XAD-2 resin is a styrene–divinylbenzene copolymer that adsorbs drugs mainly by hydrophobic and dipole–dipole interactions. The adsorbed drugs are eluted by a suitable organic solvent. Charcoal is also frequently used as an adsorptive agent. Most drugs are completely bound to a small amount of charcoal and are easily eluted by small amounts of organic solvents. Meola and Vanko developed several adsorption procedures for drugs in blood or urine using charcoal.

An automated sample extraction system (Dupont Prep I) can process several samples simultaneously. This sample processor utilizes a solid-phase column extraction technique with an inert hydrophobic sorption resin to extract the materials of interest. Extraction and washing steps are automated and twelve samples can be processed simultaneously.

## V. Detection

Barbiturates as a class possess very poor UV adsorption at 254 nm when eluted with neutral or acidic mobile phases; however, the anionic forms are strong chromophores. The reversed-phase chromatographic properties of the barbiturates have been studied extensively. The free forms of these drugs are easily separated, but little separation is observed when the pH of the mobile phase produced the anion as the dominant barbiturate species. Thus, for barbiturates, good chromatographic properties are observed in the free acid form, and good chromophoric properties are observed only in the anionic form. In addition, the use of a basic mobile phase is frequently associated with dissolution of silica-based stationary phases. Clark and Chan (*51*) reported an innovative LC method where the barbiturates are separated as the free acids by reversed-phase chromatography and then the pH of the mobile phase is changed by a post-column technique so that the drugs are detected as the chromophoric anions at 254 nm.

It is well known that many chemical compounds having little or no absorption in the near UV region may possess fairly strong absorption in the region below 200 nm. This absorption owes to the $n$–$\sigma^*$, and $\pi$–$\pi^*$, transitions, and permits detection in the region at or below 200 nm. The absorption at 195 nm displayed by the barbiturates and hypnotics allowed us to analyze these drugs without the usual sample concentration steps. Ten ng of the hypnotics could be detected.

Although sensitivity is greatly enhanced at 195 nm, there are several limitations of detection in this region. The mobile phase must be transparent at these wavelengths. A mobile phase that usually conforms to this criterion of transparency is a mixture of phosphate buffer and glass-distilled acetonitrile (UV cutoff $< 190$ nm). Other solvents, such as methanol, ethanol, tetrahydrofuran, and acetate buffer, cannot be used in the mobile phase below 200 nm. In addition, these low detection wavelengths are fairly nonspecific, hence, resolution must be adequate to separate potentially interfering substances from the analytes of interest.

## VI. Complete Analysis of Test Samples

The application of the principles detailed above may best be illustrated by a discussion of our actual sample analysis (*49*). Serum samples were analyzed for twelve common hypnotics by reversed-phase chromatography as follows:

| | |
|---|---|
| Column: | Reversed phase, $\mu$-Bondapack $C_{18}$, 30 cm × 4.0 mm |
| Mobile phase: | Acetonitrile–phosphate buffer, 21.5/78.5 parts by volume (phosphate buffer pH 4.4) |
| Flow rate: | 3 mL/min |
| Oven temp.: | 50° C |
| Detection: | 195 nm |

The concentration of acetonitrile in the mobile phase and the column temperature were selected to optimize the resolution of the hypnotic drugs. We found that 21.5 parts of acetonitrile gave the best resolution for all of twelve drugs. If the concentration of acetonitrile was reduced to 20.5 parts, the resolution between phenytoin and glutethimide was completely lost (Fig. 2). If the concentration of acetonitrile was increased to 22.5 parts, there was baseline resolution between phenytoin and glutethimide; however, the resolution between pentobarbital, amobarbital, and ethchlorvynol was lost (Fig. 3). Elevated column temperature (50° C) was necessary to resolve pentobarbital from amobarbital under these conditions. Utilizing a flow rate of 3 mL/min, in combination with the optimized mobile phase, resulted in an analysis time of approximately 22 min for all twelve hypnotics (primidone, methyprylon, phenobarbital, butabarbital, butalbital, ethchlorvynol, pentobarbital, amobarbital, phenytoin, glutethimide, secobarbital, and methaqualone) and an internal standard [5-(4-methylphenyl)-5-phenylhydantoin]. The

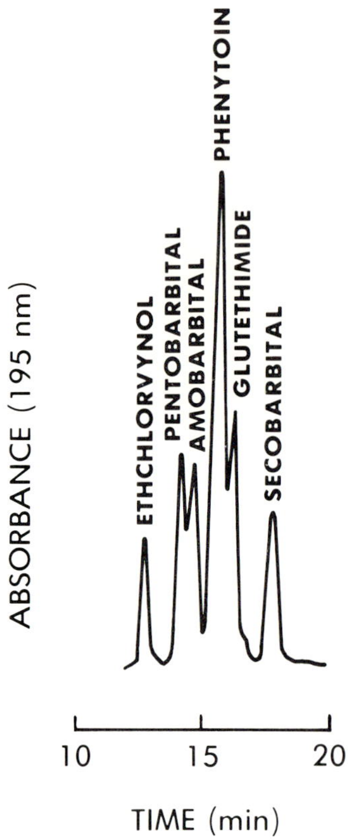


FIG. 2. Chromatography of some hypnotics on $\mu$-Bondapack $C_{18}$ with 20.5% acetonitrile and phosphate buffer.

internal standard was selected because it was minimally interfered with by sample matrix constituents. Samples and controls were prepared by one of the following methods:

*Method A: Acetonitrile precipitation.* Serum samples, 25–200 $\mu$L were transfered to Eppendorf 1.5-mL polypropylene microtubes together with an equal volume of acetonitrile containing 50 $\mu$g/mL of the internal standard. After thorough mixing, the contents were centrifuged for 1 min in an Eppendorf no. 5412 centrifuge. Twenty $\mu$L of the supernatant were injected into the chromatograph and analyzed under the chromatographic conditions outlined above.

*Method B: Extraction with ethyl acetate.* A 300 $\mu$L quantity of the serum supernatant from Method A was transferred, along with 50 $\mu$L of the glacial acetic acid, to a 12 mL glass-stoppered tube. This mixture was extracted with 3 mL of ethyl acetate for 5 min, and

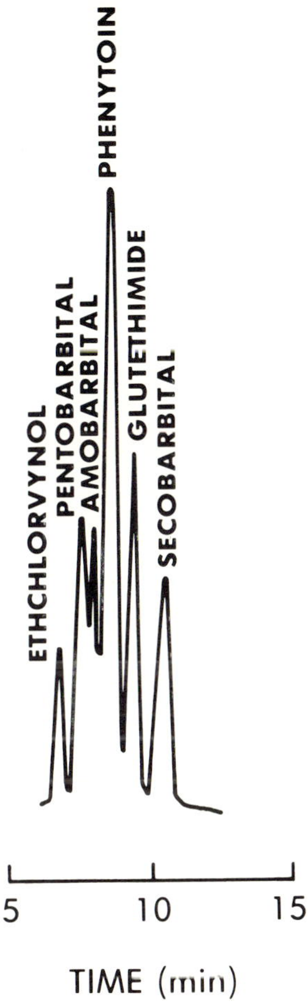


FIG. 3. Chromatography of some hypnotics on $\mu$-Bondapack $C_{18}$ with 22.5% acetonitrile and phosphate buffer.

centrifuged at 2000 rpm. The ethyl acetate layer was separated, evaporated at 70°C under reduced pressure, and the residue was dissolved in 100 $\mu$L of acetonitrile. A 10–20 $\mu$L sample of the solution was injected into the chromatograph and eluted under the same conditions as in Method A.

A serum supernatant, prepared from a serum-based control containing all of the drugs at a concentration of 20 mg/L, yielded the chromatograph in Fig. 4. The chromatograms obtained from patient sera are shown in Fig. 5. The chromatograms were compared with that obtained from the drug standards and serum-based control and drug peaks were identified by their relative retention times. Quantitation of

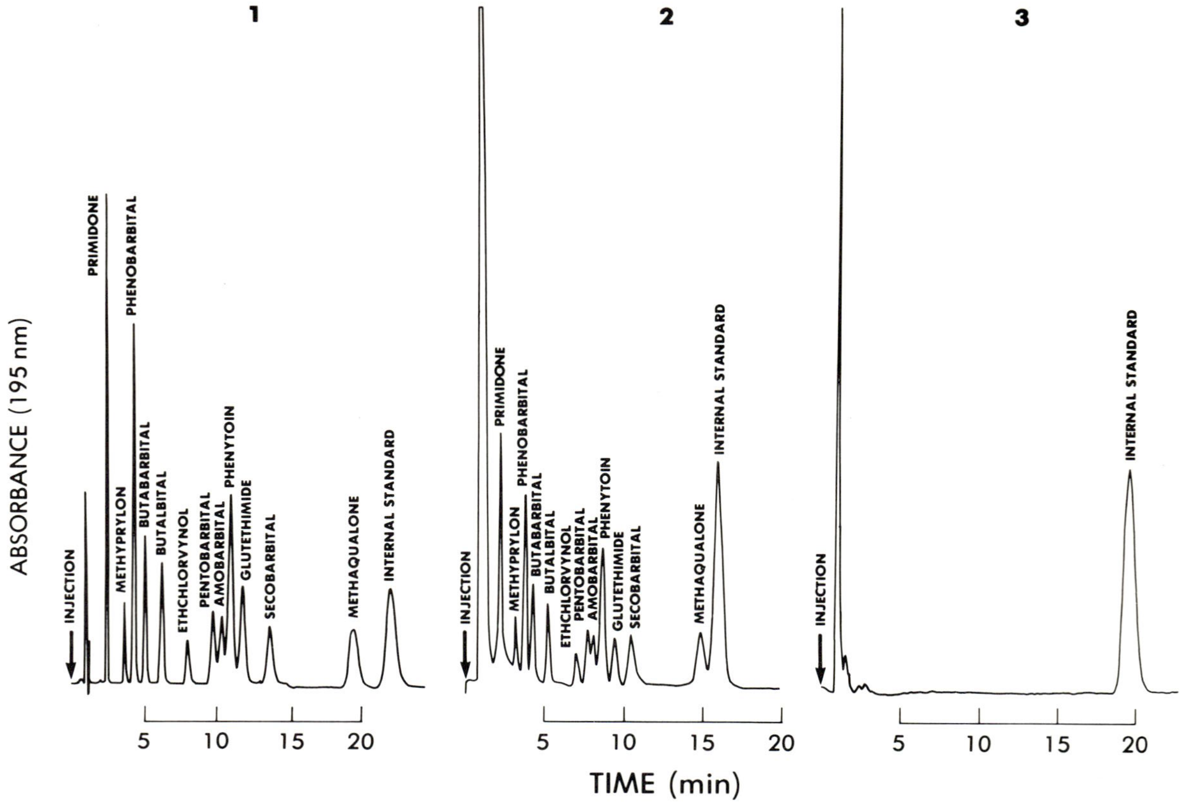


FIG. 4. Left: Chromatogram of a standard mixture of drugs. Quantity chromatographed was 250 ng, except for internal standard, which was 500 ng. Middle: Chromatogram of a serum supplemented with 200 ng of each drug, except for internal standard, which was 500 ng. The chromatograph was run at higher acetonitrile concentration. (22.5 parts). Right: Chromatogram of a drug free serum with 50 mg of added

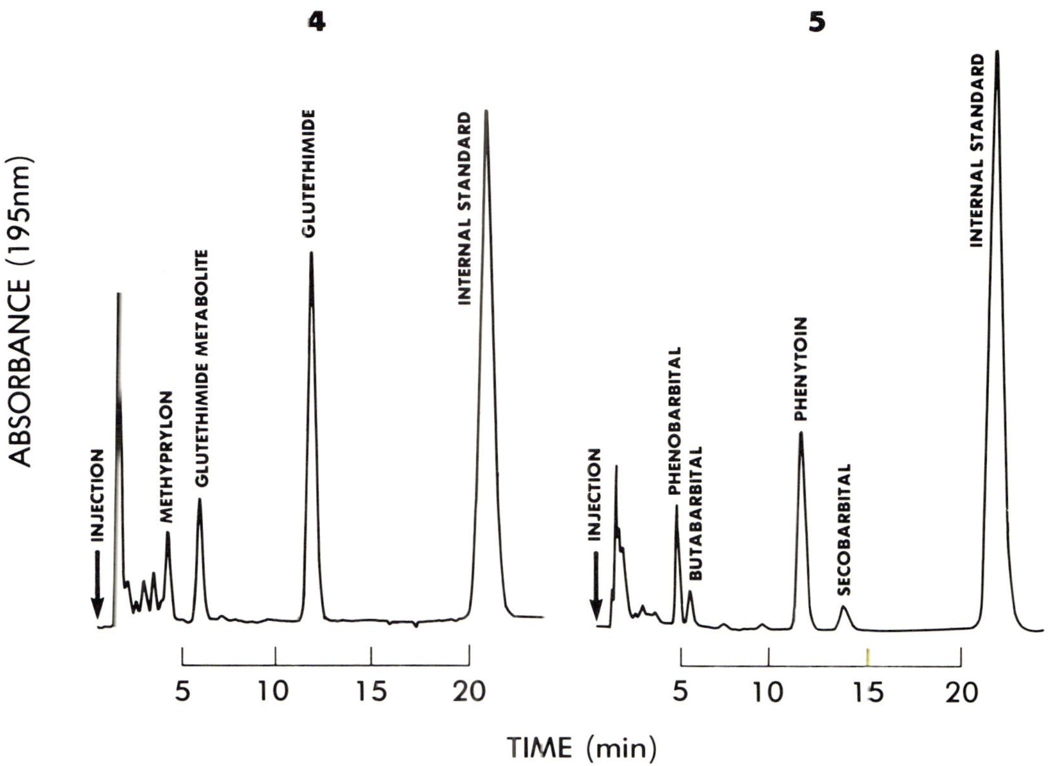


FIG. 5. Left: Chromatogram from the serum of a patient with, per liter, 13.5 mg of methyprylon and 44.0 mg of glutethimide. Peak eluting at 6 min corresponds to glutethimide metabolite. Right: Chromatogram of a patient's serum containing per liter, 4.0 mg of phenobarbital, 2.5 mg of butabarbital, 2.5 mg of secobarbital, and 8.7 mg of phenytoin. Reprinted from reference *49* by courtesy of American Association for Clinical Chemistry.

the chromatogram may be accomplished by use of peak heights or peak areas. Peak areas are best measured by an electronic integrator, or computer. The accuracy of the assay is improved by the incorporation of an internal standard which corrects for extraction and injection variations. The ratios of peak heights (or peak area) of drug to internal standard for a range of concentrations (2–100 mg/L) are plotted against concentrations. This plot is used for quantitation. All standards were made from serum and extracted as outlined above.

Any substance that chromatographs along with the test sample and has a retention time similar to any of the analytes may interfere with the assay. Interfering substances may falsely elevate the quantity of the drugs present in a sample, or decrease it by an apparent increase of an internal standard. Interfering substances may originate from two sources: those extrinsic to and those intrinsic to the sample. The first category consists of impurities introduced with reagents or glassware. A potential source of such interfering substances may be plasticizers present in plastic ware or in blood-collection tube stoppers. Interfering substances may be introduced from serum components that are co-extracted with the drug of interest. These might be other drugs ingested by the subject or endogenous constituents. Potentially interfering drugs were studied by chromatographing over 40 of these drugs. Only ethotoin and mephobarbital were found to interfere with the analysis of phenobarbital and amobarbital, respectively. Both of these drugs are rarely used. Interference from endogenous constituents was studied utilizing drug free sera.

## VII. Current Trends in LC Techniques

The introduction of a new generation of LC detectors, such as LC coupled with a mass spectrometer and a high-speed scanning ultraviolet spectrophotometer, will provide more definitive identification of drugs. Rapid scanning spectrophotometers should allow for the optical resolution of compounds that are not fully separated chromatographically. New developments in microparticulate packings and improvement in packing techniques are giving the analyst greater column efficiency and selectivity, and reduced analysis times.

## VIII. Conclusions

Most clinical laboratories use spectrophotometric methods for hypnotic screening. These methods lack both sensitivity and specificity. Spectrophotometric methods cannot differentiate accurately between long- and short-acting barbiturates. This differentiation is important in order to institute rational therapy. In addition, since alcohol and benzodiazepines are frequently ingested along with babiturates, resulting in potentiation of barbiturate activity, it is now necessary to detect these drugs at lower concentrations. Many of the problems of specificity and sensitivity have been overcome by the liquid chromatographic methods.

Since LC is a nondestructive method of analysis, the eluate from the column can be collected and further analyzed by suitable methods to confirm the presence of any drug or metabolite. In addition, the presence of a specific drug can be confirmed by the technique of absorption ratioing or UV scanning. This can easily be accomplished using fast scanning ultraviolet spectrophotometers.

The LC method described in this chapter can easily be adapted to microsamples (as little as 25 μL of serum). This eliminates the need for collection of several milliliters of blood often required for the analysis of these drugs by other screening methods. Since this method is simple and rapid (total analysis time, 40 min) it can readily be adapted for rapid screening when appropriate.

## Acknowledgments

Laurence J. Marton, M.D. is the recipient of NCI Research Career Development Award CA-00112. We thank Phil Reynolds (The Institute of Forensic Sciences, Oakland, California) for providing us with patient samples and a number of GLC and ultraviolet analyses. We also thank Jeff Wall and Brian Stafford for their excellent technical assistance, and Mary Stawski for her careful typing of this manuscript.

## References

1. Loomis, T. A., *Essentials of Toxicology,* Lea and Febiger, Philadelphia, 1968.
2. Berry, D. J., *J. Chromatogr.* **86,** 89 (1973).
3. Barret, M. J., *Clin. Chem. Newsletter* **3,** 1 (1971).
4. Meyers, F. H., Jawetz, E., and Goldfein, A., *Drug Abuse: Review of Medical Pharmacology,* Lange Medical Publications, Los Altos, Calif., 1974.
5. Low, N. C., Fales, H. M., and Milne, G. W. A., *Clin. Toxicol.* **5,** 17 (1972).

6. "Toxic Screen Cumulative Results," *DAL Newsletter,* Stanford University Hospital, **2,** 3 (1976).
7. Dupont, R., "Federal Study of Nighttime Sleeping Pills, National Institute of Drug Abuse, 1977.
8. McBay, A., *Clin. Chem.* **19,** 361 (1973).
9. Fimble, B. S., "Glutethimide," in *Methodology for Analytical Toxicology,* Sunshine, I. ed., CRC Press, Cleveland, Ohio, 1975, p. 178.
10. Berry, D. J., *J. Chromatog.* **42,** 39 (1969).
11. Davidow, B., Petri, N. L., and Quame, B., *Am. J. Clin. Pathol.* **38,** 714 (1968).
12. Cochin, J., and Daly, J. W., *J. Pharmacol. Exp. Ther.* **139,** 154 (1963).
13. Dunlop, M., and Curnow, D. H., *J. Clin. Pathol.* **20,** 204 (1967).
14. Hofmann, A. F., *Anal. Biochem.* **3,** 145 (1962).
15. Bogan, J., Rentoul, E., and Smith, H., *J. Forensic Sci. Soc.* **4,** 147 (1964).
16. Sunshine, I., TLC for Weak Acids, Neutrals, and Weak Bases, in *Handbook of Analytical Toxicology,* Sunshine, I., ed., CRC Press, Cleveland, Ohio, 1975, p. 412.
17. Mule, S. F., *J. Chromatog.* **55,** 255 (1971).
18. Jatlow, P., *Am. J. Med. Technol.* **39,** 231 (1973).
19. Goldbaum, L. R., *Anal. Chem.* **24,** 1604 (1952).
20. Williams, L. A., and Zak, B., *Clin. Chim. Acta.* **4,** 170 (1959).
21. Jatlow, P., *Am. J. Clin. Pathol.* **59,** 167 (1973).
22. Dauphinais, L. R., and McComb, R., *Am. J. Clin. Pathol.* **44,** 440 (1965).
23. Goldbaum, L. R., *Anal. Chem.* **32,** 81 (1960).
24. Bailey, D., and Jatlow, P., *Clin. Chem.* **19,** 615 (1973).
25. MacGee, J., *Clin. Chem.* **17,** 587 (1971).
26. Fiereck, E. A., and Tretz, N. W., *Clin. Chem.* **17,** 1024 (1971).
27. Brochmann-Hanssen, E., and Oke, T., *J. Pharm. Sci.* **58,** 371 (1969).
28. Flanagan, R. J., and Withers, G., *J. Clin. Pathol.* **25,** 899 (1972).
29. Sine, H. E., McKenna, M. J., Law, M. R., and Muray, M. H., *J. Chromatogr. Sci.* **10,** 297 (1972).
30. Rice, A. J., and Wilson, W. R., *Clin. Toxicol.* **6,** 59 (1973).
31. Kaufman, J. H., *Am. J. Med. Technol.* **39,** 338 (1973).
32. Levy, S. K., Schwartz, T., *Clin. Chim. Acta.* **54,** 19 (1974).
33. MacGee, J., *Anal. Chem.* **42,** 421 (1970).
34. Thoma, J., and Bondo, P., "GC for Sedative Drugs," in *Handbook of Analytical Toxicology,* Sunshine, I., ed., CRC Press, Cleveland, Ohio, 1975, p. 421.
35. Flanagan, R. J., and Berry, D. J., *J. Chromatogr.* **131,** 131 (1977).
36. Finkle, B. S., and Taylor, D. M., *J. Chromatogr. Sci.* **10,** 312 (1972).
37. Law, N. C., Aandahl, V., Fales, H. M., and Milne, G. W. A., *Clin. Chim. Acta.* **32,** 221 (1971).
38. Fales, H. M., Milne, G. W. A., and Axenrod, T., *Anal. Chem.* **42,** 1432 (1970).
39. Cleeland, R., Christenson, J., Usetegui-Gomez, M., Heveran, J., Davis, R., and Grumberg, E., *Clin. Chem.* **22,** 712 (1976).
40. Scharpe, S. L., Cooreman, W. M., Bloome, W. J., and Lakemen, G. M., *Clin. Chem.* **22,** 723 (1976).

*41.* Jain, N. C., and Cravey, R. H., *J. Chromat. Sci.* **12,** 228 (1974).
*42.* McReynolds, W. O., *J. Chromatogr. Sci.* **8,** 230 (1970).
*43.* Brochmann-Hanssen, E., and Obe, T. O., *J. Pharm. Sci.* **58,** 370 (1969).
*44.* Street, H. V., *Clin. Chim. Acta.* **34,** 357 (1971).
*45.* Rubenstein, K. E., Schneider, R. S., and Ullman, E. F., *Biochem. Biophys. Res. Comm.* **47,** 846 (1972).
*46.* Dixon, P. F., and Stoll, M. S., "The HPLC Detection of Some Drugs Taken in Overdose," in *High Pressure Liquid Chromatography in Clinical Chemistry,* Dixon, P. F., Gray, C. H., Lim, C. K., and Stoll, M. S., eds. Academic Press, New York, NY, 1976 p. 211.
*47.* Tjaden, U. R., Kraak, J. C., and Huber, J. F. K., *J. Chromatogr.* **143,** 183 (1977).
*48.* Kabra, P. M., Stafford, B. E., and Marton, L. J., *Clin. Chem.* **23,** 1284 (1977).
*49.* Kabra, P. M., Koo, H. Y., and Marton, L. J., *Clin. Chem.* **24,** 657 (1978).
*50.* Pranistis, P. A. F., Mitzoff, J. R., *J. Forensic Sci.* **19,** 917 (1974).
*51.* Clark, C. R., and Chan, J. L., *Anal. Chem.* **50,** 635 (1978).
*52.* Berry, D. J., and Grove, J., *J. Chromatogr.* **80,** 205 (1973).

# Chapter 11

## Toxicology Screening

Pokar M. Kabra, Brian E. Stafford,
and Laurence J. Marton

*Department of Laboratory Medicine*
*School of Medicine*
*University of California*
*San Francisco, California*

### I. Introduction

Rapid identification and quantitation of drugs in biofluids are helpful to the physician in managing patients with suspected drug intoxication. Higgins and O'Brien (*1*) noted that prior to 1960, drug overdoses usually consisted of a single drug. However, three years later they observed that multiple drug overdoses had risen to 13% (*2*). Law (*3*) reported that out of 240 proven drug misuse cases observed in Suburban Hospital, Bethesda, Maryland, over a four year period, 60% of the cases involved a single drug and 40% were multiple drug ingestions.

Various techniques currently employed for screening drugs include spectrophotometry, gas-liquid, thin layer and paper chromatography, enzyme multiplied immuno technique (EMIT), and gas chromatography combined with mass spectrometry.

Spectrophotometric analysis (*4*) is often time consuming, lacks specificity, and is usually applicable for only single drugs. Paper and thin layer (*5*) chromatography are valuable techniques for the detection of multiple drugs; however, they are usually time consuming and only provide semiquantitative data. Gas-liquid chromatography (GC) is an excellent method for the separation of many drugs found in gastric contents, serum, or urine. However, residues obtained from these biofluids by simple chloroform extraction often contain complex mixtures of compounds that are difficult to resolve by a single GC column, and unequivocal identification of each component cannot be obtained. For instance, glutethimide and dibutyl phthalate, a common contaminant in these fluids, are poorly separated by GC (*6*). Dual-column GC (*7*) or GC interfaced with specific and selective detectors, such as a mass spectrometer (*8*), have improved the situation significantly.

The immunological assays that have recently become available to detect drugs of abuse in biofluids are a valuable addition to current analytical methods (*9*). The principal advantages of these techniques are: high sensitivity, speed, and direct analysis of biofluids without prior extraction and concentration. A serious limitation of immunoassays is the lack of specificity for an individual drug; drugs of similar chemical structure may cross-react [e.g., codeine, a common ingredient of cough medications, reacts in the assay as well as, or better than, morphine, a compound that, when present, indicates the use of heroin (*10*)]. For this reason, all positive results must be confirmed by nonimmunological procedures.

Although the use of liquid chromatography has been extensively reported upon for the analysis of various classes of therapeutic drugs (See Section II, this volume), there are virtually no reports concerning the utilization of LC for toxic drug screening. An exception is the analysis of hypnotic drugs described in Chapter 10. This article briefly describes an LC screening method for the simultaneous analysis of twenty commonly abused drugs utilizing gradient liquid chromatography. This LC method offers several advantages over other techniques: sample manipulation prior to chromatography is minimal, several different classes of drugs can be analyzed simultaneously with good specificity, precision, and accuracy, and the column effluent can be collected for further drug identification and characterization.

## II. LC Analysis

We used a Perkin-Elmer Series 3 liquid chromatograph equipped with a variable wavelength detector (Perkin-Elmer LC55 or LC75) and a temperature controlled oven (LC 100). The reversed phase columns, either a "$\mu$ Bondapack $C_{18}$" (Water Associates, Incorporated) or an Ultrasphere-ODS 5$\mu$, (Altex) was mounted in the oven.

The sample was injected into a Rheodyne Model 7105 valve mounted on the chromatograph. A Hewlett-Packard high speed spectophotometer Model 8450 A equipped with a flow cell was interfaced with liquid chromatograph to scan the column effluent for further characterization in certain cases.

The column was eluted with acetonitrile/phosphate buffers at the rate of 3.0 mL/min using a programmed two-step gradient. The oven temperature was set at 50° C and the effluent was monitored at 210 nm. The phosphate buffer was decontaminated of the organic impurities by passing it through a preparative column 15 cm × 10 mm dry packed with 25–40 μm Lichroprep™ RP18 (E. Merck). This column was mounted between the pump and the mixing tee.

A 200-μL quantity of acetonitrile containing 10 μg of hexobarbital (as an internal standard) was added along with 25 μL of concentrated acetic acid to 200 μL of serum. The sample was vortex-mixed and centrifuged. A sample of 30 μL of the supernatant was injected into the chromatograph and eluted with an acetonitrile/phosphate gradient. The acetonitrile concentration was increased from 5 to 45% in two linear steps over a time interval of 34 min. Figure 1 illustrates the chromatographic separation achieved.

The eluted drugs were detected at 210 nm. Below 210 nm there was a nonspecific interference from the serum matrix. The minimum detection level for most drugs was approximately 5 mg/L, the benzodiazepines and methaqualone could be detected at 1 mg/L. A Hewlett-Packard high-speed spectrophotometer model 8450 A equipped with a flow cell was used as a detector for a number of analyses. This detector is capable of scanning the spectrum from 200 to 700 nm in approximately 1 s. Figure 2 illustrates a UV scan of methaqualone obtained from this detector while the peak was eluting from the column. A number of drugs that do not possess characteristic UV absorbance features can be distinguished from each other by plotting the first derivative of their spectra. This point is well illustrated for phenytoin and glutethimide in Figs. 3 and 4. Figure 3 illustrates essentially indistinguishable UV spectra for phenytoin and gluetethimide. However, a plot of the first derivative of their spectra (Fig. 4) shows striking differences between the two drugs. This may be a useful approach to the unequivocal identification of a number of coeluting compounds that cannot be identified by direct UV scanning.

The procedure was optimized for the linearity and recoveries for all of these drugs. Interference from other abused drugs, except for acetaminophen, was insignificant. Even the acetaminophen interference could be eliminated by employing a simple chloroform extraction of the serum supernatant.

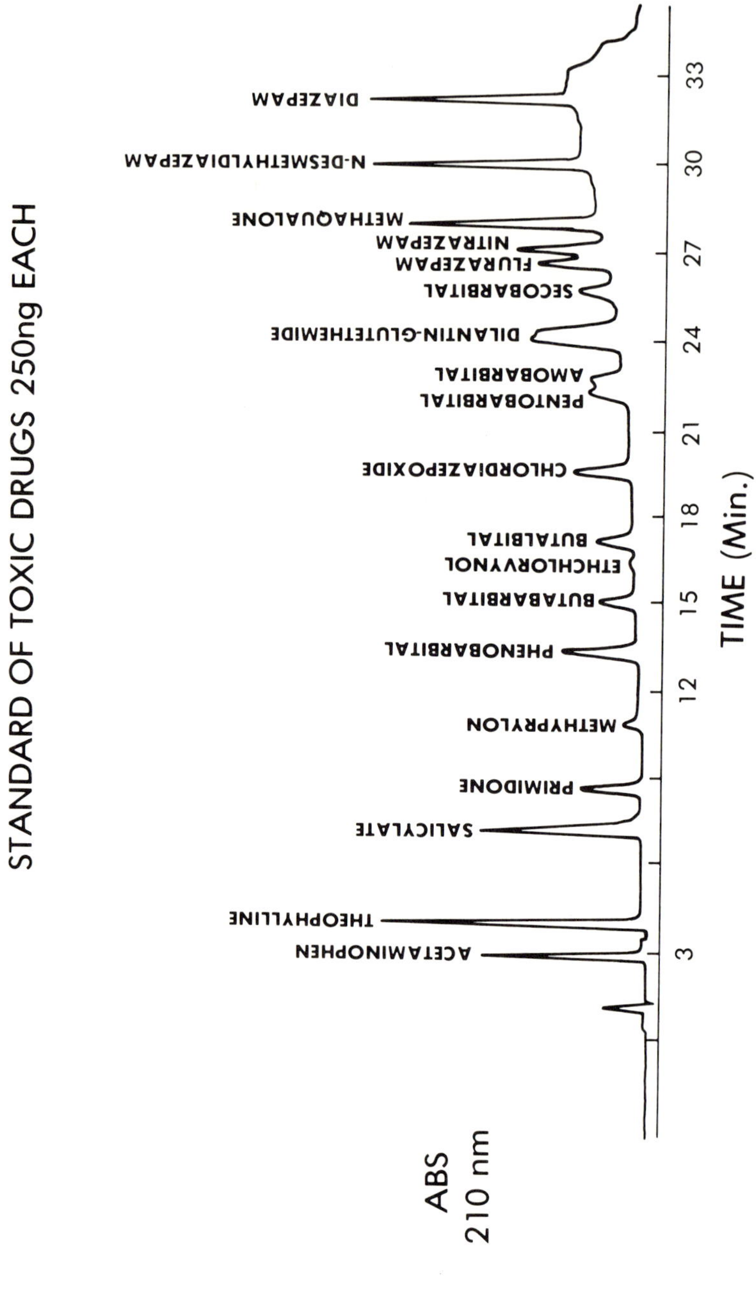


FIG. 1. Chromatogram of a standard mixture of drugs. The quantity chromatographed was 250 ng for each drug.

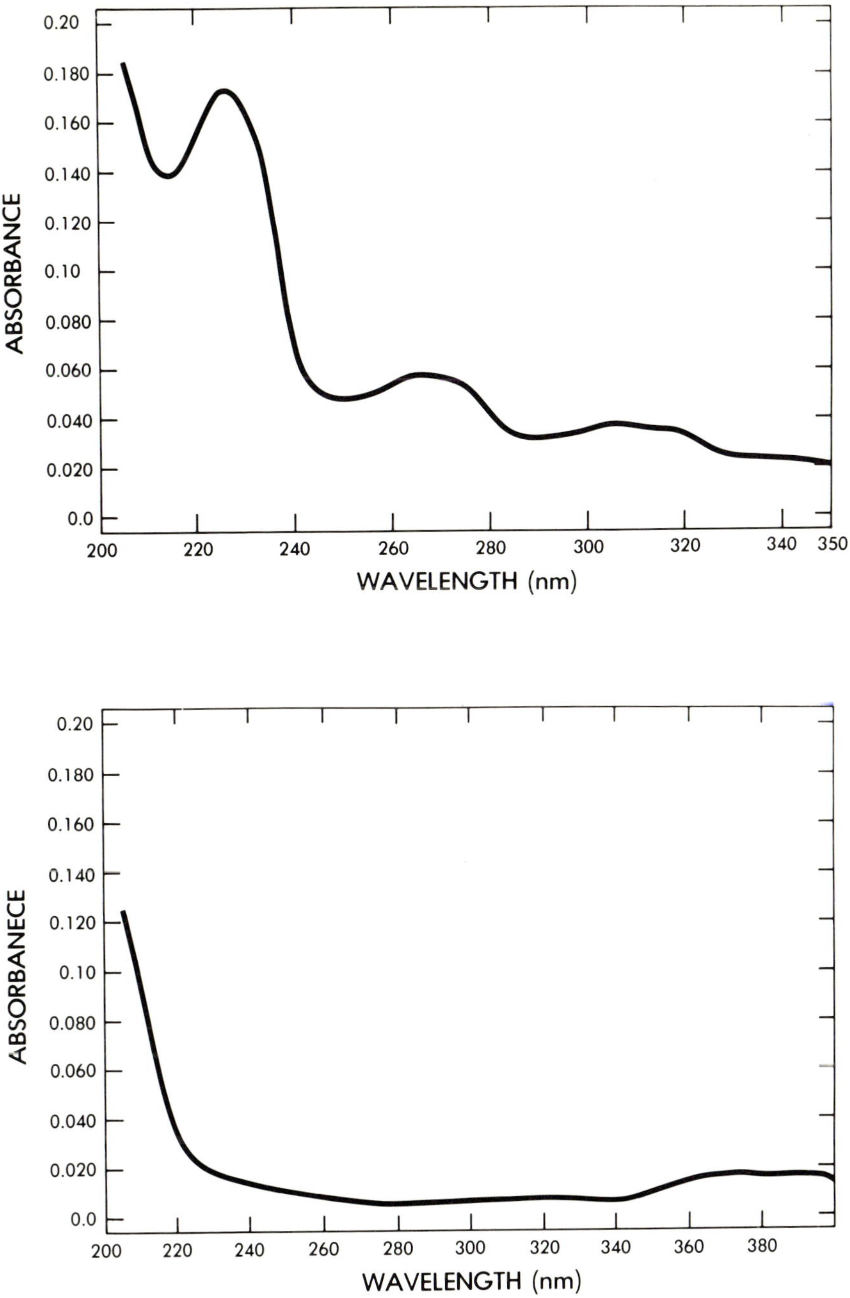


FIG. 3. Combined UV spectra of phenytoin and glutethimide in the mobile phase.

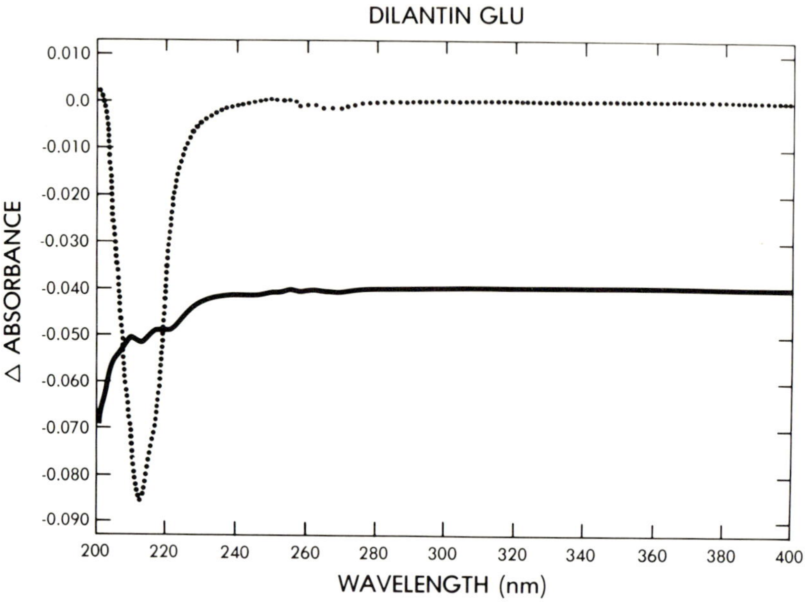


FIG. 4. A plot of first derivative spectra of phenytoin and glutethimide. Solid line denotes phenytoin, while dotted line represents glutethimide.

## III. Discussion

Poisoning accounts for an increasing proportion of emergency admissions to hospitals in the US and Europe. Thus, an increasing demand for emergency toxicology services is prompting hospitals to develop such facilities. LC methods, like the one described in this chapter, should provide a reliable, accurate, and rapid (40–45 min) analysis for a large variety of commonly abused drugs. In most laboratories the average analysis time for a comprehensive drug screen is approximately 2–3 h, requiring that a variety of techniques be employed.

Barbiturates are the most commonly prescribed drugs encountered in overdose cases. The widespread abuse of these drugs has been well documented. Barbiturates differ primarily in their duration of action. Short-acting barbiturates such as pento-, amo-, and secobarbital peak rapidly, while the blood levels of long acting ones, such as phenobarbital, may still be increasing hours after the administration of a large dose. Owing to these differences it is desirable to identify and quantitate the barbiturate present. Spectrophotometric methods lack this ability, while the LC technique is fully capable of identifying individual barbiturates. The ability to measure drugs, such as nonbarbiturate hypnotics, analgesics, chlordiazepoxide, diazepam, and other benzodiazepines, is an added advantage of the LC method.

## IV. Current Trends and Future Developments

The introduction of a new generation of LC detectors, such as mass spectrometers or high-speed scanning spectrophotometers, will provide more definitive identification of drugs. These detectors may also be able to identify compounds that are not fully separated chromatographically. New developments in microparticulate packings and improved column technology should provide greater column efficiency, selectivity, and reduced analysis time.

## Acknowledgments

Laurence J. Marton is the recipient of NCI Research Career Development Award CA-00112. We thank Mary Stawski for her careful typing of this manuscript.

## References

1. Higgins, G., and O'Brien, J. R. P., *Proc. Assn. Clin. Biochem.* **1,** 86 (1960).
2. Higgins, G., and O'Brien, J. R. P., *Proc. Assn. Clin. Biochem.* **3,** 221 (1965).
3. Law, N. C., *Am. J. Med. Technol.* **39,** 237 (1973).
4. Jatlow, P., *Am. J. Med. Technol.* **39,** 231 (1973).
5. Sunshine, I., "TLC for Weak Acids, Neutrals, and Weak Bases, in *Handbook of Analytical Toxicology,* Sunshine, I.,ed., CRC Press, Cleveland, Ohio, 1975, p. 412.
6. Law, N. C., Fales, H. M., and Milne, G. W. A., *Clin. Toxicol.* **5,** 17 (1972).
7. Thoma, J., and Bondo, P., "GC for Sedative Drugs," in *Handbook of Analytical Toxicology,* Sunshine, I, ed., CRC Press, Cleveland, Ohio, 1975, p. 421.
8. Finkle, B. S., and Taylor, D. M., *J. Chromatogr. Sci.* **10,** 312 (1972).
9. Bastiani, R. J., Phillips, R. C., Schneider, R. S., and Ullman, E. F., *Am. J. Med. Technol.* **39,** 211 (1973).
10. Brattin, W. J., and Sunshine, I., *Am. J. Med. Technol.* **39,** 223 (1973).

# Part III

# Clinical Analysis of Endogenous Human Biochemicals

# Chapter 12

## Determination of Tyrosine and Tryptophan Metabolites in Body Fluids Using Electrochemical Detection

Gregory C. Davis, † David D. Koch,‡
and Peter T. Kissinger

*Department of Chemistry, Purdue University, West Lafayette, Indiana*

and

Craig S. Bruntlett and Ronald E. Shoup

*Research Laboratory, Bioanalytical Systems, West Lafayette, Indiana*

### I. Introduction

The amino acids tyrosine and tryptophan are precursors for a number of important physiological compounds. The catecholamines, which are metabolites of tyrosine, serve as neurotransmitters in the central

†Present position: Research Chemist, Monsanto Agricultural Research Center, St. Louis, Missouri.

‡Present position: Research Associate, Clinical Chemistry Laboratories, Barnes, Hospital, St. Louis, Missouri.

and peripheral nervous systems. Serotonin, a major metabolite of tryptophan, is a potent neurotransmitter and vasoconstrictor. Without doubt, these compounds have been some of the most intensely studied molecules in the last twenty years. One of the benefits that often accrues from basic biochemical research is clinical data of diagnostic and prognostic significance. In this case, however, the results have been disappointing. In only a few instances has the measurement of metabolites of these two amino acids been shown to have real clinical significance.

This situation may be ascribed to a number of factors. One problem is the diverse function and physiological distribution of these compounds. Their measurement in physiological fluids is an integration of total bodily function. Dysfunction in one organ (e.g., the adrenal gland) can have an impact on the overall concentrations of these compounds, but the data often requires careful interpretation. Another factor is the lack of simple, yet sensitive analytical techniques needed to determine the concentration changes effected by a given disease state. There very likely could be more significant diagnostic information in the measurement of several metabolites and enzyme activities rather than the determination of one or two. This approach has not been widely adopted, owing to the lack of methodology with such wide applicability. Many of the necessary measurements could be made, but often several days of effort and considerable expense would be required to complete the work for even a single patient specimen.

In this report, the biochemical pathways of tyrosine and tryptophan will be briefly reviewed. A discussion of the clinical significance of the major metabolites and enzymes is followed by a review of the analytical methodology. A relatively new analytical technique, liquid chromatography with electrochemical detection (LCEC), has had a significant impact on the determination of these compounds. The general LCEC approach and several successful clinical methods are examined.

## II. Tyrosine Metabolism

The catecholamines (CAs) serve a number of important physiological roles in mammalian systems. In the central nervous system norepinephrine (NE) serves as a neurotransmitter, and there is much evidence supporting the neurotransmitter role of dopamine (DA) as well. In the peripheral nervous system, NE is stored in the sympathetic nerve endings and controls many sympathetically innervated organs. Epinephrine (EPI) and a small amount of NE are produced by the adrenal medulla and secreted into the bloodstream for delivery to a number of target organs. A variety of circulatory and metabolic functions are controlled by EPI and NE.

The biosynthetic pathway of the catecholamines is shown in Fig. 1. Dietary tyrosine serves as the predominant source for the neurologically important molecules, the catecholamines. The first step is the hydroxylation of tyrosine by the enzyme tyrosine hydroxylase to 3,4-dihydroxyphenylalanine (DOPA). This is the rate limiting step in catecholamine production and is subject to feedback control by all of the catecholamines. Dopa is then decarboxylated to dopamine (DA) by a nonspecific aromatic L-amino acid decarboxylase. Norepinephrine (NE) is enzymatically produced via dopamine-β-hydroxylase. This enzyme is located in the catecholamine storage granules, where the conversion takes place prior to nerve stimulation.

In the central nervous system, the catecholamine pathway essentially ends with the formation of NE, while in the adrenal medulla NE is further *N*-methylated to give epinephrine (EPI). This conversion

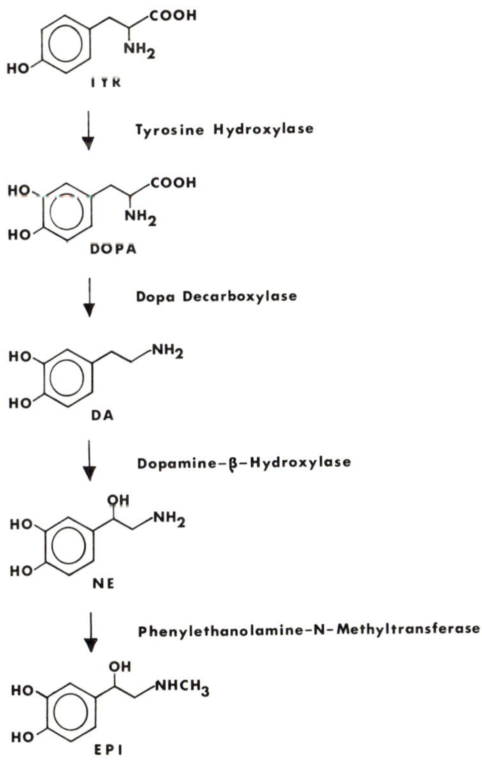


FIG. 1. Tyrosine metabolic pathway.

is accomplished by the action of phenylethanolamine-*N*-methyltransferase.

The enzymes responsible for degradation of the catecholamines are monoamine oxidase (MAO) and catechol-*O*-methyltransferase (COMT). Monoamine oxidase, located on the outer mitochondrial membrane, oxidizes the amine side chain to an aldehyde. The aldehyde is then rapidly reduced or oxidized by an aldehyde reductase or aldehyde dehydrogenase, respectively. Both the corresponding alcohol and carboxylic acid metabolites of each of the catecholamines are found to some extent.

COMT catalyzes the transfer of a methyl group from *S*-adenosyl methionine to one of the ring hydroxyl groups. Methylation occurs predominantly at the 3-position. The catecholamines may be methylated directly, giving rise to basic metabolites generally referred to as leading to a variety of hydroxymethoxyphenyl (vanillyl) derivatives. In the CNS where the catecholamines of major importance are DA and NE, *O*-methylation and oxidation of the side chain to the carboxylic acid is the predominant degradation pathway of DA leading to the catabolite homovanillic acid. For NE, however, the pathways take a different twist: *O*-methylation, oxidation of the amine, and subsequent reduction to the alcohol to give 3-methoxy-4-hydroxyphenylglycol (MHPG).

To further increase the rate of elimination of the catecholamines and their metabolites, they are conjugated with glucuronic and sulfuric acids by the corresponding transferase enzymes found primarily in the liver and also in the brain. All of the catecholamines and metabolites are excreted partially as conjugates. Figure 2 illustrates the metabolic pathways for the catecholamines. All of the metabolites shown are found in the urine. The proportions of the various metabolites vary greatly with the physiological state of the patient.

## III. Clinical Significance of Tyrosine Metabolism

### A. Urinary Catecholamines

The clinical interpretation of urinary catecholamine excretion has frequently lagged behind fundamental research in the etiology of related diseases. The reasons for these delays are multifold. Often the methodology employed in the biomedical research laboratory is too specialized, cumbersome, or time-consuming to be useful in a throughput-oriented clinical environment. For example, in well-controlled research studies where only minor differences in excretion values are

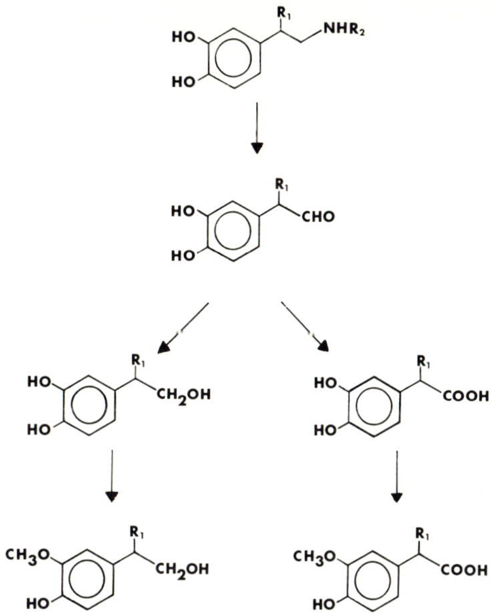


FIG. 2. General metabolic pathway for catecholamines ($R_1$, $R_2$, = H dopamine; $r_1$ = OH, $R_2$ = H norepinephrine; $R_1$ = OH, $R_2$ = $CH_3$ epinephrine).

evident, inter- and intrapatient differences caused by diet, diurnal variations, drug therapy, and/or individual metabolic variations will largely obscure any clinical meaningfulness.

Another problem with urinary measurements of catecholamines is the "melting pot" character of urine. All of the major metabolites, regardless of their source, eventually are deposited in the bladder for excretion. The major problem is identifying, for a given metabolite, the relative proportions excreted from the brain, the peripheral nervous system, the adrenal medulla, and so on. Some leeway is offered in that the metabolic pathways of a catecholamine may differ depending on the tissue examined. Norepinephrine, for example, in the brain is converted to both 3-methoxy-4-hydroxyphenylglycol and vanillyl-mandelic acid, whereas its fate in the peripheral nervous system is largely only the latter compound. These metabolic differences, however, are distinct in urine analyses only when gross changes in metabolism appear. Since the catecholamines are largely excreted as metabolites, the variations in the small percentage excreted unchanged demands sensitive and selective analytical methods.

Probably the most important diseases related to catecholamine excretion are tumors derived from the neural crest tissue (e.g., neuroblastoma) and Parkinson's disease. Parkinson's disease stems from a dopamine deficiency in the striatum (*1*); the urinary excretion of dopamine and its 3,4-dihydroxyphenylacetic acid and homovanillic acid metabolites are depressed. In addition to discerning a possible diagnosis, urinary levels of dopamine and metabolites are of therapeutic value to known Parkinson's patients undergoing L-DOPA treatments (*2, 3*).

Neuroblastoma, pheochromocytoma, ganglioneuroma, and ganglioneuroblastoma are tumors of the sympathetic nervous system derived from primitive neural crest tissue (*4, 5*). Each type of tumor is unique to a different level of nerve cell differentiation and therefore gives rise to a different profile of excreted catecholamines (*6*). Pheochromocytomas are usually benign tumors of the chromaffin tissue that can appear at all ages and in both sexes. Tissue studies have shown that they often contain epinephrine and norepinephrine at elevated levels (10–80× higher than normal). These abnormalities are reflected in an equally elevated profile of norepinephrine and epinephrine and metabolites (*7*).

Neuroblastomas are very malignant tumors of the sympathetic nervous tissue that occur most frequently in children (*8*). Urinary profiles reveal an elevation in dopamine and, in most cases, norepinephrine. In comparison to pheochromocytoma, the ratio of acidic and neutral metabolites to catecholamines is higher in neuroblastoma. Some evidence has pointed to differences in the subcellular structures of the two tumors. Primitive neuroblastoma cells lack the storage vesicles that contain norepinephrine and dopamine-$\beta$-hydroxylase, so deamination by MAO is predominant (*9*).

In each type of tumor, urinary catecholamine measurements have been used for both diagnostic and prognostic purposes (*10–12*). The suspicion of a tumor in a patient presenting vague, general symptoms may be corroborated using these clinical determinations; the immediate drop and stable resumption of normal levels after surgical removal of the tumor have generally indicated successful recoveries. For neural crest tumors, measurement of urinary catecholamines has proven useful since the tumors excrete quantities so large that the urinary profiles are distinctly different from normal limits.

Myocardial infarction is another problem of clinical interest in which the evaluation of catecholamine excretion levels has been shown useful. In one study of thirty patients (*13*), those demonstrating only moderately significant elevations in catecholamine excretion in the immediate post infarction period showed an excellent prognosis, whereas those who continued a course of abnormally high norepinephrine and epinephrine and vanillylmandelic acid proved less fortunate. The daily profiling of the excretion rates of these and related compounds for up to a week following the infarction aided in the clinical management of these patients.

In phenylketonuria (PKU), patients with decreased excretion of norepinephrine, epinephrine, and dopamine are observed (*14*). One explanation for this difference between PKU patients and normal controls lies in the competition of phenylalanine with tyrosine in the rate-limiting enzymatic ring hydroxylation step. Since phenylalanine levels are high in the PKU patient, and since phenylalanine may be converted to DOPA in a two-step process via tyrosine hydroxylase, it has been postulated that this less efficient process may account for decreased catecholamine excretion (*15*).

In essential hypertension studies, elevated urinary norepinephrine excretion has been noted in about 20% of all patients (*16*). Even though for these reasons urinary norepinephrine is not clinically useful (cf. the discussion on plasma values in section III.B), the range of values expected is important in the diagnosis of pheochromocytoma, where high norepinephrine and often epinephrine excretion are expected. Here it is imperative to differentiate the normals and hypertensives from the less usual pheochromocytoma patient. Studies involving large numbers of patients have been carried out to this end (*17*).

Outside the sympathetic nervous system, catecholamines have clinical significance in the diagnosis and management of malignant melanoma (*18, 19*). The pigment cell, or melanocyte, utilizes tyrosine in a different manner from that described previously; once DOPA is formed, the enzyme tyrosinase may convert it to a quinone, which in turn may polymerize to melanin pigments. Another possible fate is the combination of the quinone with endogenous mercaptans such as cysteine or glutathione to form ring adducts (*20*). The excretion of the latter compounds is uniquely oriented to the melanocyte and therefore serves as an appropriate marker of cell dysfunction. In malignant melanoma, DOPA and these adducts are excreted in greatly increased quantities. The confirmation of a suspected case, the evaluation of the extent of disease, and the progress of medical treatment may all be assisted by the biochemical analysis of these and related compounds.

Urinary catecholamine measurements have also been useful in understanding muscular dystrophy (*21*) and familial dysautonomia (*27*).

Analytical methods for the determination of urinary catecholamines have been based primarily on the formation of fluorescent derivatives or gas-liquid chromatography. For fluorometric analysis, norepinephrine and epinephrine may be oxidized by agents such as iodine or ferricyanide and then cyclized under alkaline conditions to form fluorescent trihydroxyindoles (*23*). Ethylenediamine may also be employed to condense the quinone and form a fluorescent two-ring derivative (*24*). Since urine is very complex, the sample must undergo sufficient cleanup routines before derivatization to avoid fluorescence contributions from endogenous interferences. High blank values limit the sensitivity of fluorescence assays. Like fluorescence, gas-liquid chromatography also requires extensive sample cleanup and derivatization. Electron capture (*25*) and mass fragmentographic detection (*26*) have been used most successfully. The complexity of urine makes the latter approach more suitable, but for most clinical laboratories the maintenance and operation of such specialized and sophisticated instrumentation is beyond reach.

## B. Serum Catecholamines

Unlike urinary catecholamine measurements, which represent an integration of sympathetic and central activity over a given period of time, serum catecholamine levels more accurately reflect the temporal changes that occur in the sympathetic nervous system in response to stress or disease.

Hypertension is thought to involve an overactivity of the sympathetic nervous system and is an example of a disease in which the measurement of serum norepinephrine has been utilized. As mentioned previously, NE in blood comes essentially from the sympathetically innervated tissues. The fact that NE values increase by 100% when a person stands (*27*) and that the urinary levels remain unchanged after adrenalectomy in man (*28*) lend further credence to this statement. The increase in NE upon standing also suggests that although sympathetic nerves innervate many tissues throughout the body, a considerable portion of this catecholamine in blood arises from sympathetic nerves that control the cardiovascular system (*27*). Examination of the sympathetic neuromuscular junction lends support to this contention. Compared to other sympathetically innervated tissues, the smooth muscles in the walls of the arteries and veins appear to have a wider junctional gap (*29*). This perhaps allows for greater diffusion into the circulatory system. The maintenance of cardiovascular tone and blood pressure by noradrenergic nerves is accomplished by a negative feedback system. The inability of this system to properly maintain the blood pressure is strongly implied in clinical disorders such as hypertension, and thus the desire to measure NE concentration. Several investigators have shown that NE is increased in the serum of hypertensives (*30–33*), while others have found little difference compared to controls (*34–36*).

Hypertension is a complex disease of apparently many origins with varying degrees of severity. The inability to properly classify patients prior to clinical studies could be responsible for some of the discrepancies (*32*). The influence of norepinephrine release may be greater in some patients than in others. A great deal of work remains to be done before serum catecholamines can be fully assessed as a diagnostic indicator for hypertension.

Plasma catecholamines have been studied in other diseases. Patients with hypothyroidism were found to have plasma norepinephrine values higher than that of normals (*37*), while patients with hyperthyroidism were found to have low to normal norepinephrine (*37, 38*). As described in the previous section, pheochromocytoma and neuroblastoma are neural crest tumors that are often characterized by an increase in urinary concentrations of catecholamines and their metabolites. The determination of plasma catecholamines can help to confirm the diagnosis of these tumors in some cases (*39*). In one such case the urinary catecholamines were high, but the blood pressure was normal and in the other, urinary free catecholamines were normal.

Plasma catecholamine measurements show great clinical potential. As has been illustrated, in any disorder involving the peripheral nervous system, serum or plasma catecholamines potentially could provide useful diagnostic information. One of the problems that has hindered the utilization of this information has been the lack of adequate analytical methods. Norepinephrine, which is found to be highest of the three in man, is present at only a few hundred picograms per milliliter of plasma or serum in healthy individuals at rest. Fluorometric, radioenzymatic, gas chromatographic, and GC/MS methods have all been employed to evaluate serum or plasma concentration.

Fluorescent methods have generally made use of derivatization since the native fluorescence of the catecholamines is insufficient for the concentrations generally found in plasma. Trihydroxyindoles (THI) formed from oxidation of NE and EPI in alkaline media have been the most popular derivatives (*40–44*) (Fig. 3). In spite of the highly fluorescent THIs, large sample volumes are still required (*5–15* mL), which is an acceptable sample size in large animals, but obviously precludes such an approach in longitudinal studies with small animals (e.g., rats). Instability of the fluorophores and interference from catechol drugs such as isoproterenol have also plagued the THI methods. Other derivatization methods have used *o*-pthalaldehyde (*45, 46*) and fluorescamine (*47*) in combination with LC. These reagents are effective in measuring primary amines, but secondary amines such as EPI must be determined by an alternate approach. It is

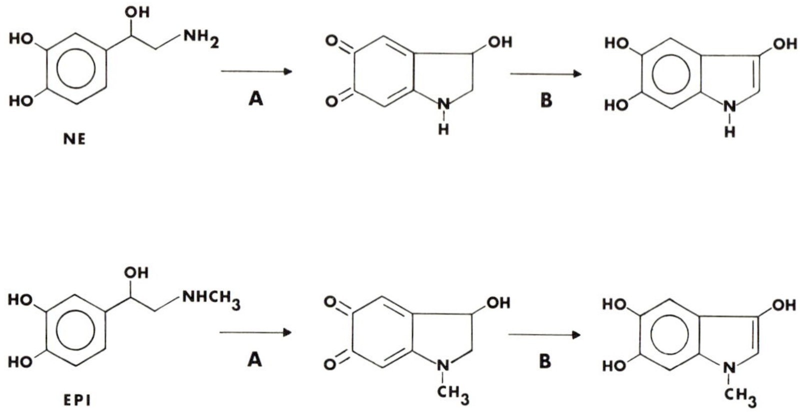


FIG. 3. Conversion of norepinephrine (NE) and epinephrine (EPI) to their corresponding fluorescent trihydroxyindoles. The reaction proceeds by (A) oxidation of the catecholamine to the orthoquinone followed by a 1,4-Michael addition closing the ring to yield an indoline. The indoline is oxidized to give the intermediate shown, which then undergoes an intramolecular rearrangement (B) to give the 3,5,6-trihydroxyindole.

obviously desirable to have a method that is applicable for all catecholamines. The use of native fluorescence with LC did not demonstrate the sensitivity required for plasma determinations (*48*) and a post-column THI procedure has met with very limited success (*46*).

Radioenzymatic techniques were introduced for the determination of circulating catecholamines to bring more specificity and sensitivity to the measurement. Early methods, such as the enzymatic isotope derivative assay (*49*), measured only total catecholamines (NE and EPI). The assay involved the addition of catechol-*O*-methyltransferase to the catecholamine sample in the presence of labeled [$^{14}$C-methyl] *S*-adenosyl-L-methionine [$^{14}$C-SAM] (Fig. 4). The *O*-methylated products, the metanephrines, were then converted to vanillin by mild periodate oxidation and the latter was extracted and

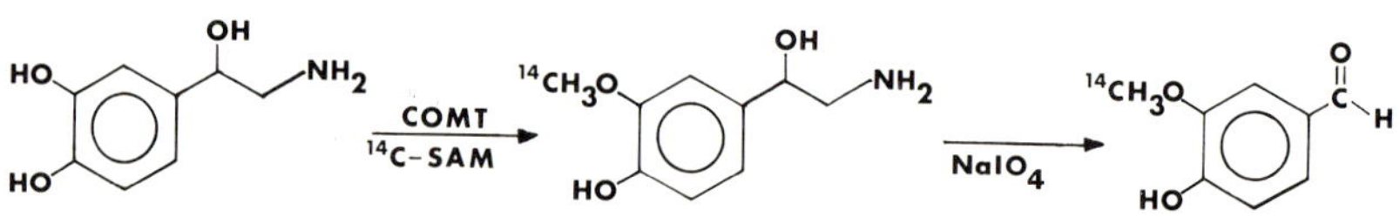


FIG. 4. Radioenzymatic determination of norepinephrine utilizing catechol-*O*-methyltransferase (COMT) and $^{14}$C-methyl *S*-adenosyl-L-methionine. The labeled product is then converted to vanillin by periodate oxidation.

counted. The assay was later modified by incorporation of a TLC separation step for the metanephrines in order to determine NE and EPI separately (*50*). Tritiated norepinephrine and epinephrine were also added as internal standards. More recently, tritiated *S*-adenosyl-L-methionine has replaced the $^{14}$C-SAM because it has a higher specific activity (*51*). The single isotope method reduced the expense of the assay somewhat by eliminating the labeled internal standards. Nevertheless, this method can exhibit good reproducibility (*52*). The analysis of small sample sizes (< 1 mL) and the attainment of extremely low detection limits (typically < 10 pg for NE) have been achieved with the radioenzymatic assay (*53, 54*). Unfortunately, these methods remain expensive and awkward to use on a routine basis.

Gas chromatography (*55, 56*) and gas chromatography/mass spectrometric techniques (*57–59*) have been developed that have demonstrated the necessary sensitivity. They have not been as popular as either the fluorometric or radioenzymatic techniques, primarily owing to the necessity of forming volatile derivatives. The introduction of extra steps, incomplete reactions, and multiple or uncharacterized products (*60*) have also played a part in discouraging adaptation of these techniques, as well as the inordinate expense.

## C. Urinary Metanephrines

One pathway of catecholamine inactivation lies in *O*-methylation via the enzyme catechol-*O*-methyltransferase to produce the usually meta *O*-methylated metabolities (*61, 62*). Normetanephrine, metanephrine, and 3-methoxytyramine are produced from norepinephrine, epinephrine, and dopamine, respectively. The three metabolites are collectively referred to as the "metanephrines."

The most important clinical application of the metanephrines has been in the diagnosis of neural crest tumors. In this regard, screening tests for elevated normetanephrine and metanephrine in pheochromocytoma have proven useful and have become standard practice. Unfortunately, most of these assays do not permit the determination of 3-methoxytyramine at all, or the measurement of normetanephrine separately from metanephrine. A number of papers have discussed the relevance of urinary metanephrines determinations in these clinical situations (*4, 63, 64*).

In addition to their value as biochemical indicators in neural crest tumors, some evidence points to their utility in psychiatric disorders. Studies have shown that both the para-*O*-methyl and the 3,4-dimethoxy metabolites have been isolated in urine from schizophrenic patients (*65, 66*). The latter compounds do show pharmacological activity and it was hypothesized that these agents might be involved in mediating the behavior of these patients.

The most common analytical techniques for the metanephrines are variations of a method devised by Pisano (*67*) for normetanephrine and metanephrine involving absorption of the compounds from hydrolyzed urine onto a cation exchange resin, elution with dilute base, conversion to vanillin by reaction with periodate, and spectrophotometric measurement of the product. The procedure affords some selectivity in that the periodate oxidation is dependent on the $\beta$-hydroxy group; however, it is severely limited in sensitivity by high blank values. Differential measurement is not possible.

None of the *O*-methylated compounds are sufficiently naturally fluorescent, but their conversion to fluorometric derivatives has been studied. Both normetanephrine and metanephrine may be oxidized to fluorescent trihydroxyindole derivatives using oxidizing agents such as iodine, periodate, or ferricyanide (*68*). Separate measurement of each compound was accomplished by oxidation at different pH values. The chief problems with fluorescence assays were unpredictable intersample quenching phenomena and the variable recoveries owing to differences in sample ionic strength.

For studies requiring simultaneous assay of all three metabolites, chromatography offers the necessary improvements. Derivatization to form a heavily fluorinated ester or amide followed by gas chromatography with sensitive electron capture detection is standard procedure. Mass fragmentography is a suitable although very costly alternative. Most assays require fairly extensive cleanup steps prior to derivatization (*26, 69*).

## D. Acid and Neutral Metabolites

The primary catecholamine degradation pathway involves essentially two enzymatic processes: *O*-methylation at the 3 position of the catechol moiety and oxidative deamination of the alkyl amine side-chain. The latter route yields an intermediate aldehyde which is either further oxidized to give the acid metabolite or reduced leaving a neutral alcohol. The enzymes responsible for the deactivation, catechol-*O*-methyltransferase (COMT) and monoamine oxidase (MAO), are widely distributed throughout the body in various tissue, glands, sympathetic and parasympathetic nerves, ganglia, blood vessels, and all areas of the brain. The highest COMT activity is found in the liver, whereas heart and brain tissue contain the highest MAO activity. Evidence indicates that neither process is solely associated with either the central or peripheral system or even that one process dominates the other because all metabolites, as depicted in Fig. 2, can be found to some extent in tissues and fluids associated with catecholamine activation.

The acidic and neutral catabolites of clinical interest are 3-methoxy-4-hydroxyphenylacetic acid (homovanillic acid, HVA) derived from dopamine, and 3-methoxy-4-hydroxymandelic acid (vanilmandelic acid, VMA) and 3-methoxy-4-hydroxyphenylglycol (MHPG), the acidic and neutral metabolic products of epinephrine and norepinephrine. MHPG is generally found as the sulfate or glucuronide conjugates. The remainder of the catabolites not mentioned here can be found in the physiological fluids commonly assayed, namely cerebrospinal fluid, serum or plasma, and urine, but at much lower concentrations. The clinical significance of these compounds has not been shown to be a more reliable measure of patient status than the terminal metabolic products, VMA, HVA, and MHPG.

Altered concentrations of catecholamine metabolites in urine are indicative of a number of physiological and pathological conditions associated with the neuronal and hormonal properties of these compounds. Because of the varied sources of the metabolites, care must be exercised in relating body fluid concentration changes to a specific site of origin or clinical state. Nevertheless, there are several disorders that can be directly linked with a high percentage accuracy to urinary concentrations of these metabolites.

The determination of greatest diagnostic value has been obtained by monitoring urinary levels of VMA in suspected diseases involving neural crest tumors, primarily pheochromocytoma (*70–72*) and neuroblastoma (*71–74*). Increased concentrations of VMA associated with pheochromocytoma differentiates these patients from those with essential hypertension. VMA is also increased in children afflicted with neuroblastoma. Here VMA is not only used as a diagnostic criteria, but, in combination with HVA to determine a HVA/VMA ratio, is reported to be of prognostic value following treatment (*73, 75*). HVA determinations are also important in confirming Parkinsonism, as well as in monitoring therapeutic response to L-dopa treatment because of the direct relationship to central nervous system dopamine metabolism (*72, 76, 77*). As would be expected, gaging drug regimens for maximal efficacy by following metabolism and drug disposition in neurological disorders is an important ongoing research area. One subject of extreme interest and controversy in recent years is the relationship between urinary concentrations of MHPG and norepinephrine metabolism in the CNS (*78–80*). The "catecholamine hypothesis" suggests that NE pathways are disrupted in persons afflicted with a number of neurological disorders (manic depression, schizophrenia), and because the end product of NE catabolism in the brain is predominantly MHPG (specifically MHPG-$SO_4$), the concentration in urine may be a marker of CNS NE activity. At this time no definitive conclusion is possible, although it appears that urinary MHPG does not derive solely from central NE metabolism, but also to a large degree from the peripheral NE pool. Nevertheless, decreased concentrations have been shown to occur in certain depressive disorders (*79, 81*). Other evidence indicates that in humans urinary MHPG arises predominantly from the sympathetic nervous system NE (*82*) and may provide a better clinical evaluation of associated sympathetic diseases (e.g., hypertension).

Urinary levels of MHPG alone may not discriminate between depressive disorders, but in combination with concentrations of other CA metabolites found in urine, subgroups within this group can be defined (*83*). This would, one hopes, lead to better classification of these individuals and suggest the most appropriate therapy.

Measurement of urinary acid and neutral CA metabolites have been shown to be of clinical importance in the neural crest tumors, Parkinsonism and, to some extent, psychiatric disorders, as noted above, but they may also prove to be significant in other disease states. These include melanoma (*84*), hypertension (*82*), degree of neurological damage following lead poisoning (*85*), cirrhosis (*86*), heart disease (*87*), and movement disorders (*88*) other than Parkinsonism. More definitive progress in these areas will require analytical methods for the acidic and neutral metabolites that are fast, reliable, and can be used on a routine basis for larger patient populations than have been studied to date.

The most commonly performed CA metabolite determination in the clinical laboratory is VMA in suspected pheochromocytoma and neuroblastoma patients. In these cases, where concentrations can be many times basal values, the colorimetric methods derived from Pisano's original work are most often employed (*89, 90*). These methods are not specific in the reactions employed and are subject to interferences. For determining normal or depressed concentrations, more selective methods are definitely needed. The same is true for HVA determinations where the most common methods measure HVA colorimetrically (*91*) or fluorimetrically (*92*). Greater specificity in these techniques can be obtained by using further purification steps such as paper (*93*), thin-layer (*94*), ion-exchange (*95*), or other column (*96*) chromatographic techniques. Earlier methods for MHPG determination also relied on formation of a colored product via periodate oxidation (*97*) or production of a fluorescing adduct (*98*) following isolation of the free compound. Since most of the urinary MHPG exists as the sulfate or glucuronide conjugate, and MHPG is acid labile, all methods employ a rather lengthy enzymatic hydrolysis step prior to the final isolation.

The recent instrumental methods for determination of urinary HVA, VMA and MHPG take advantage of the selectivity that can be obtained using gas-liquid chromatography coupled to a sensitive detector. Flame ionization and mass spectrometric detectors have been used for MHPG (*71, 99–101*) as well as the simultaneous determination of HVA and VMA (*74, 102–105*). All of these methods necessitate the derivatization of these compounds to volatile adducts that require procedures generally too complex for for routine clinical use. Even with the separation power afforded by GC, interferences can still occur because of the large number of similar compounds undergoing the derivatization reaction. Recent advances in liquid chromatography may circumvent many of the problems associated with earlier methods and come closer to meeting all of the criteria for routine clinical laboratory practice.

## E. Dopamine-β-Hydroxylase

Dopamine-β-hydroxylase (DβH) is the enzyme responsible for the biosynthesis of norepinephrine from dopamine. The enzyme is a mixed function oxidase located in the synaptic storage vesicles of noradrenergic nerve terminals. These vesicles serve as the site of β-hydroxylation as well as neurotransmitter storage compartments. The release of the neurotransmitter following nerve stimulation is described by a mechanism known as exocytosis (*106–108*). The process involves the migration of the vesicles to the presynaptic membrane, whereupon the vesicular and presynaptic membranes temporarily fuse. This is closely followed by the formation of an opening in the membrane wall that allows the release of the neurotransmitter into the synapse. The opening is large enough to permit soluble vesicular proteins such as DβH to diffuse into the synaptic gap (*106, 109*). As described earlier, inactivation of the neurotransmitter can take place by reuptake across the membrane or by enzymatic degradation. DβH with a molecular weight of 290,000 (*110*) is too large to be reabsorbed. As a result, the enzyme diffuses from the junctional gap entering the circulatory system possibly via the lymph (*111, 112*).

DβH is also located in the chromaffin cells of the adrenal medulla (*113*). If the measurement of serum DβH is to serve as an index of the sympathetic nervous system as some have suggested, then the contribution from the adrenal glands needed to be evaluated. In a study using rats, treatment with 6-hydroxydopamine (which is known to destroy sympathetic neurons) decreased the serum enzyme activity by 25% (*114*). Removal of the adrenal glands did not affect the serum enzyme activity. This data suggests that sympathetic neurons provide the majority of circulating enzyme. Recent reports have contested this statement. In certain extreme forms of stress, such as hemorrhagic hypertension, the adrenal glands made a significant contribution to plasma DβH (*115, 116*). Nevertheless, the most important factor appears to be the sympathetic contribution. With confirmation of DβH activity in serum and the release of the enzyme into circulation, a number of studies have investigated its use as a biochemical marker. The enzyme half-life in circulation is much longer than that of norepinephrine (*117*) and as such, might be more convenient to measure.

One area of intense study has been serum DβH and its relation to hypertension and stress. Higher than normal DβH could be an indicator of overactivity of the sympathetic nerves owing to stress. Several studies have reported that serum DβH is a good indicator of sympathetic nerve activity (*118–122*) and have shown a positive correlation with blood pressure (*118*). Others have found little value in DβH as an index of peripheral noradrenergic activity (*123–126*) with some demonstrating no relationship to blood pressure (*123*–27). The disparity in views may in part be explained by the difficulty in interpreting small changes in DβH which are superimposed on the high circulating levels normally found in man. Physiological factors such as organ perfusion, plasma volume, and so on could contribute significantly to the interpretation of data (*117*).

Another controversial area is the role DβH plays in behavioral disorders such as schizophrenia (*128–134*) and manic depression (*135*). Wise and Stein have reported that DβH is significantly reduced in post-mortem brain regions of schizophrenics in comparison to controls (*128*). This tends to support the theory that schizophrenic behavior results from a breakdown of the noradrenergic reward system. Attempts to reproduce the data have generally been unsuccessful (*132, 133*) and the subject is by no means closed. The use of serum DβH activity measurements for diagnostic purposes was suggested even though the primary contribution is from the peripheral and not the central noradrenergic activity. It was thought that if the same genetic loci govern the peripheral as well as ithe central nervous system, then one would have an easily accessible (albeit indirect) index of central activity for studying behavioral disorders. The utility of serum DβH for this purpose has met with both positive (*131*) and negative (*134*) results. Possible reasons for disagreement include improper clinical criteria for patient diagnosis and/or the enzyme assay methodology.

Since the measurement of serum DβH reflects peripheral activity, evidence is growing that cerebrospinal fluid (CSF) DβH reflects central activity. A recent study found that depressed patients treated with monoamine oxidase inhibitors (that are known to decrease central noradrenergic activity) caused a significant decrease in CSF DβH (*136*). These and other results support the use of DβH in CSF as a clinical marker for central noradrenergic activity (*137*).

Although CSF DβH measurements show great promise, at the present time the consensus regarding the measurement of serum DβH is that it is of limited diagnostic value. Even though the premise that circulating DβH reflects sympathetic activity is generally accepted, owing to the wide range of normal values and the high degree of overlap with disease state levels, interpretation of the data remains controversial. One bright spot is the use of DβH in longitudinal studies. Intraindividual DβH is known to remain fairly constant on day-to-day basis (*117, 119, 123*). Long term studies could provide more conclusive data during the course of a disease or in the evaluation of a particular drug therapy.

The analytical methodology used to measure DβH activity has been cited as being responsible for some of the disparity of opinion. The most commonly employed methods include radioenzymatic (*138–142*), fluorometric (*143, 146*) and spectrophotometric (*147, 148*) techniques. The double-enzyme radiochemical assay is one of the first techniques sufficiently sensitive to measure the low DβH activities found in animal tissues (*138*) (Fig. 5). Substrate, phenylethylamine or tyramine, is β-hydroxylated by DβH present in the sample. Phenylethanolamine-*N*-methyltransferase (PNMT), which is added to the sample, *N*-methylates the hydroxylated amine with a $^{14}$C-methyl group provided by labeled *S*-adenosyl-L-methionine. The labeled product is subsequently extracted and counted. Although this approach is sensitive, the method is not without problems (*150*). The substrate, tyramine or phenylethylamine, inhibits the second enzyme PNMT at the concentration necessary to saturate DβH. As a result the substrate concentrations are lowered considerably and the DβH

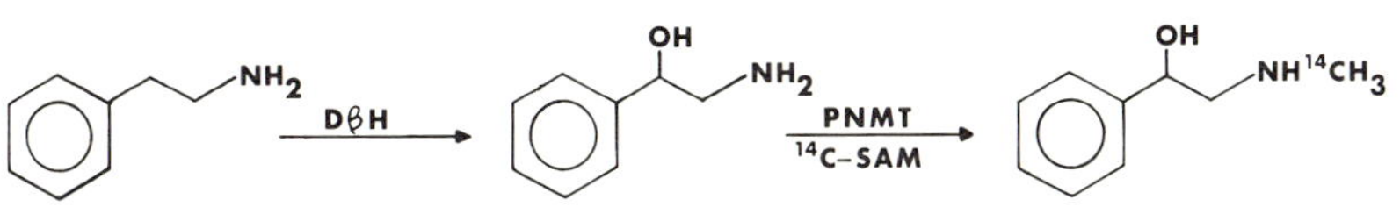


FIG. 5. Double-enzyme radiochemical determination of dopamine-β-hydroxylase.

reaction is not operated at $V_{max}$. The data obtained from the use of this assay can only be compared to data collected under identical experimental conditions because of this. Copper (II) ions and/or *N*-ethylmaleimide, which are usually added to the incubation mixture to prevent the inhibition of the DβH reaction (presumably by endogenous sulfhydryl compounds) will also inhibit the PNMT reaction (*149*). The addition of EDTA and/or dithiothreitol after the DβH step is necessary to overcome this problem. The double-enzyme radioassay also suffers from a narrow linear range and blank problems associated with PNMT specificity and purity (*150*).

Single step radioenzymatic techniques have attempted to overcome the difficulties associated with the use of PNMT by employing a labeled substrate (e.g., [2-$^{14}$C]tyramine) (*140–142*). The labeled substrate is β-hydroxylated by DβH and the product isolated by ion exchange chromatography (*140, 141*). The product is oxidized to benzaldehyde by the addition of periodate. It is then extracted into ether and counted. An essential step in this approach has been the need to purify the substrate solution prior to the assay. This substantially reduces the blank to acceptable levels and increases the sensitivity (typ. < 10 pmol). However, these assays have not been operated under substrate saturating conditions either, presumably to avoid reducing the specific activity of the labeled substrate by dilution with cold substrate (*140*).

Fluorometric methods have been developed as an alternative to the expensive and complex radiochemical assays (*138, 139*). Although they can easily measure DβH activity in human serum, the determination of the enzyme activity in CSF or animal sera is still a challenge. The introduction of LC (or TLC) has improved the specificity of the detection (usually by derivatization or native fluorescence), but the sensitivity still lies at approximately 50 pmol (*144–146*).

Spectrophotometric methods are based on the oxidation of the B-hydroxylated product (e.g., octopamine) by periodate to a benzaldehyde that can be monitored by UV (*147–149*). This has been a very successful and procedurally simple method for the measurement of DβH activity in human serum (*147*). The use of dual wavelength spectrophotometry has improved the sensitivity by reducing the contribution from the blank (*148, 149*). The sensitivity is still not quite competitive with either the fluorometric or radiochemical techniques.

## F. Catechol-*O*-Methyltransferase (COMT)

As the enzyme responsible for the *O*-methylation of the catecholamines, COMT represents a primary pathway for the termination of neuronal action following neurotransmitter release. The enzyme is present in many tissues, with principle activity located in the liver and kidney. COMT is also present in human erythrocytes and has been the object of considerable study. The thought was that if this enzyme was similar biochemically to the enzyme found in brain synaptosomes, then a readily accessible enzyme model was available to evaluate central enzyme activity (*151*). This would be very important for studying diseases in which central enzyme dysfunction is strongly implied. Under this premise human erythrocyte COMT has been examined in patients suffering from autism (*151*), schizophrenia (*152–154*), and affective illnesses (*155–158*). In some instances the results indicate that COMT could provide a clinical index of neurochemical disorders (*152, 153, 155, 157*), while other studies suggest the opposite is true (*154, 156, 158*).

Some of the factors that could be responsible for the discrepancies in the data include differences in patient diagnosis and severity of disease, environmental effects on the patients (e.g., length of hospitalization), and assay methodology (particularly in the choice of substrate). Studies investigating one or more of these factors have found their influence to be insignificant (*157, 158*). The subject remains a matter of controversy.

Another potential diagnostic use of erythrocyte COMT is that of evaluating system response to catechol drugs. Isoproterenol, a catechol drug used to control asthma, is metabolized by pulmonary COMT (*9*) and is known to have wide variability in individual drug sensitivity. Weinshilboum successfully demonstrated that human eythrocyte COMT activity correlated with lung and kidney activity in the same individuals (*160*). Using the erythrocyte enzyme as a model could help predict patient response to the drug.

The measurement of COMT has been accomplished with spectrophotometric (*161, 162*), fluorometric (*163, 164*), radioenzymatic (*165–167*), and GC/MS (*164*) methods. Spectrophotometric enzyme assays have used synthetic substrates (e.g., nitrocatechol or 3,4-dihydroxyacetophenone) selected for their chromogenic properties or have followed the change in absorbance of the incubation mixture. Although procedurally simple, they have lacked the sensitivity required to determine erythrocyte activity and are only applicable to tissues with high enzyme activity (e.g., liver). Fluorometric techniques are more sensitive, but are procedurally more involved, requiring extensive extractions (*163*) or derivatization (*164*). High fluorescent blanks have also been a problem when using crude tissue preparations. Radioenzymatic techniques are by far the most sensitive to date, and can easily measure the erythrocyte COMT activity. The typical radioenzymatic approach is to use labeled SAM

($^{14}C$ or $^{3}H$) and measure the *O*-methyl product. This approach has been called into question because of the presence of SAM in human red blood cells. This introduces the possibility of low results owing to endogenous unlabeled substrate, but a method using $^{3}H$-norepinephrine had very high blanks (*169*). Sophisticated GC/MS methods can overcome most of these problems; however, the procedures are often complicated by derivatization steps or the use of volatile synthetic substrates. Besides the obvious expense, GC/MS has yet to be proven as a routine instrument for clinical analysis.

## IV. LCEC Methods for Tyrosine Metabolism

### A. Overview

The first practical electrochemical detector for liquid chromatography was developed for the purpose of monitoring the catecholamines in brain tissue (*170*). Following this initial publication, liquid chromatography with electrochemical detection (LCEC) has been applied to a wide variety of problems involving the trace determination of easily oxidized or reduced substances (*171*).

Figure 6 illustrates the detection process in LCEC. In this example, an oxidizable component eluting from the column is passing through a thin-layer amperometric detector cell. A fluorocarbon gasket (typ. 50–125 μm) that defines the channel is sandwiched

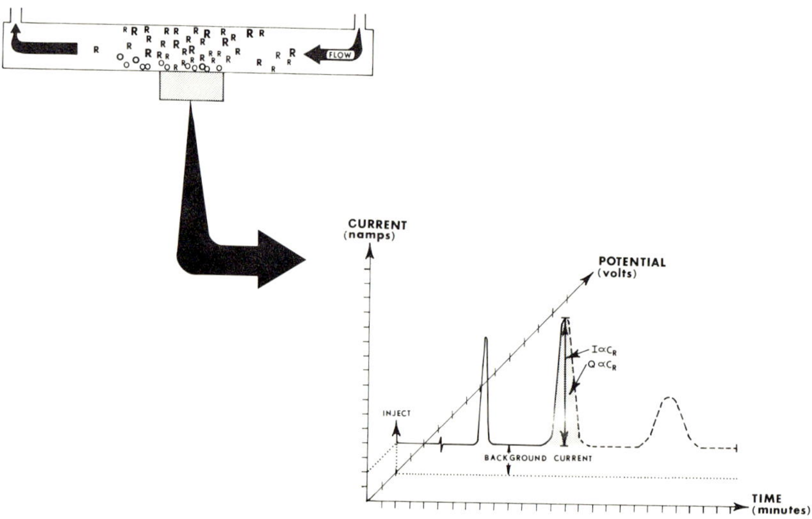


FIG. 6. Principles of electrochemical detection (courtesy of Bioanalytical Systems Inc.).

between two blocks machined from plexiglas, Kel-F, or similar engineering plastics. The electrode(s) is sealed flush with the channel wall. The electrode may be constructed from a variety of materials (gold, platinum, glassy carbon); however, carbon paste, which is an admixture of finely divided graphite and mineral oil, is an excellent choice for most neurochemical methods requiring oxidative detection.

The working electrode is held at a fixed operating potential so that if R comes in contact with the electrode surface it is immediately oxidized to O. As the figure suggests, only a small percent (typ. 5%) of the reactive molecules passing through the channel undergo an electrochemical conversion. Thus EC detection is amperometric and not coulometric (100% conversion). The merits of amperometry vs coulometry have been discussed at length elsewhere (*172*). Contrary to what one might expect, 100% conversion does not result in the most favorable signal-to-noise ratio (i.e., the lowest detection limit).

Selecting the operating potential is very important (analogous to selecting the wavelength for an optical absorbance detector). The ultimate utility and applicability of electrochemical detection depends upon the voltammetric behavior of the compound(s) of interest. One means of determining and evaluating the electrochemical characteristics is to examine current voltage curves. Figure 7 illustrates these so-called hydrodynamic voltammograms for two classes of natural

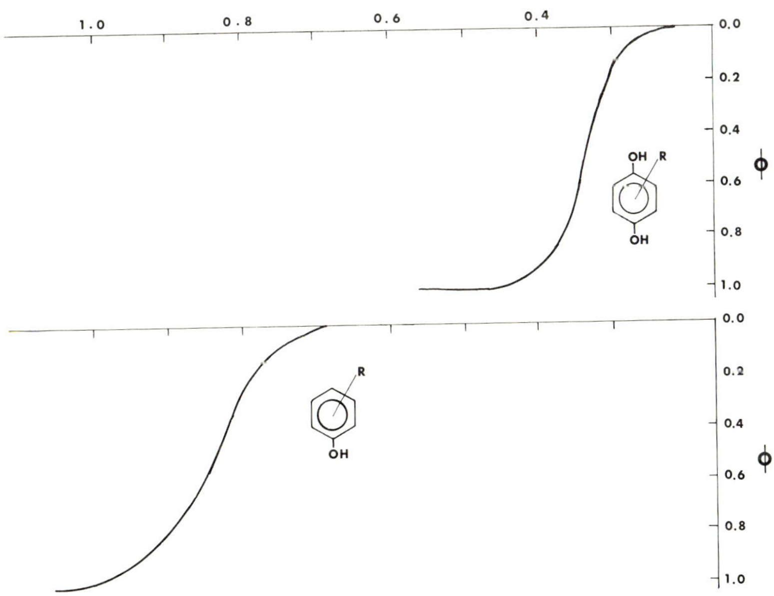


FIG. 7. Comparison of current–voltage curves (hydrodynamic voltammograms) for two classes of natural phenols.

phenols. Hydrodynamic voltammograms are generated by scanning the potential (either stepwise or with a very slow ramp) in a positive or negative direction. As long as the flow rate and the concentration of the reactant are constant, the current at each potential will be independent of time. The current response is proportional to the analyte concentration at every point along the S-shaped voltammogram. In practice, the amperometric detector for LC is usually operated on the flat part of the curve (the mass transport limited "plateau" region). This is done so that if there is a minor shift in the electrode potential owing to reference electrode drift or *iR* drop, the current response changes very little. This would not be the case if the potential were chosen on the rising portion of the curve, although measurements are often taken here to improve selectivity if necessary.

Figure 7 also illustrates the selectivity that can be obtained between electrochemically active molecules using this detector. For example, a potential can be chosen sufficient to oxidize the hydroquinone, yet, insufficient to obtain any response from the phenol. Thus hydroquinone can be selectively determined in the presence of any monophenolic substance. Conversely, if a potential chosen is sufficiently positive to detect the phenol, it would also be more than adequate for hydroquinone oxidation. In this case, then, the selectivity would have to be obtained by means other than that based solely on the electrochemistry. Hence, the need for the LC separation prior to EC detection, particularly in cases where high oxidizing potentials are required. In addition to the reduced selectivity, high oxidation or reduced potentials generally reduce the sensitivity of the determination. As the potential required for the electrochemical reaction to take place becomes large, the greater the background current owing to oxidation of the solvent (mobile phase) and/or supportiong electrolyte. Accompanying the higher background current is greater noise and ultimately a loss in sensitivity. In addition, note how the voltammetric wave of the hydroquinone is a sharper curve in comparison to the phenol, indicating faster electron transfer kinetics (*172*). In most instances, the kinetics of the electrochemical reaction are not important with regard to actual detection of the particular analyte.

The application of electrochemical detection to the quantitation of aromatic metabolites relies on the very facile reactions that occur at carbon electrodes. Figure 8 illustrates the two electrochemical reactions that occur at the electrode when a catecholamine, or essentially any one of the known unconjugated metabolites, is oxidized. The mechanism of the electrochemical oxidation of tryptophan metabolites is unknown, although in most cases the

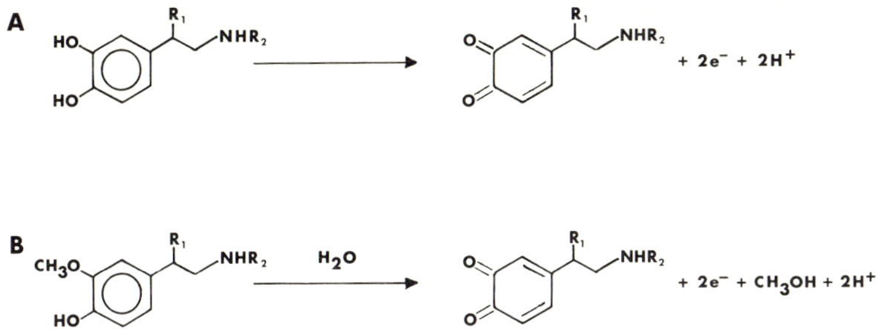


FIG. 8. Oxidation of (A) catechol and (B) vanil compounds to the orthoquinones. (A: $R_1$, $R_2$ = H dopamine; $R_1$ = OH, $R_2$ = H norepinephrine, $R_1$ = OH, $R_2$ = $CH_3$ epinephrine. B: $R_1$, $R_2$ = H 3-methoxytyramine; $R_1$ = OH, $R_2$ = H normetanephrine; $R_1$ = OH, $R_2$ = $CH_3$ metanephrine).

products are polymeric. Liquid chromatography (LC) has many advantages for the trace determination of polar organic substances. The instrumentation and reagents are quite inexpensive and the number of sample manipulations can often be reduced when compared to gas-phase, fluorescence, or radioenzymatic methods. We do not claim that our approach has overcome all (or even most) of the difficulties. The primary disadvantages include (1) the fact that samples must be processed in series for the final quantitation and (2) that the reliability of the instrumentation (including columns) is not perfect. Although the latter problem has been dramatically improved in the last few years, there remains considerable room for further progress, particularly with respect to columns. The first difficulty is made less troublesome by the improved speed of separation, the use of column switching techniques, and the fact that the LC instrument can be easily automated. The fact that a liquid chromatograph can be constructed or purchased very inexpensively makes it possible to utilize several instruments in parallel. In our laboratory we normally work with individual instruments dedicated to the catecholamines (norepinephrine, epinephrine, dopamine), metanephrines (metanephrine, normetanephrine, 3-methoxytyramine), the acid metabolites (homovanillic, vanilmandelic, and dihydroxyphenylacetic acids), and tryptophan metabolites (tryptophan, 5-hydroxytryptamine, and 5-hydroxyindoleacetic acid).

The goal of any bioanalytical method is to selectively extract out information from a complex sample matrix containing a great deal of uninteresting "noise." At some point the method will most likely have to differentiate between compounds of very similar structure. This is particularly true if one is to successfully examine the metabolites of

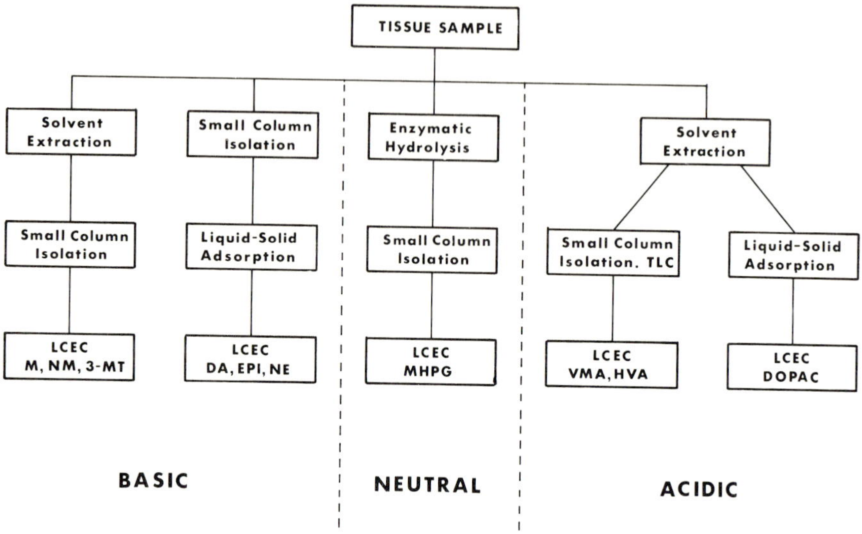


FIG. 9. Flowchart describing the analytical schemes available for the measurement of tyrosine metabolites.

tyrosine. Figure 9 illustrates the strategy used in the determination of these compounds. They can be conveniently classified as bases, neutrals, and acids. A simple pH adjustment followed by solvent extraction or liquid–solid adsorption will permit isolation of these classes prior to individual determinations.

## B. LCEC Methods for Urinary Catecholamines

As mentioned in the previous section, the preliminary fractionation of a complex sample often pays for itself many times over in sample throughput during instrumental analysis. A good case in point is the determination of urinary catecholamines first proposed by us in 1977 (*173*) and later embellished (*174–176*). A preliminary pass of diluted urine—usually 1–3 mL—over miniature cation exchange columns at pH 6.5 retains all amines, while the rest of the sample is eluted to waste. The amines are then eluted with a few milliliters of acidic $(NH_4)_2SO_4$. The pH of the collected eluant is raised to 8.5 with concentrated Tris buffer and the sample then mixed at once in a conical screw-top vial with 100 mg of acid washed alumina. The catecholamines are adsorbed onto the alumina, the solid adsorbent is allowed to settle, and the supernatant is aspirated to waste. A small volume of 0.2 $M$ $HClO_4$ (typically 300–500 $\mu$L) is used to desorb the catecholamines from the alumina. The supernatant after filtration is ready for injection onto the LC.

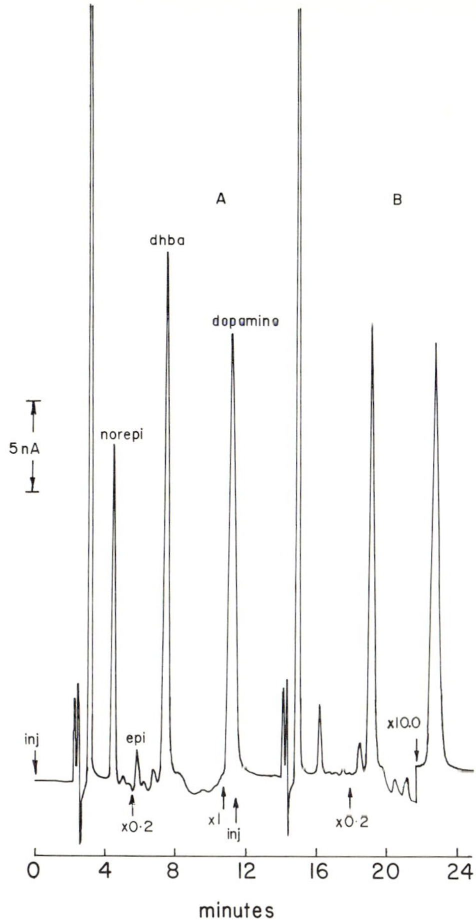


FIG. 10. Chromatograms of urinary catecholamines from healthy (A) and diseased (B) individuals. The dopamine levels differ by a factor of ten. Conditions: stationary phase, Whatman Partisil ODS 5; mobile phase, 0.1 *M*, pH 2.7, phosphate buffer, with 5% methanol and 25 mg/L octyl sodium sulfate; flow rate, 1.2 mL/min; detector, EC, +720 mV vs Ag/AgCl, CP–O electrode.

Typical chromatograms of urinary catecholamines from patients with neuroblastoma are shown in Fig. 10. The sample on the left represents a profile from an apparently healthy control. In addition to norepinephrine, epinephrine, and dopamine, each sample is supplemented with 50 ng/mL of the internal standard 3-methoxy-4-hydroxybenzylamine to insure a complete account of sample recovery from one specimen to the next. Samples can be injected approximately every 10 min.

When a selective and highly sensitive detector is used with high resolution chromatography, one might question why any cleanup procedures for the catecholamines should be necessary. For some brain or CSF samples, a simple "homogenize and inject" philosophy often works. For urine, however, a quick extraction has several advantages. First, only amines with catechol functional groups will remain in the final extract. With the moderately acidic mobile phases used, both neutral and acidic metabolites, if present in the final extract, would take 1 h or longer to elute. The preconcentration afforded in the sample cleanup step is also advantageous. A three milliliter sample may be concentrated as much as 10-fold with the above procedure by judiciously placing the alumina extraction as the final step.

Reverse-phase chromatography is ideal for the separation. Dilute pH 3 phosphate buffers containing 0.1–0.5 m*M* sodium octyl sulfate are satisfactory for the mobile phase. The long-chain detergent transforms the reverse phase $C_{18}$ bonded silica into a dynamic ion exchange column. The separation may be tuned to the desired effect by modifying the detergent concentration (*177*).

## C. LCEC Methods for Serum Catecholamines

The determination of serum or plasma catecholamines by LCEC was not successfully accomplished until very recently (*178–184*). This owed primarily to two factors: First, the continuing improvement in the efficiency of liquid chromatography columns. As the efficiency of the columns increase, the volume of mobile phase necessary to elute the compounds from the column bed decreases. The analyte is contained within a smaller volume and the effective concentration of the compound as it passes through the detector is higher. Second, a better understanding of the electrochemical detector and the parameters that effect the signal-to-noise (S/N) ratio. Detection limits for the catecholamines in clinical samples have been extended to as low as 25 pg with a S/N ratio of 5/1.

The separations in the early publications (*178, 179*) were performed on cation exchange columns. The results for norepinephrine were elevated somewhat compared to values obtained by other techniques. This may owe to less than adequate selectivity of the cation exchange columns. Our experience with plasma or serum has shown that even with the specific alumina isolation procedure (described below), many oxidizable contaminants can be detected at the sensitivity setting necessary for low level catecholamine determinations. Reverse-phase chromatography has been employed to improve the selectivity of the separation (*180, 184*). The procedure is a modification of the LCEC method for urinary catecholamines described in the previous section. Short cation exchange isolation columns in combination with alumina liquid/solid adsorption provides an improvement in sample cleanup superior to alumina alone. The cation exchange columns will first isolate cationic species (e.g., protonated amines) from the biological sample. Following their elution from the columns, the compounds can be isolated further by making use of the reactivity of the catechol moiety with alumina.

Sample cleanup is an important factor often glossed over. The use of the "most sensitive" detector does not solve every trace analysis problem. Sample concentration and cleanup often allow one to choose a more easily implemented and less costly assay method. Electrochemical detection, one of the three standard LC detection schemes for trace determinations (*185*), is an attractive alternative to the costly and tedious radioenzymatic procedures described earlier. In fact, a recent study demonstrated that the LCEC method gave excellent agreement when compared to a popular radioenzymatic technique (*183*). The LCEC methods are still at a disadvantage in sample size requirements (2 mL) compared to radioenzymatic assays (typically $< 100$ μL). Figure 11 illustrates the determination of NE

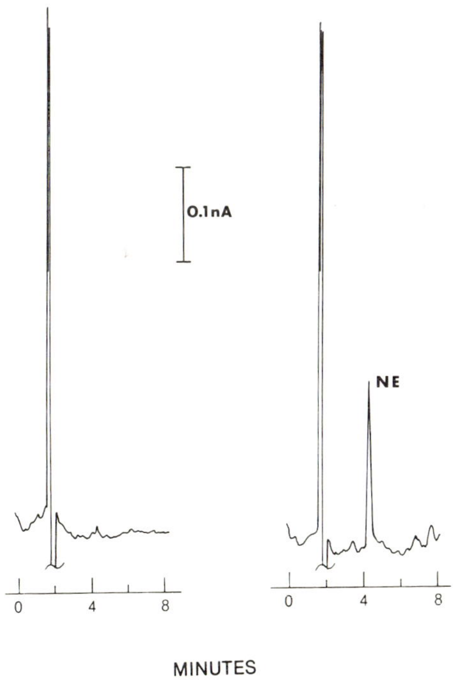


FIG. 11. Norepinephrine isolated from human serum. Spherisorb 5 μm C-18 column (25 cm × 4.0 mm), applied potential +0.750 V vs Ag/AgCl, CP-W electrode material. The chromatogram on the left is for a reagent blank. (Reprinted from ref. *184*, courtesy of American Chemical Society.)

isolated from a human serum sample. The values obtained using this assay are in good agreement with the literature.

## D. LCEC Methods for Urinary Metanephrines

The metanephrines differ from the catecholamines by only one methyl group. This small difference, however, affects both the electrochemistry and chromatography. The vanillyl (3-methoxy-4-hydroxyphenyl) functional group is slightly more difficult to oxidize than the orthohydroquinone backbone of the catecholamines. For a given pH, the additional potential required is approximately +150 mV. The oxidation reaction is described in Fig. 8. The followup hydrolysis step is very rapid.

With regard to the chromatography, the additional methyl group reduces the polarity of each substance so as to enhance their adsorption onto the nonpolar stationary phase. In comparison to the catecholamines, the metanephrines demonstrate larger capacity factors under the same chromatographic conditions. For this reason, the use of ion pairing agents may be avoided. Simple buffers can be used, with rapid equilibration times.

Since the metanephrines are not adsorbed onto alumina, the sample extraction procedure used for urinary catecholamines is not workable. Alternatively, we opted for a preliminary separation on small cation exchange columns followed by solvent extraction. The metanephrines are extracted as a group on the isolation columns; the solvent extractions offer preconcentration capability (*186*).

After a preliminary acid hydrolysis at 95° C to cleave sulfate and glucuronide conjugates, 3–5 mL samples of urine diluted with phosphate buffer (pH 6.5) are passed through the isolation columns. The columns are then washed with dilute boric acid solution to remove catecholamines. The metanephrines are eluted with 4 mL of pH 10.0 ammonium hydroxide solution and extracted twice with an ethylacetate/acetone solvent mixture. The combined organic phases are collected, evaporated under nitrogen, and brought up in 500 $\mu$L of acetic acid. The concentrate is ready for injection onto the liquid chromatograph.

Figure 12 demonstrates a representative chromatogram obtained under the instrumental conditions and extraction procedures described. In addition to normetanephrine, metanephrine, and 3-methoxytyramine, both tyramine and serotonin (5-hydroxytryptamine) have been identified in the chromatogram. Under the conditions used in this assay, however, tyramine did not give a linear response (the detector potential was less than the oxidation peak potential) and the serotonin value was irreproducible (presumably

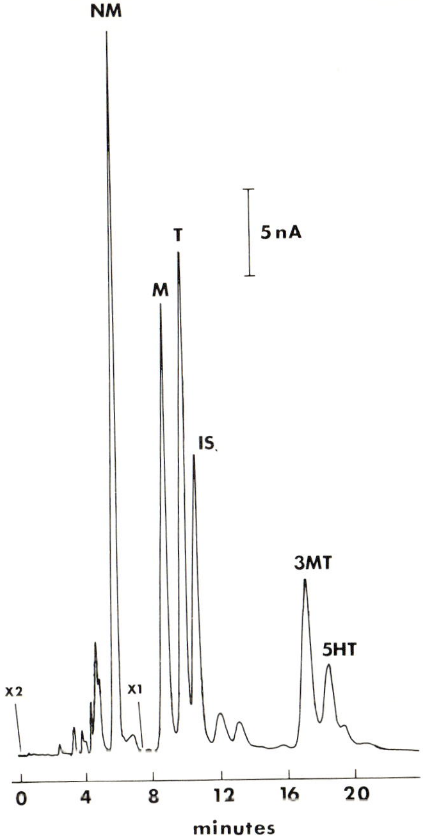


FIG. 12. Typical chromatogram of an actual urine sample from a healthy individual. Chromatographic conditions per ref. *186*. Abbreviations: NM, normetanephrine, M, metanephrine; T, tyramine; IS, 3-methoxy-4-hydroxybenzylamine (internal standard); 3MT, 3-methoxytryamine; 5HT, serotonin. (Reprinted from ref. *186*, courtesy of the American Association for Clinical Chemistry.)

owing to uncontrolled decomposition at high pH). These factors precluded the accurate quantitation of these substances. An internal standard added to all urine samples, 3-methoxy-4-hydroxybenzylamine, is eluted after tyramine, about midway through the chromatogram.

Approximately 1000 samples from healthy persons as well as from patients with hypertension, pheochromocytoma, and neuroblastoma have been assayed with the procedure and we have encountered no major difficulties. With respect to previous spectrophotometric and fluorometric assays, this approach eliminates cumbersome chemical differentiation steps for the individual quantitation of each substance. Instead, this task is performed by the power of modern liquid chromatography and supplemented by the sensitivity of the amperometric detector.

## E. LCEC Methods for the Acidic and Neutral Metabolites

The first liquid chromatographic methods for the determination of HVA, VMA, and MHPG of practical interest were carried out using columns packed with pellicular anion exchange or reverse-phase materials (*187–190*). For the most part these have been replaced by the higher efficiency microparticulate reverse-phase materials (*191–199*). The UV and electrochemical (EC) detectors have been the LC detectors of choice for the determination of these compounds in urine. Although the UV detector appears to be sufficiently sensitive for monitoring elevated levels of these components (*1, 193, 194*), one is hard-pressed to determine serum or CSF levels without exorbitant amounts of sample. Also, because of the large number of UV active compounds found in urine, rapid sample throughput can not be achieved without extensive pre-LC cleanup or expensive LC instrumentation to generate appropriate mobile-phase gradients. EC detectors, on the other hand, have two to three orders of magnitude greater sensitivity for these molecules, making them ideal for trace analysis from not only urine, but also from serum and CSF (*200*). The selectivity offered by this detector, because of the electrochemical activity requirement, precludes the necessity for arduous sample cleanup. For screening purposes, simple dilute and inject methodology is satisfactory (*197, 198*). In cases where lower concentations are expected, and as interferences become more prominent, a simple pre-LC TLC isolation procedure is all that is required (*189, 196*).

A third LC detector recently introduced, the periodate oxidative monitor (*199*), takes an autoanalyzer approach to determine VMA. From the initial report, the method appears promising for urinary analysis of normal and elevated amounts and, with modification, may be applicable to other physiological samples, as well as to other compounds that can be oxidized by periodate to vanillin (e.g., MHPG, metanephrine, normetanephrine) and measured spectrophometrically.

One of the drawbacks of LC in the clinical laboratory has been the serial nature by which samples must be analyzed. Many times all of the pertinent information is contained within the first few minutes of the chromotogram and everything following is unwanted "chemical noise" that must elute before the next sample can be injected. There are two instrumental approaches to eliminating or reducing the time required to remove the chemical noise from the system. One manipulates the mobile phase, the other manipulates the stationary phase.

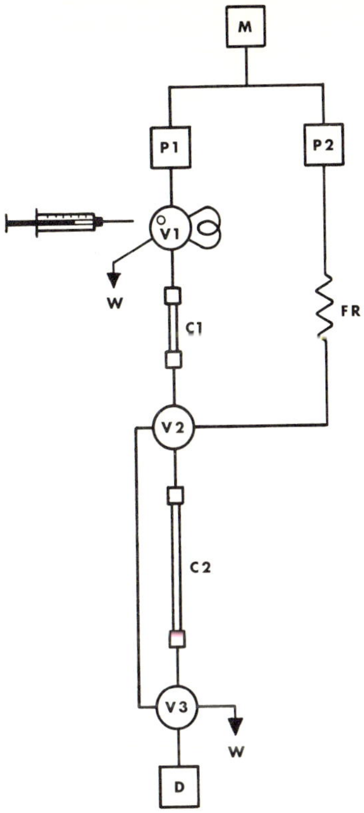


FIG. 13. Dual-pump split-column chromatograph: M is the mobile phase reservoir, $P_1$ and $P_2$ are the pumps, $V_1$ is a 6-port injection valve, $C_1$ and $C_2$ are analytical LC columns, $V_2$ and $V_3$ are 4-way valves, D is the electrochemical detector, and W is a waste receptacle. (Courtesy of Bioanalytical Systems Inc.)

As noted earlier, analysis time can be greatly reduced by programming the mobile phase from a solvent of low elution strength to one of high eluting power. This can be done by any number of continuous gradients generated by a solvent programmer and gradient mixer, or it can be done abruptly as a *pulse* gradient. The former approach can be very expensive and the reproducibility of mixing systems are questionable. The latter is simple, easy to implement, of low cost, and, generally, accomplishes the same goal as the former with greater reproducibility.

A second approach for speeding analysis time is to program the stationary phase. This technique, called split-column or column switching chromatography (*192*), reduces the number of strongly retained components of no interest in the analysis. Figure 13 illustrates this technique. As the analytes of interest pass through column C1 and

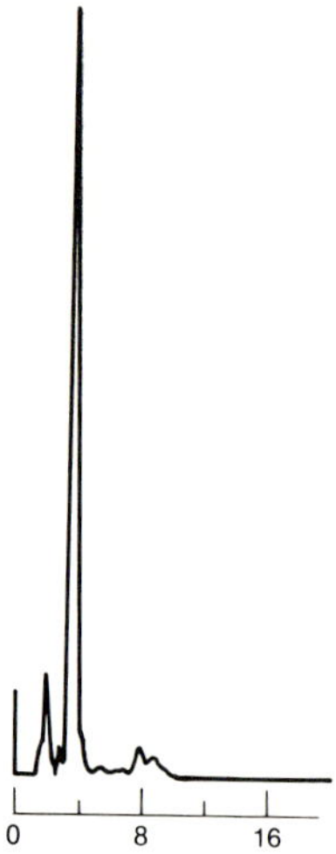


FIG. 14. Typical chromatogram for the determination of VMA using column-switching following a TLC isolation step. LC conditions: mobile phase, pH 4.0 McIlvaine buffer; flow rate, 2.0 mL/min; C1, 10 cm MPLC 18-MP; C2, 15 cm BIOPHASE ODS (C1 and C2 from Bioanalytical Systems Inc., West Lafayette, IN.); column switching time, 35 s following injection; electrode operating potential, 0.80 volt vs silver/silver chloride reference electrode. TLC conditions: silica gel pre-adsorbant, pre-channeled 250 $\mu$m plates (Whatman Inc., Clifton, NJ); developing solvent, upper layer of 2/3/1, benzene/acetic acid/water.

valve V2 onto column C2, V2 is switched to divert all the remaining components on C1 to waste, and yet continue elution of the desired compounds on C2 to the detector. When all of the components on C1 have been washed off, the valve can be switched back to its original position and the next sample can be injected. Figure 14 illustrates the determination of VMA using this technique.

## F. LCEC Methods for Serum D$\beta$H

In studying tyrosine metabolism, the assay of the major enzymes in the metabolic pathway would complement the metabolite determinations. Enzyme activity measurements have potential diagnostic use in revealing key trouble points within a given pathway. LCEC has been a powerful analytical technique for the determination of catecholamines and their metabolites. The extension of this approach to enzyme activity studies is a natural one.

The assay for serum DβH developed in our laboratory used dopamine, the natural substrate of DβH, instead of the frequently employed tyramine (*201*). The enzymatic product, norepinephrine, can be easily isolated by the alumina procedure described earlier. However, a problem arose in this method that is fairly typical of enzyme activity assays. Dopamine is used under substrate-saturating conditions to ensure that the rate of the reaction is strictly dependent upon the activity of the enzyme. Since the product isolation procedure will necessarily isolate dopamine (which is also a catechol), one is faced with having to quantitate the enzyme product in the presence of a large amount of substrate. Fortunately, norepinephrine and dopamine can easily be resolved using ion-pair reverse-phase LC. Capacity factors (which are a measure of the affinity a compound has for a stationary phase) for norepinephrine and dopamine are 2.5 and 21 respectively. This difference owes mainly to the β-hydroxyl group on the side chain of norepinephrine that reduces the hydrophobic character of the molecule. Hydrophobic interactions between the compound and stationary phase are thought to be the basis for the separation in reverse-phase chromatography (*202*). Unfortunately, these same conditions, which are chosen and optimized to resolve norepinephrine from the void volume components, dictate a high capacity factor for dopamine. This lengthens the analysis time considerably. Essentially all of the desired analytical information is available after the elution of norepinephrine (typically less than 5 min). One must wait for the elution of dopamine (usually > 20 min) before the next injection can be made. In addition, the amount of dopamine isolated and injected easily overloads the column, producing a broad tailing peak that saturates the output of the detector. This can be deleterious to the useful lifetime of the electrode surface.

Column switching provides an attractive remedy for this problem. Instead of using a single column for the separation, two shorter columns are employed (Fig. 15). The chromatographic conditions are adjusted slightly so that norepinephrine passes through both the columns and detector just as dopamine is eluting from column 1. The pneumatically actuated 4-way valve (controlled by a digital programmer) diverts the dopamine to waste. The dual-pump split-column chromatograph increases the sample throughput and avoids the detrimental effects of detector saturation.

As an alternative one might suggest the use of a step gradient to wash the dopamine from the column. With the use of ion-pair reagent, it is not possible to instantly re-equilibrate the stationary phase because of the large capacity factor of the modifier. Any savings in time would certainly be offset by the time required to re-establish the original conditions.

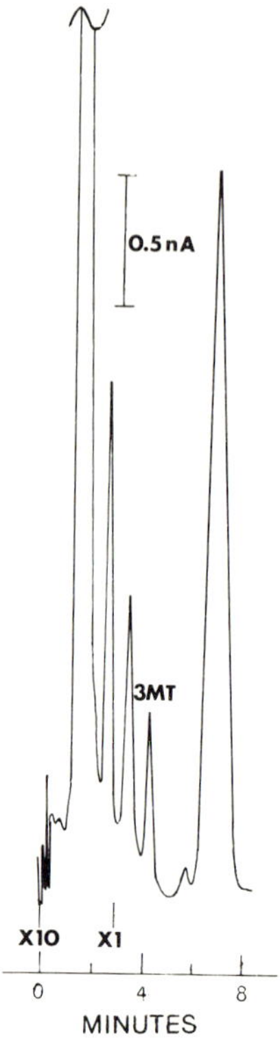


FIG. 15. Chromatogram of a quenched COMT incubation mixture from human erythrocytes. The internal standard peak precedes 3-methoxytyramine. (Reprinted from ref. *203,* courtesy of American Chemical Society.)

## G. LCEC Methods for COMT

The approach taken for the assay of COMT by LCEC is similar to that of DBH. Dopamine is used as the substrate under saturating conditions with the necessary cofactors (*203*). The source of the enzyme is lysed red blood cells (RBC). The reaction is stopped by the addition of acid and the *O*-methyl products are isolated on small cation exchange columns. Dopamine is a protonated amine under these conditions and as such will also be isolated. Boric acid, which forms a complex similar to alumina with catechols, can be passed through the column eluting the substrate dopamine. The amount of residual dopamine finally eluted with the *O*-methyl enzyme products does not pose the same problems as described in the D$\beta$H assay. This eliminates the need for split column chromatography in this assay. In fact, using a combination of ion-pair reagent and methanol (10% or less), the aqueous mobile phase conditions can easily be adjusted such that dopamine will have a much smaller capacity factor than the metanephrines. By playing these two parameters against one another the compounds are easily resolved. A typical chromatogram for COMT activity in RBC is shown in Fig. 15. The calculated activity compares favorably to literature values.

Enzyme activity assays for the other major enzymes in the tyrosine metabolic pathway have been developed using LCEC. Tyrosine hydroxylase (*204, 205*), dopa decarboxylase (*206*), and phenylethanolamine-*N*-methyltransferase (*207*) were measured by this technique. An early LCEC method for COMT was described by Borchardt using a less selective cation exchange column (*208*).

## V. Tryptophan Metabolism

Tryptophan is metabolized along several pathways into more than fifty compounds, many of which are thought to be physiologically significant. The key pathway from a neurochemical standpoint is that leading to 5-hydroxytryptamine (5-HT, serotonin), which is illustrated in Fig. 16. Tryptophan (Trp) is converted by tryptophan hydroxylase

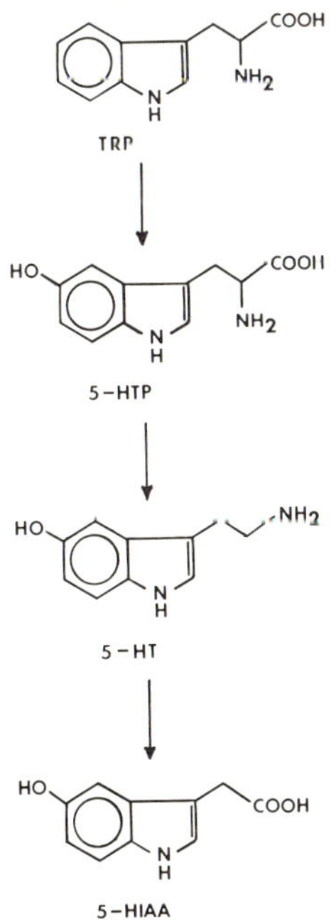


FIG. 16. Tryptophan metabolic pathway.

into 5-hydroxytryptophan (5-HTP), which is then decarboxylated to serotonin (5-HT). Serotonin is primarily metabolized to 5-hydroxyindoleacetic acid (5-HIAA). Synthesis of serotonin in the brain is thought to be regulated by brain levels of its precursor, tryptophan, which are in turn highly dependent on the plasma concentrations of the amino acid (*209, 210*). The 5-hydroxytryptophan levels in the brain are exceedingly small because of the nearly ubiquitous presence of aromatic L-amino acid decarboxylase (*211*). Thus, tryptophan hydroxylase is the critical enzyme for serotonin synthesis. It generally is present in nerve endings containing serotonin activity. Serotonin is known to be localized in the raphe nuclei, hypothalamus, amygdala, striatum, and pineal gland, among others (*212, 213*). Useful information about 5-HT activity is provided through the measurement of its major metabolite, 5-HIAA (*214*). The most advantageous choice, if possible, is to measure 5-HT simultaneously with 5-HIAA and perhaps tryptophan.

## VI. Clinical Significance of Tryptophan Metabolism

### A. Tryptophan

Clinical studies of tryptophan metabolism have generally focused on the metabolites (5-HT, 5-HIAA) rather than the amino acid itself. The clinical significance of tryptophan measurements alone has yet to be determined. One area of investigation is that of brain 5-HT synthesis, which is affected by the relative concentration of the amino acid in blood. Tryptophan competes with eight other amino acids for the same transport protein, which enables them to cross the blood–brain barrier, so that more serotonin will be synthesized in the brain if the relative concentration of tryptophan increases. It has been shown in rats that ingestion of a carbohydrate diet causes insulin to be secreted, facilitating uptake of all the neutral amino acids except tryptophan (*215*). As the relative concentration of tryptophan goes up, more reaches the brain, and serotonin synthesis increases. Still, a great deal of controversy exists regarding control of 5-HT synthesis, whether it be limited by the availability of tryptophan or enzymatic feedback control. Studies of tryptophan levels in circulation can aid in this regard. The determination of 5-HT and 5-HIAA in the circulation in combination may also provide clues to this important metabolic process.

The most popular method for tryptophan is based on fluorescence, but involves a tedious workup and reaction with formaldehyde (*216*). The same reaction was utilized in a kinetic method for tryptophan in foodstuffs (*217*), and also in a procedure that used ultrafiltration to differentiate the free from albumin-bound plasma tryptophan (*218*). Separation techniques have been employed to improve the selectivity in the measurement. Gravity-fed ion exchange columns have been successfully applied and several concurrent methods for Trp, 5-HT, and 5-HIAA using radioenzymatic techniques have been described (*219, 220*). Reverse-phase LC far and away has had the largest impact on the improvement in analysis not only for tryptophan and its metabolites, but in trace organic analysis in general. The efficiency of modern LC columns for the simultaneous determination of tryptophan metabolites as well as those of tyrosine is well established (*221*).

## B. Serotonin and 5-Hydroxyindoleacetic Acid

Serotonin has been implicated in a wealth of behavioral maladies. Often the precise 5-HT contribution is impossible to define owing to the complex neural circuitry, but the major role now hypothesized is to dampen or inhibit neuronal activity (*222*). This is not unusual; many neurotransmitters exhibit inhibition as their primary action. Also not exceptional is the inability to explain all the facets of a disease state based on observations of one particular transmitter such as serotonin. Commonly, two or three neurochemicals are interacting to produce the end result. In particular, because of the morphological relationships between 5-HT and norepinephrine structures, it is thought that their functions are interrelated. This complication points out the importance of assaying several compounds and their metabolites simultaneously.

One of the most interesting, and at the same time the most controversial, hypotheses for serotonin involves the psychoses schizophrenia and depression (*223*). In essence, the theory, the "serotonin hypothesis" of mental disease, states that schizophrenia may result when brain serotonin is in excess, while depression is possible when brain serotonin is deficient. It is clear that the measurement of 5-HT and 5-HIAA in urine, serum, and cerebrospinal fluid will, in most cases, help to refute or substantiate these conjectures.

The situation is clearer with the psychoses known collectively as the "affective illnesses," or depression. Patients show definitely lower levels of both serotonin and 5-HIAA in the cerebrospinal fluid and urine. Treatment with tryptophan and 5-hydroxytryptophan has therapeutic value, although the values of 5-HIAA do not reach the values noted in healthy patients.

A major portion of serotonin is metabolized to 5-hydroxyindoleacetic acid by monoamine oxidase. The levels of 5-HIAA in body fluids, as discussed earlier, can be indicative of the use of the serotonin pathway. Urinary excretion greatly increases in patients exhibiting Hartnup's disease or those with malignant carcinoid. Excision of the cancer lowers the 5-HIAA back to normal, unless metastasis has occurred. Excretion of 5-HIAA decreases during vitamin $B_6$ deficiency (*227*). Small increases occur during ovulation, pregnancy and stress. Moderate increases are seen in patients with Whipple's disease (*228*).

The interest in serotonin and its metabolites has paralleled the development of adequate methods for their analysis. As the methods became available or were improved, remarkable advances in understanding the role of these substances followed. Serotonin was first determined by bioassay techniques, taking advantage of its ability to contract animal tissue, e.g., rat uterus (*229*). The bioassays are exceedingly sensitive and are probably specific for the molecule of interest, but they are not suitable for a clinical laboratory.

Other methods for 5-HT include a colorimetric assay made possible by the red color generated when aqueous solutions of the metabolite are treated with nitrous and sulfuric acids followed by 1-nitroso-2-naphthol (*230*). All of the hydroxyindole compounds undergo the same coupling reaction. The sensitivity is not generally adequate for biological fluids. Fluorometric methods for serotonin and related molecules have been reviewed by Maickel (*231*). The fluorescence methods are relatively fast, sensitive, and applicable to many of the important metabolites. However, interferences are a constant problem and the detection limits, although good, are often insufficient for accurate, reproducible results. A novel automatic analysis for serotonin and 5-HIAA uses the fluorescence enhanced by *o*-phthalaldehyde derivatization, but the method was apparently inadequate for urine (*232*).

The requirement for specificity has led to many methods using a separation of some kind. Gas chromatographic methods for several tryptophan derivatives provide sensitive detection and efficient separation, but quantitative analysis is often difficult and derivatization is required. GC procedures using mass spectrometric detection have been developed for 5-HT and 5-HIAA (*233, 234*). Mass fragmentography (MF) has been used for 5-HIAA (*235, 236*), but exceedingly long sample preparations were required.

As mentioned earlier in the discussion of tryptophan analysis, cleanup of samples using column chromatography prior to fluorescence has dramatically improved selectivity. Bio-Rex 70 columns have been employed for a simultaneous assay of 5-HT and the catecholamines (*237*), and in tandem with other resins for some of the metabolites (*238*). These columns have also been used for 5-HT with Sephadex G-10 for 5-HIAA (*239*). Serotonin from plasma was concentrated on Amberlite CG-50 before measurement of native fluorescence in HCl (*240*), but the background was rather high.

Cation-exchange column chromatography was applied to the separation of 5-hydroxyindoles from blood, although long elution times were required (*241*). Six urinary indoles were separated in 20 min using reverse-phase LC with fluorescence detection (LCF) (*242*). About 5–15 ng could be detected and five of the six indoles were successfully analyzed after a single urine-deproteinization step. However, a large unknown background peak appeared in the middle of the other peaks that was unexplained and apparently could not be avoided. Other methods based on reverse-phase LC with UV detection (LCUV) for 5-HIAA (*243*), tryptophan, and several kynurenine and serotonin metabolites (*244*) suffer from inadequate detection limits. A recent report described the use of reverse-phase ion-pair LCUV to the analysis of tryptophan, 5-HPT, 5-HT, 5-HIAA, and kynurenine from plasma, saliva, and urine (*245*), but again, the sensitivity was poor. Detection using native fluorescence is fairly routine, but in a method for several tryptophan metabolites from serum (*246*), no 5-HT or 5-HIAA was seen. A similar report demonstrated greater sensitivity, but interferences are a problem (*247*). Derivatization with *o*-phthalaldehyde prior to separation is a promising means of improving the sensitivity, but this can be awkward (*248*). Several attempts involving the use of post-column reactions appear to be even less successful (*249, 250*).

## VII. LCEC Methods for Tryptophan Metabolites

### A. Tryptophan

The method we have found to be most useful for Trp, 5-HT, and 5-HIAA involves isolation on small, gravity-fed extraction columns prior to use of LCEC (*251–253*). The resin selected depends on the compound preferred. Each compound can be determined independently, although a desirable feature of this approach is the ability to assay several concurrently. A strong cation exchange resin (Dowex AG-50) is employed for liquid–solid isolation of tryptophan. Elution is followed by injection onto a reverse-phase liquid chromatograph. Amperometric detection is employed at a potential of +1.0 V.

Tryptophan from cerebrospinal fluid, plasma, or urine can be determined (*251*).

## B. Serotonin and 5-Hydroxyindoleacetic Acid

Isolation of 5-HT occurs on a weak cation exchange resin (Amberlite CG-50). The eluate is injected onto a similar chromatograph as for tryptophan, but at a detector potential of only +0.50 V. Applicability to tissue, serum, CSF, and urine has been established (*251*). The three-step procedure, combining liquid–solid extraction, chromatographic separation, and electrochemical detection, provide this method with selectivity such that serotonin is typically the only compound that will oxidize at the chosen potential.

A gel filtration resin (Sephadex G-10) selectively absorbs 5-HIAA. The assay for tissue samples generally functions best when a protectant such as cysteine is added. Injection onto the same chromatograph as for 5-HT is made, again at +0.50 V. Urine, serum, and CSF can be assayed as well as tissue samples (*251*).

When desired, two or more of the above metabolites can be determined in a sequential process. In all cases, the Amberlite step comes before the Dowex step, which precedes Sephadex isolation. When an investigation requires determination of 5-HT and 5-HIAA, the method begins with the Amberlite isolation. The effluents from these columns are applied directly to the Sephadex resin, and adsorption of 5-HIAA takes place at the same time the elution of 5-HT is carried out. Other combinations would proceed similarly. If for some reason 5-HPT and Trp are desired simultaneously, several options are available. The same chromatograph could determine both, but a potential of +1.0 V would be necessary. Another approach would be to employ two separate instruments optimized for each compound.

Other examples of the application of LCEC to tryptophan and its metabolites include work that demonstrates the superiority of electrochemical detection over UV (*254*). LCF (with derivatization) provides nearly the same sensitivity.

## C. Precolumn Sample Enrichment of Serum or Plasma Serotonin

The method described above for the tryptophan metabolites is well suited for routine service in the analysis of urine or large regions of brain tissue. The technique is relatively straightforward and the technology is rather inexpensive.

The need to determine serotonin in serum or plasma requires a method with extremely good sensitivity. The small, gravity-fed extraction columns used for the cleanup step in the methods above decrease the overall time required for the assay and are quite convenient, but do not interface well with liquid chromatography. The minimum volume needed to elute the compound(s) of interest is much larger than the usual LC injection volume of ca. 20 $\mu$L. Without some modifications, better than 90% of the isolated compound is wasted, decreasing the overall sensitivity.

One approach would be to simply increase the injection volume. Normally this would destroy the chromatographic efficiency, but will work in this case since the capacity factor for serotonin in purely aqueous mobile phases is very large. The serotonin from a large volume aqueous injection (e.g., a 2 mL loop) will effectively be enriched at the top of the column and will not chromatograph until the mobile phase that contains an organic modifier (e.g. methanol) penetrates the column. In this conventional trace enrichment scheme, two problems arise. The trace contaminants are also enriched, and detector baseline fluctuations may occur owing to the large amount of water from the sample injection disturbing the mobile phase composition. A precolumn that will isolate the compound but would not enrich the weakly retained interferences is one solution to this problem. Unfortunately this approach leads to a loss in efficiency and the precise delivery of the sample. Incorporating a second injection valve and a second pump into the apparatus provides the enrichment scheme with the best capabilities. This sample enrichment system is outlined in Fig. 17. Precise injection of 2.0 mL is accomplished with valve V1 and it is delivered to the precolumn with an aqueous mobile phase which allows strong retention and enrichment of the compound. Back-flushing the precolumn (valve V2) with the analytical mobile phase containing methanol initiates the reverse-phase separation.

A typical chromatogram for the determination of 5-HT in 100 $\mu$L and 1 mL of plasma is shown in Fig. 18. The values found in plasma and serum, 3.31 $\pm$ 0.24 and 72.1 $\pm$ 4.9 ng/mL, respectively, compare well with recently published data (*255–257*).

The performance of the sample enrichment chromatograph was checked and found to be linear from 5 pg/mL to 5 ng/mL. The linearity would be expected to continue well above 50 ng/mL, but at these concentrations the sample enrichment feature would not be necessary. Detection limits for 5-HT at a signal-to-noise ratio of 5 have been calculated to be 1.1 pg/mL, or 6.5 pM.

The method described here can be extended to 5-HIAA, as presented earlier (*251*), simply by adding the Sephadex G-10 step to the

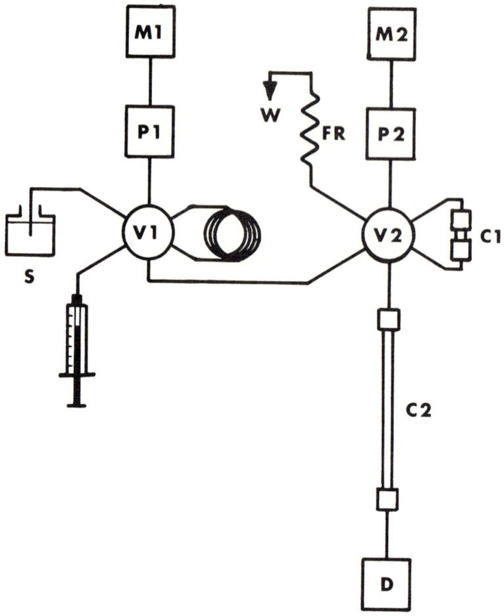


FIG. 17. Schematic of sample enrichment for liquid chromatography. Mobile phase 1 (M1, enrichment mobile phase) is pumped by pump 1 (P1) through valve 1 (V1) to column 1 (C1, precolumn). The sample (S) is injected by V1 and delivered to valve 2 (V2) where it is enriched by C1. When V2 is rotated, mobile phase 2 (M2, analytical mobile phase) from P2 back-flushes the contents of C1 onto C2 (analytical column) and to the detector D. (Courtesy of Bioanalytical Systems Inc.)

isolation. This method moreover is applicable to other samples such as cerebrospinal fluid. Other compounds have been determined using this sample enrichment/column switching apparatus with electrochemical detection, including environmental samples (*258*) and NADH (*259*).

## VIII. Conclusions

Liquid chromatography with thin-layer electrochemical detection is rapidly becoming the methodology of choice for the determination of the neurologically active biogenic amines and their metabolites in both tissue and body fluid specimens. Although significant progress has been made since the inception of LCEC eight years ago, there is much room for improvement. Simplified sample workup procedures and more reliable columns are two major areas where further work is needed. Recent developments in stationary phase programming

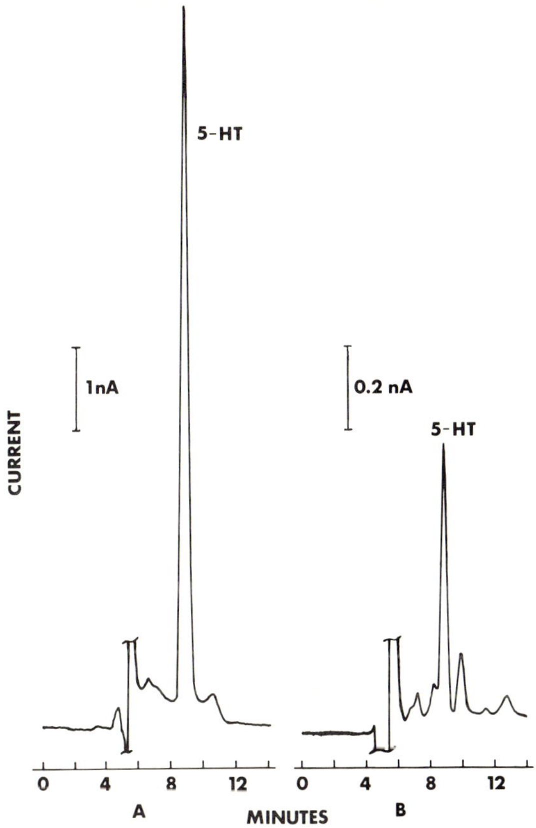


FIG. 18. Determination of serotonin in human plasma using the sample enrichment system. (A) 1 mL of plasma. Chromatographic conditions: (1) enrichment branch: 3 cm × 4.6 mm Merck RP-18; 0.42M, pH 5.1 ammonium acetate; flow rate 1.8 mL/min; 2.0 mL injection loop. (2) analytical branch: 30 cm × 4 mm Waters $\mu$-Bondapack $C_{18}$; 0.5 *M,* pH 5.1 ammonium acetate, 15% MeOH; flow rate 1 mL/min; 0.50 V electrode potential vs Ag/AgCl. (B) 100 $\mu$L of plasma. Conditions as in A. (Reprinted from ref. *252,* courtesy of American Chemical Society.)

(especially the split column and on-line trace enrichment schemes described above) point the way toward a number of improved procedures in the near future. The electrochemical detection of many tyrosine and tryptophan metabolites is now routine for injection of picomole amounts isolated from biological samples. Quantitation at lower levels has been achieved in some laboratories and further improvements in the minimum detectable quantity can be expected.

Although the present article has emphasized the application of LCEC to endogenous compounds of neurological interest, there have been many other applications to biomedical problems. Thiols (e.g., glutathione, cysteine, and pencillamine), phenothiazines, ascorbic acid, uric acid, methylxanthines, nitroglycerine, and a number of other compounds have been measured in biological samples using LCEC. Recently various pre- and post-column reaction schemes have been devised that extend electrochemical detection to compounds that are themselves not electroactive at easily accessible potentials. Amino acids, fatty acids, and unsaturated lipids are among those classes of compounds that are now detectable with good sensitivity using indirect amperometric methods (*260*).

In general, the use of LC for determination of endogenous metabolites in clinical samples is a far more difficult chore than is the therapeutic drug monitoring described in earlier chapters of this book. It is especially important to recognize that although LC is versatile, it is rare that one instrument can be used for several different assays concurrently. The diversity of endogenous metabolites normally requires that a separate instrument be dedicated to each assay. It is therefore prudent to use relatively inexpensive modular LC systems whenever feasible until the sample load becomes such that automated systems are cost effective.

## References


*1.* Hornykiewicz, O., *Brit. Med. Bull.* **29,** 172 (1973).
*2.* Bianchine, J. R., *New Engl. J. Med.* **295,** 814 (1976).
*3.* Riggin, R. M., Alcorn, A. L., and Kissinger, P. T., *Clin. Chem.* **22,** 782 (1976).
*4.* Gitlow, S. E., Mendlowitz, M., and Bertani, L. M., *Amer. J. Cardiol.* **26,** 270 (1970).
*5.* Gitlow, S. E., Bertani, L. M., Rausen, A., Gribetz, D., and Dziedzic, S. W., *Cancer,* **25,** 1377 (1970).
*6.* Pochedly, C., and Strauss, L., Histogenesis and Pathology of Neuroblastoma, in *Neuroblastoma,* Pochedly, C., ed., Publishing Sciences Group, Acton, Mass., 1978, p. 155.
*7.* Spiegel, H. E., *Ann. Clin. Lab. Sci.* **4,** 174 (1974).
*8.* Pochedly, C., ed., *Neuroblastoma,* Publishing Sciences Group, Acton, Mass., 1978.
*9.* Itoh, T., and Ohmori, K., *J. Lab. Clin. Med.* **81,** 887 (1972).
*10.* Gitlow, S. E., Pertsemlidis, D., and Bertani, L. M., *Am. Heart J.* **82,** 557 (1971).
*11.* Kaser, H., *Pharm. Rev.* **18,** 659 (1966).

12. Gitlow, S. E., Dziedzic, L. B., Strauss, L., Greenwood, S. M., and Dziedzic, S. W., *Cancer* **32,** 898 (1973).
13. Hingerty, D., and O'Boyle, A., *Clinical Chemistry of the Adrenal Medulla,* Thomas, Springfield, Ill., 1972.
14. Hoeldtke, R.,*Metabolism* **23,** 663 (1974).
15. Ikeda, M., Levitt, M., and Udenfriend, S., *Biochem. Biophys. Res. Commun.* **18,** 482 (1965).
16. Nestel, P. J., and Esler, M. D., *Circ. Res.* **26** (Supp II), 75 (1970).
17. Moyer, T. P., Jiang, N. S., Tyce, G. M., and Sheps, S. G., *Clin. Chem.* **25,** 256 (1979).
18. Blois, M. S., and Banda, P. W., *Cancer Res.* **36,** 3317 (1976).
19. Voorhess, M. L., *Cancer* **26,** 146 (1970).
20. Hansson, C., Edholm, L. E., Agrup, G., Rorsman, H., and Rosengren, A. M., *Clin. Chim. Acta* **88,** 419 (1978).
21. Kabara, J. J., Riggin, R. M., and Kissinger, P. T., *Proc. Soc. Exp. Biol. Med.* **151,** 168 (1976).
22. Goodall, M., Gitlow, S. E., and Alton, H., *J. Clin. Invest.* **50,** 2734 (1971).
23. Sandhu, R. S., and Freed, R. M., Catecholamines and Associated Metabolites in Human Urine, in *Standard Methods of Clinical Chemistry,* Vol. 7, Cooper, G. R., ed., New York, Academic Press, 1972, p. 231.
24. Seki, T., *J. Chromatogr.* **155,** 415 (1978).
25. Wong, K. P., Ruthven, C. R. T., and Sandler, M., *Clin. Chim. Acta* **47,** 215 (1973).
26. Muskiet, F. A. J., Thomasson, C. G., Gerding, A. M., Fremouw-Ottevangers, D. C., Nagel, G. T., and Wolthers, B. G., *Clin. Chem.* **25,** 453 (1979).
27. Kopin, I. J., Lake, R. C., and Ziegler, M., *Ann. Intern. Med.* **88,** 671 (1978).
28. Euler, U. S. von, Franksson, C., and Hellström, J., *Acta Physiol. Scand.* **31,** 1 (1954).
29. Bevan, J. A., *Fed. Proc.* **36,** 2439 (1977).
30. DeChamplain, J. Farley, L., Cousineau, D., and Ameringen, M. -R., van, *Circ. Res.* **38,** 109 (1976).
31. Louis, W. J., Doyle, A. E., and Anavekar, S., *New. Engl. J. Med.* **288,** 599 (1973).
32. Brecht, H. M., and Schoeppe, W., *Clin. Sci. Mol. Med.* **55,** 81s (1978).
33. Philipp, T., Distler, A., Cordes, U., and Wolff, H. P., *Clin. Sci. Mol. Med.* **55,** 61s (1978).
34. Lake, C. R., Ziegler, M. G., Coleman, M. D., and Kopin, I. J., *New Engl. J. Med.* **296,** 208 (1977).
35. Jones, D. H., Hamilton, C. A., and Reid, J. L., *Clin. Sci. Mol. Med.* **55,** 73s (1978).
36. Franco-Morselli, R., Bandouin-Legros, M., and Meyer, P., *Clin. Sci. Mol. Med.* **55,** 97s (1978).
37. Coulombe, P., Dussault, J. H., and Walker, P., *Metabolism* **25,** 973 (1976).

38. Stoffer, S. S., Jiang, N., Gorman, C. A., and Pikler, G. M., *J. Clin. Endocrinol. Metab.* **36,** 587 (1973).
39. Leung, F. Y., and Griffiths, J., *CMA Journal* **111,** 1321 (1974).
40. O'Hanlon, J. F., Jr., Campuzano, H. C., and Horvath, S. M., *Anal. Biochem.* **34,** 568 (1970).
41. Griffiths, J. C., Leung, F. Y. T., and McDonald, T. J., *Clin. Chim. Acta* **30,** 395 (1970).
42. Valori, C., Renzini, V., Brunori, C. A., Porcellati, C., and Corea, L., *Ital. J. Biochem.* **18,** 394 (1969).
43. Renzini, V., Brunori, C. A., and Valori, C., *Clin. Chim. Acta* **30,** 587 (1970).
44. Miura, Y., Campese, V., DeQuattro, V., and Meijer, D., *J. Lab. Clin. Med.* **89,** 421 (1977).
45. Davis, T. P., Gehrke, C. W., Gehrke, C. W. Jr., Cunningham, T. D., Kuo, K. C., Gerhardt, K. O., Johnson, H. D., and Williams, C. H., *Clin. Chem.* **24,** 1317 (1978).
46. Okamoto, K. -I., Ishida, Y., and Asai, K., *J. Chromatogr.* **167,** 205 (1978).
47. Imai, K., *J. Chromatogr.* **105,** 135 (1975).
48. Krstulovic, A. M., and Powell, A. M., *J. Chromatogr.* **171,** 345 (1979).
49. Engelman, K., Portnoy, B., and Lovenberg, W., *Am. J. Med. Sci.* **255,** 259 (1968).
50. Engelman, K., and Portnoy, B., *Circ. Res.* **26,** 53 (1970).
51. Cryer, P. E., *Diabetes* **25,** 1071 (1976).
52. Cryer, P. E., Santiago, J. V., and Shah, S. D., *J. Clin. Endocrinol. Metab.* **39,** 1025 (1974).
53. DaPrada, M., and Zürcher, G., *Life Sci.* **19,** 1161 (1976).
54. Peuler, J. D., and Johnson, G. A., *Life Sci.* **21,** 625 (1977).
55. Lovelady, H. G., and Foster, L. L., *J. Chromatogr.* **108,** 43 (1975).
56. Imai, K., Wang, M. -T., Yoshiue, S., and Tamura, Z., *Clin. Chim. Acta* **43,** 145 (1973).
57. Wang, M. -T., Imai, K., Yoshioka, M., and Tamura, Z., *Clin. Chim. Acta* **63,** 13 (1975).
58. Jacob, K., Vogt, W., Knedel, M., and Schwertfeger, G., *J. Chromatogr.* **146,** 221 (1978).
59. Ehrhardt, J. D., and Schwartz, J., *Clin. Chim. Acta* **88,** 71 (1978).
60. Boutagy, J., and Benedict, C., *J. Chromatogr.* **151,** 263 (1978).
61. Axelrod, J., and Tomchick, R., *J. Biol. Chem.* **233,** 702 (1958).
62. Nagatsu, T., *Biochemistry of the Catecholamines: The Biochemical Method,* University of Tokyo Press, Tokyo, 1973.
63. Liebner, E. J., and Rosenthal, I. M., *Cancer* **32,** 623 (1973).
64. Bravo, E. L., Tarazi, R. C., Gifford, R. W., and Stewart, B. H., *New Engl. J. Med.* **301,** 682 (1979).
65. Senoh, S., Daly, J., Axelrod, J., and Witkop, B., *J. Am. Chem. Soc.* **81,** 6240 (1959).
66. Keuhl, F. A., Hichens, M., Ormond, R. E., Meisinger, M. A. P., Gale, P. H., Cirillo, V. J., and Brink, N. G., *Nature* **203,** 154 (1964).
67. Pisano, J. J., *Clin. Chim. Acta* **5,** 406 (1960).
68. Weil-Malherbe, H., and Smith, E. R. B., *Pharmacol. Rev.* **18,** 331 (1966).

69. Nelson, L. M., Bubb, F. A., Lax, P. M., Weg, M. W., and Sandler, M., *Clin. Chim. Acta* **92,** 235 (1979).
70. Bertani-Dziedzic, L. M., Krstulovic, A. M., Ciriello, S., and Gitlow, S. E., *J. Chromatogr.* **164,** 345 (1979).
71. Muskiet, F. A. J., Fremouw-Ottevangers, D. C., Wolthers, B. G., and deVries, J. A., *Clin. Chem.* **23,** 863 (1977).
72. Wadman, S. K., Ketting, D., and Voute, P. A., *Clin. Chim. Acta* **72,** 49 (1976).
73. Gitlow, S. E., Bertani-Dziedzic, L. M., and Dziezic, S. W., in *Neuroblastoma,* Prochedly, C., ed., Publishing Science Group, Acton, MA, 1976, pp. 115–154.
74. Brewster, M. A., Berry, D. H., and Moriarty, M., *Clin. Chem.* **23,** 2247 (1977).
75. Laug, W. E., Siegel, S. E., Shaw, K. N., Landing, B., Baptista, J., and Gutenstein, M., *Pediatrics* **62,** 77 (1978).
76. Mitchell J., and Coscia, C. J., *J. Chromatogr.* **145,** 295 (1978).
77. Lakke, J. P. W. F., Nienhuis, R. J., Schut, T., and Korf, J., *Current Concepts in Treatment of Parkinsonism,* Yahr, M. D., ed., Raven Press, New York, NY, 1974, pp. 37–60.
78. Maas, J. W., *Ann. Intern. Med.* **88,** 556 (1978).
79. Pickar, D., Sweeney, D. R., Maas, J. W., and Heninger, G. R., *Arch. Gen. Psychiatry* **35,** 1378 (1978).
80. Hollister, L. E., Davis, K. L., Overall, J. E., and Anderson, T., *Arch. Gen. Psychiatry* **35,** 1410 (1978).
81. Schildkraut, J. J., Orsulak, P. J., Schatzberg, A. F., Gudeman, J. E., Cole, J. O., Rohde, W. A., and LaBrie, R. A., *Arch. Gen. Psychiatry* **35,** 1427 (1978).
82. Ixxo, J. L., Horwitz, D., and Keiser, H. R., *Life Sci.* **24,** 1403 (1979).
83. Schildkraut, J. J., Orsulak, P. J., LaBrie, R. A., Schatzberg, A. F., Gudeman, J. E., Cole, J. O., and Rohde, W. A., *Arch. Gen. Psychiatry* **35,** 1436 (1978).
84. Trapeznikov, N. N., Raushenbakh, M. O., Ivanova, V. D., and Yavorsky, V., *Cancer* **36,** 2064 (1975).
85. Silbergeld, E. K., and Chisolm, J. J. Jr., *Science* **192,** 153 (1976).
86. Zoghbi, F., Emerit, J., Fermanian, J., Bousquet, O., and Legrand, J. C., *Biomedicine* **27,** 37 (1977).
87. Teodorescu, O., Cosmatchi, R., and Capalna, S., *Physiologie* **15,** 113 (1978).
88. Tabaddor, K., Wolfson, L. I., and Sharpless, N. S., *Neurology (Minneap.)* **28,** 1249 (1978).
89. Pisano, J. J., Crout, J. R., and Abraham, D., *Clin. Chim. Acta* **7,** 285 (1962).
90. Gumbolt, G., *Clin. Chem.* **23,** 1949 (1977).
91. Knight, J. A., and Haymond, R. E., *Clin. Chem.* **23,** 2007 (1977).
92. Kahane, Z., and Vestergaard, P., *Clin. Chim. Acta* **47,** 453 (1973).
93. Stott, A. W., Lindsay Smith, J. R., Hanson, P., and Robinson, R., *Clin. Chim. Acta* **63,** 7 (1975).
94. Badella, M., Routh, M. W., Gump, B. H., *Clin. Chem.* **22,** 2046 (1976).

*95.* Shibuya, H., Watanabe, S., and Toru, M., *Experientia* **31,** 623 (1975).
*96.* Messiha, F. S., Bakutus, E., and Frankos, V., *Clin. Chim. Acta* **45,** 159 (1973).
*97.* Ruthven, C. R. J., and Sandler, M., *Clin. Chim. Acta* **12,** 318 (1965).
*98.* Zawad, J. S., and Brown, F. C., *Clin. Chim. Acta* **73,** 187 (1976).
*99.* Fellows, L., Riederer, P., and Sandler, M., *Clin. Chim. Acta* **59,** 255 (1975).
*100.* Biondi, P. A., Cagnasso, M., and Secchi, C., *J. Chromatogr.* **143,** 513 (1977).
*101.* Murray, S., Baillie, T. A., and Davies, D. S., *J. Chromatogr.* **143,** 541 (1977).
*102.* Melchert, H. U., and Hoffmeister, H., *J. Clin. Chem. Clin. Biochem.* **15,** 81 (1977).
*103.* Takahashi, S., Yoshioka, M., Yoshiue, S., and Tamura, Z., *J. Chromatogr.* **145,** 1 (1978).
*104.* Roginsky, M. S., Gordon, R. D., and Bennett, M. J., *Clin. Chim. Acta* **56,** 261 (1974).
*105.* Narasimhachari, N., Leiner, K., and Brown, C., *Clin. Chim. Acta* **62,** 245 (1975).
*106.* Weinshilboum, R. M., Thoa, N. B., Johnson, D. G., Kopin, I. J., and Axelrod, J., *Science* **174,** 1349 (1971).
*107.* Axelrod, J., *Pharmacol. Rev.* **24,** 233 (1972).
*108.* Algate, D. R., and Leach, G. D. H., *J. Pharm. Pharmac.* **30,** 162 (1978).
*109.* Gewirtz, G. P., and Kopin, I. J., *Nature (London)* **227,** 406 (1970).
*110.* Geffen, L., *Life Sci.* **14,** 1593 (1974).
*111.* Ngai, S. H., Dairman, W., Marchelle, M., and Spector, S., *Life Sci.* **14,** 2431 (1974).
*112.* Åberg, H. E., Hansson, H. E., Wetterberg, L., Ross, S. B., and Frödin, Ö., *Life Sci.* **14,** 65 (1974).
*113.* Viveros, O. H., Arqueros, L., and Kirshner, N., *Life Sci.* **7,** 609 (1968).
*114.* Weinshilboum, R., and Axelrod, J., *Science* **173,** 931 (1971).
*115.* Pinardi, G., Talmaciu, R. K., Santiago, E., and Cubeddu, L. X., *J. Pharmacol. Exp. Ther.* **209,** 176 (1979).
*116.* Cubeddu, L. X., Santiago, E., Talmaciu, R., and Pinardi, G., *J. Pharmacol. Exp. Ther.* **203,** 587 (1977).
*117.* Schanberg, S. M., and Kirshner, N., *Biochem. Pharmac.* **25,** 617 (1976).
*118.* Geffen, L. B., Rush, R. A., Louis, W. J., and Doyle, A. E., *Clin. Sci.* **44,** 617 (1973).
*119.* Wooten, G. F., and Cardon, P. V., *Arch. Neurol.* **28,** 103 (1973).
*120.* Planz, G., Wiethold, G., Appel, E., Böhmer, D., Palm, D., and Grobecker, H., *Europ. J. Clin. Pharmacol.* **8,** 181 (1975).
*121.* Alexander, N., McClaskey, J., and Maronde, R. F., *Life Sci.* **18,** 655 (1976).
*122.* Nagatsu, T., Kato, T., Numata, S. Y., Ikuta, K., Sano, M., Nagatsu, I., Takeuchi, T., Matsuzaki, M., and Umezawa, H., *Experentia* **33,** 581 (1977).
*123.* Horwitz, D., Alexander, R. W., Lovenberg, W., and Keiser, H. R., *Circ. Res.* **32,** 594 (1973).

*124.* DeQuattro, V., Campese, V., Lurvey, A., Yen, G., and Kypridakis, G., *Biochem. Med.* **15,** 1 (1976).
*125.* Lake, C. R., Ziegler, M. G., Coleman, M., and Kopin, I. J., *Circ. Res.* **41,** 865 (1977).
*126.* Hörtnagl, H., Stadler-Wolffersgrün, R., Brücke, Th., Hammerle, A. F., and Hackl, J. M., *Nauyn-Schmiedeberg's Arch. Pharmacol.* **303,** 235 (1978).
*127.* Aoki, K., Tazumi, K., Yoshida, T., Kato, S., Sato, I., and Takikawa, K., *Clin. Sci. Mol. Med.* **51,** 465s (1976).
*128.* Wise, C. D., and Stein, L., *Science* **181,** 344 (1973).
*129.* Wise, C. D., Belluzzi, J. D., and Stein, L., *Pharmac. Biochem. Behav.* **7,** 549 (1977).
*130.* Okada, T., Ohta, T., Shinoda, T., Kato, T., Ikuta, K., and Nagatsu, T., *Neuropsychobiology* **2,** 139 (1976).
*131.* Fujita, K., Ito, T., Maruta, K., Teradaira, R., Beppu, H., Nakagami, Y., Kato, Y., Nagatsu, T., and Kato, T., *J. Neurochem.* **30,** 1569 (1978).
*132.* Cross, A. J., Crow, T. J., Killpack, W. S., Longden, A., Owen, F., and Riley, G. J., *Psychopharmacology* **59,** 117 (1978).
*133.* Wyatt, R. J., Schwarz, M. A., Erdelyi, E., and Barchas, J. D., *Science* **187** 368 (1975).
*134.* Shopsin, B., Freedman, L. S., Goldstein, M., and Gershon, S., *Psychopharmacologia* **27,** 11 (1972).
*135.* VanCauter, E., and Mendlewicz, J., *Life Sci.* **22,** 147 (1978).
*136.* Lerner, P., Major, L. F., Murphy, D. L., Lipper, S., Lake, C. R., and Lovenberg, W., *Neuropharmacology* **18,** 423 (1979).
*137.* DePotter, W. P., DePotter, R. W., DeSmet, F. H., and Fraeyman, N. N., *Arch. Int. Pharmacodyn.* **232,** 334 (1978).
*138.* Molinoff, P. B., Weinshilboum, R., and Axelrod, J., *J. Pharmacol. Exp. Ther.* **178,** 425 (1971).
*139.* Henry, D. P., Johnson, D. G., Starman, B. J., and Williams, R. H., *Life Sci.* **17,** 1179 (1975).
*140.* Nagatsu, T., Thomas, P., Rush, R., and Udenfriend, S., *Anal. Biochem.* **55,** 615 (1973).
*141.* Wise, C. D., *J. Neurochem.* **27,** 883 (1976).
*142.* Bouclier, M., Mandel, P., and Aunis, D., *Pharmacol. Res. Comm.* **9,** 743 (1977).
*143.* Ohkura, Y., Ohtsubo, K., Kohashi, K., and Zaitsu, K., *Chem. Pharm. Bull.* **25,** 748 (1977).
*144.* Fujita, K., Nagatsu, T., Maruta, K., Teradaira, R., Beppu, H., Tsuji, Y., and Kato, T., *Anal. Biochem.* **82,** 130 (1977).
*145.* Flatmark, T., Skotland, T., Ljones, T., and Ingebretsen, O. C., *J. Chromatogr.* **146,** 433 (1978).
*146.* Kopun, M., and Herschel, M., *Anal. Biochem.* **85,** 556 (1978).
*147.* Nagatsu, T., and Udenfriend, S., *Clin. Chem.* **18,** 980 (1972).
*148.* Kato, T., Hiroshi, K., and Nagatsu, T., *Biochem. Med.* **10,** 320 (1974).
*149.* Kato, T., Wakui, Y., Nagatsu, T., and Ohnishi, T., *Biochem. Pharmacol.* **27,** 829 (1978).
*150.* Laduron, P., *Biochem. Pharmac.* **24,** 557 (1975).

*151.* O'Brien, R. A., Semenuk, G., Coleman, M., and Spector, S., *Clin. Exp. Pharmacol. Physiol.* **3,** 9 (1976).
*152.* Poitou, P., Assicot, M., and Bohuon, C., *Biomedicine* **21,** 91 (1974).
*153.* Walker, H. A., Danielson, E., and Levitt, M., *J. Autism Childhood Schiz.* **6,** 263 (1976).
*154.* Ebstein, R., Belmaker, R. H., Benbenisty, D., and Rimon, R., *Biol. Psychiat.* **11,** 613 (1976).
*155.* Briggs, M. H., and Briggs, M., *Experientia* **29,** 278 (1973).
*156.* Mattsson, B., Mjorndal, T., Oreland, L., and Perris, C., *Acta Psychiat. Scand. Suppl.* **255,** 187 (1974).
*157.* Dunner, D. L., Cohn, C. K., Gershon, E. S., and Goodwin, F. K., *Arch. Gen. Psychiat.* **25,** 348 (1971).
*158.* Dunner, D. L., Levitt, M., Kumbaraci, T., and Fieve, R. R., *Biol. Psychiat.* **12,** 237 (1977).
*159.* Blackwell, E. W., Briant, R. H., Conolly, M. E., Davies, D. S., and Dollery, C. T., *Brit. J. Pharmacol.* **50,** 587 (1974).
*160.* Weinshilboum, R. M., *Life Sci.* **22,** 625 (1978).
*161.* Coward, J. K., and Wu, F. Y., *Anal. Biochem.* **55,** 406 (1973).
*162.* Borchardt, R. T., *Anal. Biochem.* **58,** 382 (1974).
*163.* Axelrod, J., and Tomchick, R., *J. Biol. Chem.* **233,** 702 (1958).
*164.* Lin, R. -L., and Narasimhachari, N., *Anal. Biochem.* **57,** 46 (1974).
*165.* Gulliver, P. A., and Tipton, K. F., *Biochem. Pharmacol.* **27,** 773 (1978).
*166.* Jonas, W. Z., and Gershon, E. S., *Clin. Chim. Acta* **54,** 391 (1974).
*167.* Griffiths, J., and Linklater, H., *Clin. Chim. Acta* **39,** 383 (1972).
*168.* Hoberg, E., and Hempel, K., *Clin. Chim. Acta* **90,** 107 (1978).
*169.* Bade, P., Christ, W., and Rakow, D., *Nauyn-Schmiedebergs Arch. Pharmacol. Suppl.* **282,** R3 (1974).
*170.* Kissinger, P. T., Refshauge, C. J., Dreiling, R., and Adams, R. N., *Anal. Lett.* **6,** 465 (1973).
*171.* *Bibliography of Recent Reports on Electrochemical Detection,* Bioanalytical Systems, West Lafayette, Indiana.
*172.* Shoup, R. E., Bruntlett, C. S., Bratin, K. B., and Kissinger, P. T., *Principles and Practice of Liquid Chromatography with Electrochemical Detection,* BAS Press, West Lafayette, Indiana, 1980.
*173.* Kissinger, P. T., Riggin, R. M., Alcorn, A. L., and Rau, L. D., *Biochem. Med.* **13,** 299 (1975).
*174.* Riggin, R. M., and Kissinger, P. T., *Anal. Chem.* **49,** 2109 (1977).
*175.* Hansson, C., Agrup, G., Rorsman, H., Rosengren, A. M., Rosengren, E., and Edholm, L. E., *J. Chromatogr.* **162,** 7 (1979).
*176.* Moyer, T. P., Jiang, N. S., Tyce, G. M., and Sheps, S. G., *Clin. Chem.* **25,** 256 (1979).
*177.* Shoup, R. E., PhD Dissertation, Purdue University, 1980, p. 58.
*178.* Hallman, H., Farnebo, L. O., Hamberger, B., and Jonsson, G., *Life Sci.* **23,** 1049 (1978).
*179.* Allenmark, S., and Hedman, L., *J. Liq. Chromatogr.* **2,** 277 (1979).
*180.* Davis, G. C., PhD dissertation, Purdue University, 1980.
*181.* Riggin, R. M., and Kissinger, P. T., *Anal. Chem.* **49,** 2109 (1977).

182. Kissinger, P. T., *Anal. Chem.* **49,** 447A (1977).
183. Hjemdahl, P., Daleskog, M., and Kahan, T., *Life Sci.* **25,** 131 (1979).
184. Davis, G. C., Kissinger, P. T., and Shoup, R. E., *Anal. Chem.* **53,** (1981).
185. Kissinger, P. T., Felice, L. J., Miner, D. J., Preddy, C. R., and Shoup, R. E., "Detectors for Trace Organic Analysis by Liquid Chromatography: Principles and Applications," in *Contemporary Topics in Analytical and Clinical Chemistry,* Vol. 2, Hercules, D. M., Hieftje, G. M., Snyder, L. R., Evenson, M. A., eds., Plenum Press, New York, 1978, pp. 55–175.
186. Shoup, R. E., and Kissinger, P. T., *Clin. Chem.* **23,** 1268 (1977).
187. Mrochek, J. E., Dinsmore, S. R., and Ohrt, D. W., *Clin. Chem.* **19,** 927 (1973).
188. Yoshida, A., Yoshioka, M., Tanimura, T., and Tamura, Z., *J. Chromatogr.* **116,** 240 (1976).
189. Felice, L. J., and Kissinger, P. T., *Anal. Chem.* **48,** 794 (1976).
190. Felice, L. J., and Kissinger, P. T., *Clin. Chim. Acta* **76,** 317 (1977).
191. Molnár, I., and Horvath, C., *Clin. Chem.* **22,** 1497 (1976).
192. Kissinger, P. T., Bruntlett, C. S., Davis, G. C., Felice, L. J., Riggin, R. M., and Shoup, R. E., *Clin. Chem.* **23,** 1449 (1977).
193. Molnár, I., and Horvath, C., *Chromatographia* **11,** 260 (1978).
194. Mitchell, J., and Coscia, C. J., *J. Chromatogr.* **145,** 295 (1978).
195. Buchanan, D. N., Fucek, F. R., and Domino, E. F., *J. Chromatogr.* **162,** 394 (1979).
196. Felice, L. J., Bruntlett, C. S., Shoup, R. E., Kissinger, P. T., *Trace Organic Analysis,* N. B. S. Special Publ. **519,** 391 (1979).
197. Morrisey, J. L., and Shihabi, Z. K., *Clin. Chem.* **25,** 2043 (1979).
198. Morrisey, J. L., and Shihabi, Z. K., *Clin. Chem.* **25,** 2045 (1979).
199. Rosano, T. B., and Brown, H. H., *Clin. Chem.* **25,** 550 (1979).
200. Wightman, R. M., Plotsky, P. M., Strope, E., Delcore, R. Jr., and Adams, R. N., *Brain Res.* **131,** 345 (1977).
201. Davis, G. C., and Kissinger, P. T., *Anal. Chem.* **51,** 1960 (1979).
202. Horvath, C., Melander, W., and Molnar, I., *J. Chromatogr.* **125,** 129 (1976).
203. Shoup, R. E., Davis, G. C., and Kissinger, P. T., *Anal. Chem.* **52,** 483 (1980).
204. Blank, C. L., and Pike, R., *Life Sci.* **18,** 859 (1976).
205. Nagatsu, T., Oka, K., and Kato, T., *J. Chromatogr.* **163,** 247 (1979).
206. Christensen, H. D., and Blank, C. L., The Determination of Neurochemicals in Tissue Samples at Subpicomole Levels, in *Biological/Biomedical Applications of Liquid Chromatography, Vol. 3,* Hawks, G. L., ed., Dekker, New York, 1979.
207. Borchardt, R. T., Vincek, W. C., and Grunewald, G. L., *Anal. Biochem.* **82,** 149 (1977).
208. Borchardt, R. T., Hegazi, M. F., and Schowen, R. L., *J. Chromatogr.* **152,** 255 (1978).
209. Young, S. N., and Sourkes, T. L., *Adv. Neurochem.* **2,** 133 (1977).
210. Fernstrom, J. D., *Metabolism* **26,** 207 (1977).
211. Green, A. R., and Grahame-Smith, D. G., 5-Hydroxytryptamine and

Other Indoles in the Central Nervous System, in *Handbook of Psychopharmacology,* Iverson, L. L., Iverson, S. D., and Snyder, S. H., eds., Plenum Press, New York, NY, 1975, p. 169.

*212.* O'Brien, R. A., Cerebral Distribution of Serotonin, in *Serotonin in Mental Abnormalities,* Boullin, D. J., ed., Chichester, England, 1978, p. 41.

*213.* Saavedra, J. H., *Federation Proc.* **36,** 2134 (1977).

*214.* Reinhard, J. R., Jr., and Wurtman, R. J., *Life Sci.* **21,** 1741 (1977).

*215.* Lovenberg, W., Besselaar, G. H., Bensuiger, R. E., and Jackson, R. L., Physiologic and Drug-Induced Regulation of Serotonin Synthesis, in *Serotonin and Behavior,* Barchas, J., and Usdin, E., eds., Academic Press, New York, NY, 1973, p. 49.

*216.* Denckla, W. D., and Dewey, H. K., *J. Lab. Clin. Med.* **69,** 160 (1967).

*217.* Steinhart, H., *Anal. Chem.* **51,** 1012 (1979).

*218.* Bloxam, D. L., Hutson, P. H., and Curzon, G., *Anal. Biochem.* **83,** 130 (1977).

*219.* Hery, F., Rouer, E., and Glowinski, J., *Brain Res.* **43,** 445 (1972).

*220.* Gaudin-Chazal, G., Daszuta, A., Faudon, M., and Ternaux, J. P., *Brain Res.* **160,** 281 (1979).

*221.* Davis, T. P., Gehrke, C. W., Gehrke, C. W., Jr., Cunningham, T. D., Kuo, K. C., Gerhardt, K. O., Johnson, H. D., and Williams, C. H., *Clin. Chem.* **24,** 1317 (1978).

*222.* Marczynski, T. J., Serotonin and the Central Nervous System, in *Chemical Transmission in the Mammalian Central Nervous System,* Hockman, C. H., and Bieger, D., eds., University Park Press, Baltimore, MD, 1976, Ch. 7.

*223.* Woolley, D. W., *The Biochemical Basis of Psychoses,* Wiley, New York, NY, 1962, Ch. 4.

*224.* Goodwin, F. K., Post, R. M., Dunner, P. L., and Gordon, E. K., *Amer. J. Psychiat.* **130,** 73 (1973).

*225.* Van Praay, H. M., Krof, J., and Puite, J., *Nature* **225,** 1259 (1970).

*226.* Gayford, J. J., Parker, A. L., Phillips, E. M., and Raswell, A. R., *Br. J. Psychiat.* **122,** 597 (1973).

*227.* Yess, N., Price, J. M., Brown, R. R., and Swan, P. B., *J. Nutr.* **84,** 229 (1964).

*228.* Goldenburg, H., *Clin. Chem.* **19,** 38, (1973).

*229.* Erspamer, V., *Nature* **170** 281 (1952).

*230.* Undenfriend, S., Weissbach, H., and Clark, C. T., *J. Biol. Chem.* **215,** 337 (1955).

*231.* Maickel, R. P., Fluorometric Analysis of 5-Hydroxytryptamine and Related Compounds, in *Methods of Neurochemistry,* Vol. 2, Fried, R., ed., Dekker, New York, NY, 1972.

*232.* Korf, J., *Anal. Biochem.* **53,** 146 (1973).

*233.* Beck, O., Wiesal, F. A., and Sedvall, B., *J. Chromatogr.* **134,** 407 (1977).

*234.* Godse, D. D., Warsh, J. J., and Stance, H. C., *Anal. Chem.* **49,** 915 (1977).

*235.* Sedvall, G., Bjerkenstedt, L., Swahn, C. G., Wiesel, F. -A., Wode-Helgodt, B., *Adv. Biochem. Psychopharmacol.* **16,** 343 (1977).

*236.* Sjoquist, B., and Johansson, B., *J. Neurochem.* **31,** 621 (1978).
*237.* Barchas, J., Erdelyi, E., and Angwin, P., *Anal. Biochem.* **50,** 1 (1972).
*238.* Smith, J. E., Lane, J. D., Shea, P. A., and Aprison, M. A., *Anal. Biochem.* **64,** 149 (1975).
*239.* Kemerer, V. F., Lichtenfeld, K. M., and Koch, T. R., *Clin. Chim. Acta.* **92,** 81 (1979).
*240.* Frattini, P., Cucchi, H. L., Santagostino, G., and Corona, G. L., *Clin. Chim. Acta.* **92,** 353 (1979).
*241.* Guibault, G. G., and Froelich, D. M., *Clin. Chem.* **20,** 812 (1974).
*242.* Graffeo, A. P., and Karger, B. L., *Clin. Chem.* **22,** 184 (1976).
*243.* Fornstedt, N., *Anal. Chem.* **50,** 1342 (1978).
*244.* Yong, S., and Lan, S., *J. Chromatogr.* **175,** 343 (1979).
*245.* Riley, C. M., Tomlinson, E., Jeffries, T. M., and Redfern, P. H., *J. Chromatogr.* **162,** 153 (1979).
*246.* Krstulovic, A. M., and Matzura, C., *J. Chromatogr.* **163,** 72 (1979).
*247.* Anderson, G. M., and Purdy, W. C., *Anal. Chem.* **51,** 283 (1979).
*248.* Gehrke, C. W., Gehrke, C. W., Jr., Cunningham, T. D., Kuo, K. C., Gerhardt, K. O., Johnson, H. D., and Williams, C. H., *J. Chromagtogr.* **162,** 293 (1979).
*249.* Krstulovic, A. M., and Powell, A. M., *J. Chromatogr.* **171,** 345 (1979).
*250.* Garnier, J. P., Bousquet, B., and Droux, C., *J. Liq. Chromatogr.* **2,** 539 (1979).
*251.* Koch, D. D., and Kissinger, P. T., *J. Chromatogr.* **164,** 441 (1979).
*252.* Koch, D. D., and Kissinger, P. T., *Anal. Chem.* **52,** 27 (1979).
*253.* Koch, D. D., and Kissinger, P. T., *Life Sci.* **26,** 1099 (1980).
*254.* Richards, D. A., *J. Chromatogr.* **175,** 293 (1979).
*255.* Sasa, S., Blank, L., Wenke, D. C., and Sczupak, A., *Clin. Chem.* **24,** 1509 (1978).
*256.* Frattini, P., Cucchi, M. L., Santagostino, G., and Corona, G. L., *Clin. Chim. Acta* **92,** 353 (1979).
*257.* Joseph, M. H., and Baker, H. F., *Clin. Chim. Acta* **72,** 125 (1976).
*258.* Rice, J. R., and Kissinger, P. T., manuscript in preparation.
*259.* Davis, G. C., Poon, M. J., and Kissinger, P. T., manuscript in preparation.
*260.* King, W. P., and Kissinger, P. T., *Clin. Chem.* **26,** 1484 (1980).

# Chapter 13

## Steroids

Felix J. Frey, Brigitte M. Frey, and Leslie Z. Benet

*Division of Clinical Pharmacology*
*Department of Medicine and Department of Pharmacy*
*University of California*
*San Francisco, California*

### I. Introduction

If a radioimmunoassay, a protein binding method, or a colorimetric assay for the assessment of a steroid level is replaced by high performance liquid chromatography (HPLC), the cost for the determination of a steroid level increases at least initially because one must acquire the new HPLC equipment. Therefore, if an older method provides the same results as the new, "advanced" HPLC method, the only advantage resulting from the introduction of a high performance chromatographic assay is that gained by the manufacturer in terms of greater sales. Thus, justification for the assessment of steroids by HPLC is only obtained if the quality and/or quantity of information gained is significantly increased as compared to that provided by the conventional methods. But this evidential relation, that more and better information justifies a higher price in any case, is no longer true in health care, with the birth some years ago of the categoric imperative for the reduction of costs in the medical sector. That is, each new technology introduced for health maintenance should demonstrate at least a stabilizing impact on total medical expenditures. Therefore, after reviewing the presently available HPLC methods for the clinically important steroids, we will consider whether HPLC analyses for these steroids can be recommended without violating this vox populi.

## II. Glucocorticoids

Of all the steroids determined for clinical purposes, the glucocorticoids are the most often requested. The determination of cortisol or its metabolites is indispensable for the elucidation of disease states such as hypocorticism, hypercorticism, and congenital adrenal hyperplasia (*1, 2*). Recently the cortisol/cortisone ratio in the amniotic fluid was identified as an important predictor of a respiratory distress syndrome, since the capacity of the fetal lung to convert cortisone to cortisol increases with advancing gestational age (*3*). Furthermore, in the last few years, interest in the assessment of exogenous and endogenous glucocorticoids increased when it was shown that the absorption and metabolism of exogenous steroids and the secretion of endogenous glucocorticoids may vary considerably from person to person in patients taking exogenous steroids, and that this variability may be predictive of steroid efficacy and side effects (*4–7*).

Endogenous glucocorticoid production is usually estimated by colorimetric reactions of urinary steroid metabolites, assessing either the 17-hydroxycorticoids, the 17-ketosteroids, or the 17-ketogenic steroids (*8–10*). All these methods are deficient, however, in that they either do not measure all glucocorticoid metabolites, and/or they measure steroids not related to the endogenous glucocorticoid pathway. It seems, therefore, obvious—although not proven for all situations—that the specific and sensitive quantification of particular steroids or metabolites in urine or plasma may be of greater diagnostic value (*1*).

Older techniques for the assessment of cortisol are not completely specific. The presence of corticosterone, deoxycorticosterone, and certain drugs limits the use of fluorometric assays (*11*). The detection limit for RIA, approximately 1 ng/mL, is lower than that obtained with HPLC, which is in the order of 5–10 ng/mL. However, RIA antibodies exhibit various degrees of cross-reactivity with other endogenous steroids such as cortisone, 11-deoxycortisone, 17-hydroxyprogesterone, corticosterone, and deoxycorticosterone (*12–14*). Furthermore, the commercially available antibodies cannot distinguish between endogenous and exogenous glucocorticoids. Similar cross-reactivity is reported for the competitive protein binding techniques (*12, 15*). This cross-reactivity, which may be unimportant in many clinical situations, limits the value of results obtained by RIA or protein binding methods in neonates, pregnant women, patients with adrenal hyperplasia, patients receiving metyrapone, and in most situations where exogenous glucocorticoids have been administered. In these situations, a gas chromatographic–mass spectrometric method offers both increased sensitivity and specificity, but the costly equipment and the complicated preparation of the samples make the clinical use of the GC–MS less attractive (*16*). HPLC methods are less expensive, quite specific, and sufficiently sensitive for the assessment of endogenous and exogenous glucocorticoids.

In the first high performance liquid chromatographic techniques reported, cortisol was not separated from cortisone or cortisone from the solvent front of plasma extracts (*17–20*). Interference by exogenous glucocorticoids also limited the application of one such assay (*18*), and the solvent mixture of a second procedure is poorly miscible (*17*). In order to overcome the above-mentioned deficiencies, we developed a new HPLC assay capable of separating cortisol, cortisone, dexamethasone, prednisone, prednisolone, and methyl prednisolone simultaneously (*21*).

The steroids are extracted from 1 mL of plasma with methylene chloride/ether, washed with acid and base, and separated isocratically on a normal-phase silica column with a mobile phase consisting of methylene chloride/tetrahydrofuran/methanol/glacial acetic acid (96.5/1/2.1/0.05 by volume). For retention times to be reproducible, the water content of the column is maintained constant by water-saturating 300 mL/L of the methylene chloride used in the solvent system. The presence of the $\alpha,\beta$-unsaturated ketone with an absorption maximum at 250 nm allows one to quantify concentrations of the steroids with a UV detector. Retention times range from 6 to 20 min (Figs. 1A and 1B). In contrast to other steroid assays for which the range of linearity studied was relatively limited (*17–19*), use of the above-mentioned assay with measurement quantified on a double pen recorder showed that 1000-fold concentrations in plasma could be measured without dilution of the sample. The latter is a prerequisite especially for pharmacokinetic studies, where concentrations may range up to several thousand micrograms per liter after intravenous administration. The lower detection limit for all the analyzed steroids is 10 ng/mL. The intraday variability is 1–10% and the interday variability, 3–11%. Of 26 drugs and 20 steroids tested, only theophylline presents an interference problem (*21*).

In our opinion, the greatest advantage of this HPLC method compared to RIA and/or protein binding assays is that exogenous and endogenous steroids may be measured simultaneously and specifically.

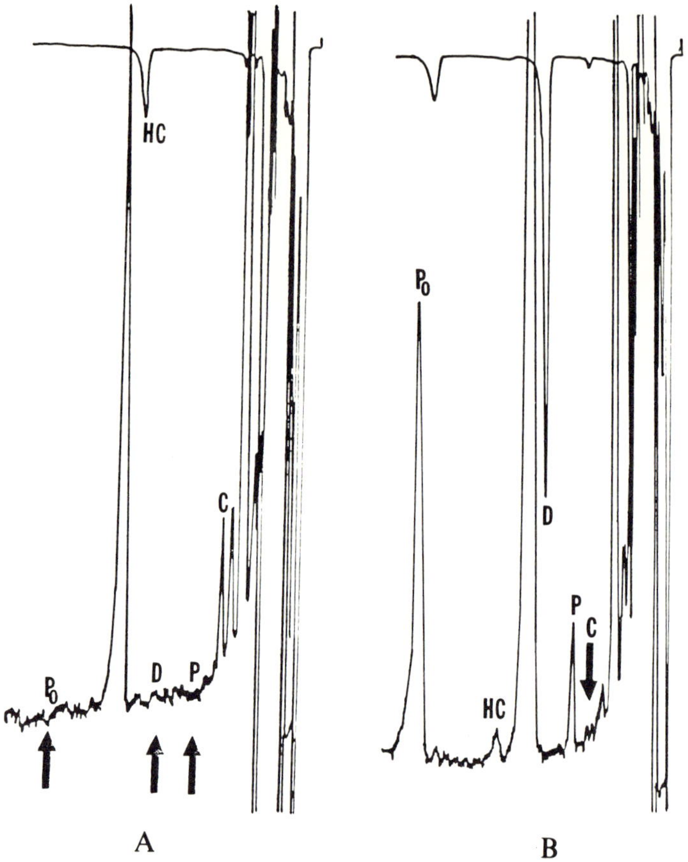


FIG. 1. A: Dual pen recording of chromatogram for blank human plasma extract. $P_0$ = prednisolone, HC = cortisol, D = dexamethasone, P = prednisone, C = cortisone. The attenuation of the upper pen is 10 times that for the lower pen recording. Addition of internal standard to this sample indicated that concentrations of HC and C were 204 and 17 ng/mL, respectively (*21*). B: Dual pen recording of chromatogram for plasma sample from the same subject. This sample was obtained 7 h after a 50-mg oral prednisone dose. $P_0$ = 265 ng/mL, P = 25.7 ng/mL, HC $<$ 10 ng/mL. The attenuation of the upper pen is 10 times that for the lower pen recording (*21*). Reproduced with permission of the publisher.

This allows the investigator to examine the interrelationship between exogenous and endogenous glucocorticoids, as demonstrated in Fig. 2 for a patient receiving daily 15-mg oral doses of prednisone in the treatment of Erythema multiforme. The data plotted in Fig. 2 (which is representative of other patients we have studied) clearly demonstrates

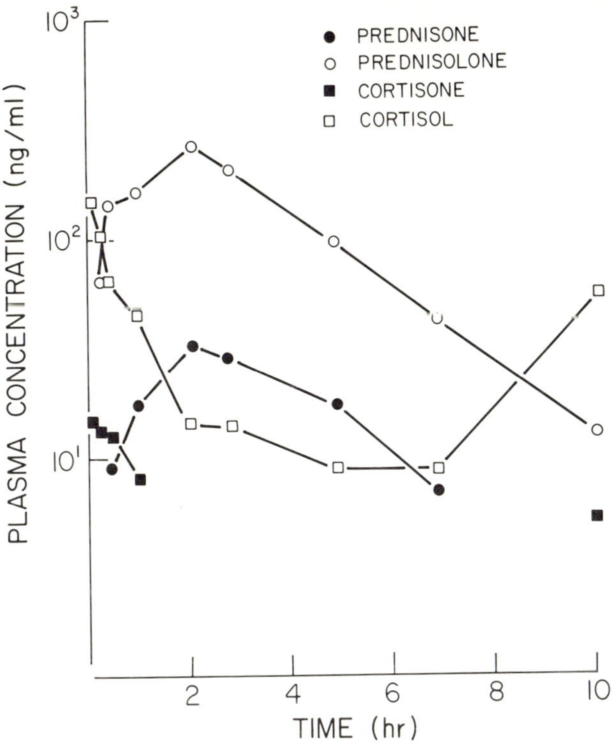


FIG. 2. Semilogarithmic plot of plasma concentrations of glucocorticoids in a patient under chronic prednisone treatment (15 mg/d) after an oral dose of prednisone.

the release of endogenous cortisol and cortisone, while prednisone and prednisolone clearance measurable in plasma (*22*). Measurements of prednisolone clearance and volume of distribution, the availability of prednisolone following oral prednisone doses, and the levels of endogenous cortisol and cortisone appear to account at least partially for the variability and occurrence of cushingoid side effects (*6*).

## III. Aldosterone

The assessment of aldosterone levels is critical for the diagnostic workup of hyper- or hypomineralocorticoid disease states such as primary and secondary hyperaldosteronism, Bartter's syndrome, hyporeninemic hypoaldosteronism, Addison's disease, and certain forms of congenital adrenal hyperplasia (*2, 23–25*). In contrast to these diseases, however, the pathogenetic importance of aldosterone in essential hypertension is still a matter of debate (*26*). The lack of knowledge concerning the importance of mineralocorticoids in essential hypertension owes mostly to the complex interrelationship between body fluids, aldosterone, and other hormonal systems. However, part of the uncertainty may be attributed to methodological problems with the aldosterone assay, since even the experienced investigators who developed the aldosterone RIA report a day-to-day variability of more than 14% (*23, 27–29*). Further, the RIA antibodies used exhibit significant and variable cross-reactivity with the primary metabolite, tetrahydroaldosterone, which appears in the urine at a concentration that is about five times higher than the concentration of the 18-aldosterone glucuronide. A series of thin-layer chomotographic purification steps have recently been proposed to improve the specificity of the antibodies (*29*). However, the improved specificity does not resolve one problem inherent to all urinary aldosterone RIA measurements. That is, assessed by a specific RIA method, the 18-aldosterone glucuronide represents only about 7–15% of the total endogenous aldosterone production. If measurement of this metabolite is taken as a reflection of actual plasma aldosterone levels, one must assume a constant ratio between this glucuronide and all other metabolites. In this case, an HPLC method that measures more than one steroid specifically, as we have demonstrated for glucocorticoids (*22*), would be helpful.

Until recently, little development of HPLC methods for the quantification of aldosterone had been undertaken. Since aldosterone plasma concentrations are of the order of 0.1–0.5 ng/mL in normal individuals on a daily sodium and potassium intake of about 100 mEq and 60–100 mEq, respectively, when plasma samples are obtained in the supine position, HPLC is not suitable unless a prior derivatization is carried out to enhance sensitivity. In the urine, the concentrations of 18-aldosterone glucuronide are much higher, since approximately 2–10 μg of this metabolite are excreted per day. These concentrations can be assessed by HPLC, as described by DeVries and coworkers (*30*).

The urine is adjusted to pH 1 with sulfuric acid, and the hydrolysis reaction is allowed to proceed at room temperature for 24 h. The hydrolyzed steroids are extracted with methylene chloride three times. This organic layer is washed once with 1% and once with 0.1% sodium hydroxide, and twice with a sodium chloride saturated aqueous solution. The organic layer is then dried over anhydrous sodium sulfate, filtered, and evaporated to dryness. The residue is dissolved in 70% methanol, washed three times with toluene, and evaporated to dryness. The residue is then purified by TLC using three different solvent systems. Prednisolone is added as an internal standard and quantification at 254 nm is carried out on HPLC using a 10-$\mu$ silica column and a solvent system consisting of chloroform and methanol (98.5/1.5). A recovery of 63% is reported (*30*). To correct for the loss of material occurring during the thin layer separation steps, aldosterone standard was added to a part of each urine extract (*30*). However, no internal standard was used in the initial hydrolysis and extraction steps, which may explain in part the high coefficient of variation (16%) observed. The reported detection limit of about 1.5 ng/mL urine would make the assay suitable for clinical purposes, but the extraction and purification procedure is somewhat too complicated.

The purification could possibly be shortened by the use of a $\beta$-glucuronidase (EC 3.2.1.31), which hydrolyzes most of the urinary glucuronides, but not 18-aldosterone glucuronide. The hydrolyzed, water-insoluble steriods could then be extracted with methylene chloride prior to the acid hydrolysis of the 18-aldosterone glucuronide (*31*). It should be possible to measure tetrahydroaldosterone using HPLC. The simultaneous detection of this metabolite and the hydrolyzed parent compound, however, would require two detectors set at different wavelengths, since tetrahydroaldosterone does not contain the $\alpha,\beta$-unsaturated ketone of aldosterone that provides the absorption maximum at 254 nm.

## IV. Estrogens

Many obstetricians believe that the measurement of estrogens is of some value in the assessment of fetal well-being. In this regard, the measurement of estriols is superior to estradiols (*32*). However, it is unclear whether conjugated, unconjugated, or total estriol in plasma or in urine is the best parameter for monitoring fetal–placental function (*32, 33*). Estriol determination is mandatory for the diagnosis of adrenal feminization. In the future, estrogen determination may prove to be relevant for the dynamic testing of hypothalamic–pituitary function in patients with sterility (*34*).

In nonpregnant women, plasma levels of estradiol are less than 0.2 ng/mL and estriol less than 0.4 ng/mL. An assay useful for placental monitoring would require accurate measurement of unconjugated estriol plasma levels as low as 4 ng/mL, since 100% fetal deaths were reported when estriol concentrations in plasma fell below this level after the 32nd week of gestation (*33*). In contrast, urinary estrogens in pregnant women may be measured in the milligram range. Thus, most investigators have tried initially to analyze estrogen concentrations in the urine.

Estrogens have two absorption maxima: 220 and 280 nm. At 280 nm at least 20 ng of estrogen is required to give a measurable UV signal (*36*), an amount that is much higher than what is usually present in 1 mL of plasma. Therefore, estrogen measurements are still carried out using modifications of a nonspecific colorimetric reaction first described by Kober in 1931 for measuring estrogens in urine (*37*). More recently, radioimmunoassays became available for determination of both unconjugated and total plasma estriol (*35, 38*). The antibodies for unconjugated estriol are reported to have only minor cross-reactivity with other steroids (*38*). Gas–liquid chromatographic procedures have also been described for the determination of urinary placental estriol (*39, 40*). These time-consuming procedures require derivatization, and interference by other steroids demands a more complicated extraction procedure than is necessary when HPLC techniques are used.

Dolphin described the first HPLC assay for estrogen determination (*36, 42*). However, this method is not suitable for routine clinical work, since it lacks an internal standard and the recovery reported is only 27%. Gotelli et al. introduced a new assay (*41*) that uses carbamezepine as the internal standard and reported an analytical recovery of 80% even though several steps precede the final separation. Briefly, the method is as follows: Estrogens, excreted in urine as sulfate and glucuronide conjugates, are hydrolyzed using $\beta$-glucuronidase and acid. The solution is saturated with sodium carbonate and sodium chloride and estriol extracted into a mixture of diethyletherpetroleum ether (60/40 by volume). This separation step is somehow critical, since a pH higher than 10 in the aqueous phase leads to a loss of more than 60% of the estrogens and a low pH increases the amount of unwanted acidic components in the organic layer (*43*). The organic phase is evaporated to dryness and reconstituted in methanol.

The estriol is chromatographed isocratically on a $\mu$ Bondapak $C_{18}$ column using a mobile phase consisting of an acetonitrile/phosphate buffer (pH 4.4) mixture (22/78 by volume) at a flow rate of 3 mL/min (*41*). The steroid is detected at 280 nm. The assay is linear from 5 to 100 mg/L of estriol. The inter- and intraday variabilities are reported to be low (< 3.7%). The results obtained are comparable with those obtained using gas–liquid chromatography (*40*). The method does not distinguish 16-epiestriol and carbamazepine nor 16,17-epiestriol and estriol. This interference between the epiestriols and the internal standard as well as unchanged estriol is inconsequential when the assay is used for routine clinical analysis; however, the interference may be disturbing when one is carrying out more accurate metabolic studies.

There is no study available demonstrating a clear clinical advantage for any one of the three methods currently used for estrogen determination, i.e., nonspecific colorimetry, RIA, and HPLC. The advantage of the RIA method is the sensitivity obtained, allowing the detection of estrogen in the plasma of nonpregnant women. A method combining the sensitivity of RIA and the specificity of HPLC would be helpful in investigating certain clinical problems. It may also be possible to improve the sensitivity of the HPLC method by dansylating the estrogens, as has been done in a method described for measurements of estrogens in pharmaceutical preparations (*44*). Dansylated estrogens may be measured by fluorescent spectrophotometry at concentrations as low as 0.04 ng/mL. Such a method would be clinically useful, allowing the clinician to obtain specific plasma concentration measurements. A preliminary report indicates that the use of native fluorescence of the estrogens may simplify the detection of estrogens (*74*). Using excitation and emission wavelengths of 220 and 608 nm, respectively, was said to detect as little as 0.1 ng/mL estriol (*74*). Such methods would be clinically useful, allowing the clinician to obtain specific plasma concentration measurements.

## V. Vitamin D

Vitamin $D_2$, ergocalciferol, and vitamin $D_3$, cholecalciferol, must be hydroxylated to become biologically active (*45*). Hydroxylation at the 25 position occurs in the liver, while hydroxylation at positions 1 and 24 occurs in the kidney (*45*). Disease states affecting the liver, such as cirrhosis, have been found to be associated with low plasma concentrations of the 25-hydroxymetabolites, while in patients with chronic renal failure the total concentrations of 1,25- and 24,25-dihydroxy vitamin D are low (*45, 46*). There is some evidence that the 1,25-metabolites are reduced in post-menopausal women, in patients with hypoparathyroidism and in the presence of excess glucocorticoids, while during chronic anticonvulsant and glucocorticoid therapy, 25-hydroxylated vitamin D was claimed to be low (*47–50*).

The characteristic *cis*-triene ultraviolet spectrum of the vitamin D metabolites, with an absorption maximum at 265 nm, is a favorable precondition for HPLC analysis (*51, 54*). Plasma concentrations of unmetabolized vitamin $D_2$ (~ 1.2 ng/mL), $D_3$ (~ 2.3 ng/mL), 25-hydroxy vitamin $D_2$ (~ 3.9 ng/mL), and 25-hydroxy vitamin $D_3$ (~ 27 ng/mL) have been reported to be sufficiently high for assessment by HPLC using a UV detector at 254 nm (*51–53*). The concentrations of total 1,25-dihydroxy vitamin D (~ 0.04 ng/mL) and total 24,25-hydroxy vitamin D (~ 3.5 ng/mL), however, are too low to be quantitated by their UV absorbance (*53*). For these latter metabolites, competitive protein-binding assays are required (*52, 53*). A normal-phase HPLC procedure (*52*) is used to isolate 1,25-dihydroxy vitamin D before the protein-binding assay is carried out.

The 25-hydroxyl metabolites can be assessed directly by means of an HPLC assay separately as $D_2$ and $D_3$ molecules (*51, 52*). The $D_2$ and $D_3$ molecules differ by only one double bond in the side chain. The procedure for the separation of these two 25-hydroxy vitamin D metabolites includes three steps: First, a lipid extraction with chloroform/methanol is performed on 2 mL of plasma spiked with [$^3$H]25-hydroxy vitamin $D_3$ (*51*). Following that extraction, a prepurification of the lipids is necessary. For this purpose, three different fractionation procedures were investigated to separate the parent compounds from their metabolites. A liquid chromatographic method using an isopropanol/hexane solvent system on a Zorbax-SIL column was found to be superior to an "open column" system employing either Sephadex LH 20 (3 or 16 g) or HAPS column chromatography. Retention times on the Zorbax-SIL column were shorter than 15 min. In the third step, 25-hydroxy vitamin $D_2$ is separated from 25-hydroxy vitamin $D_3$ in a reversed phase HPLC assay using a Zorbax-ODS column and a methanol/water solvent system (91/9) (*51*). Using a slightly different solvent system for the third step, i.e., 98.5% methanol and 1.5% water, a separation of the parent $D_2$ and $D_3$ compounds was claimed to be possible (*51*). However, for both the parent $D_2$ and $D_3$ compounds, and the above mentioned 25-hydroxy vitamin $D_2$ and $D_3$, base line separation was not achieved (*51*). Detection limits and possible assay interference by drugs are not reported. The day-to-day variability was about 16% (*51*). A comparison of measurements of total 25-hydroxy vitamin D, assessed using a protein binding assay (which cannot separate 25-hydroxy vitamin $D_2$ and 25-hydroxy vitamin $D_3$), with HPLC measurements indicated a good correlation for total metabolites (*51, 55*).

An extension and improvement of this method was recently described in which the parent compounds $D_2$ and $D_3$ and the metabolites 25-hydroxy vitamin $D_2$ and $D_3$ are measured in 5 mL of plasma directly by HPLC, while total 24,25-dihydroxy vitamin D and total 1,25-dihydroxy vitamin D are assessed using a competitive protein-binding assay following separation by HPLC (*53*). The

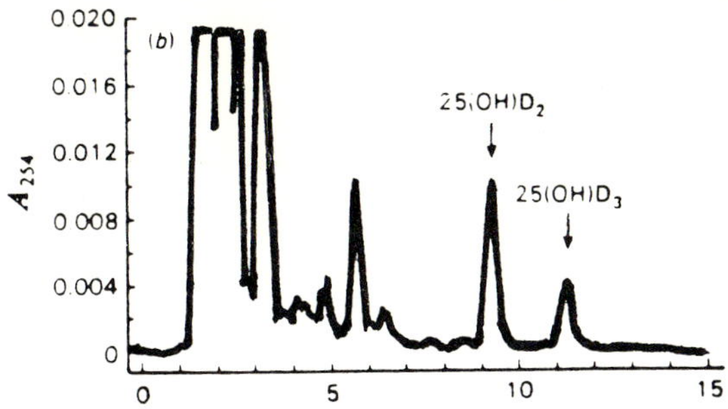


FIG. 3. HPLC with UV detection at 254 nm of $25(OH)D_2$ and $25(OH)D_3$ isolated from human plasma (*53*). Reproduced with permission of the publisher.

method requires batch elution chromatography on Sephadex LH-20 and Lipidex 5000 followed by one or two high performance liquid chromatographic separations. With this method, the 25-hydroxy metabolites are completely separated (Fig. 3) (*53*). However, vitamin $D_2$ and $D_3$ are not separated completely and the concentrations that are claimed to be measurable directly using a UV detector may be too low to be assessed by this method. The published standard curves (*53*) depict only levels above 10 ng, although detection was claimed to be accurate at lower concentrations. The intra-assay coefficients of variation were 8–17% and the inter-assay coefficients of variation were 10–26% (*53*). Potential interference by drugs was not tested (*53*).

## VI. Bile Acids

The primary bile acids, cholic acid and chenodeoxycholic acid, are formed in the liver (*56*). These primary acids are transformed by anaerobic intestinal bacteria into the secondary bile acids, deoxycholic acid and lithocholic acid. Ninety-five percent of the bile acids that pass down the small intestine are absorbed through the terminal ileum (*57*). Some of the secondary bile acids are then further transformed by the liver into tertiary bile acids (*56*). All bile acids are present in the body either as conjugates of glycine or taurine or in their unconjugated form. Conjugation is carried out in the liver, while deconjugation is carried out by intestinal bacteria (*56*). Gastroenterologists attempt to use the conjugation–unconjugation processes of the bile acids in their diagnostic workup of both liver and intestinal diseases (*58, 59*). For instance, in severe acute hepatitis and cirrhosis, total fasting bile acid levels in serum are elevated with a predominance of chenodeoxycholic acid, while in obstructive liver disease, cholic acid usually is the major component (*60*). Patients with a stagnant loop syndrome may have abnormally high serum values of unconjugated bile acids and those with regional ileitis absorb less bile acids than normal (*61*).

For most clinical situations, however, it has not yet been established which bile acid has the greatest diagnostic discriminatory value; it is also not known whether measurement of bile acids adds anything new to the conventional laboratory repertoire (*58, 59, 62*). To answer such questions, assays would have to be available to assess 12 steroids at concentrations lower than 1 $\mu$g/mL (*63*), if, indeed, one considers only the primary and secondary conjugated and unconjugated bile acids.

In a recent preliminary report, separation of the taurine and glycine conjugates of cholic acid, deoxycholic acid, chenodeoxycholic acid, and lithocholic acid was described (*64*). The solvent system consisted of 0.0005 *M* tetrabutylammonium in methanol/$H_2O$/isopropanol (65/31/4) at pH 4. A $C_{18}$ column was used and the steroids were detected at 210 nm. A lower detection limit of 250 ng was reported. The four abovementioned bile acids, in the unconjugated form, can be separated on a similar reversed phase system, but without ion pairing, in a solvent system consisting of variable proportions of 2-propanol and 10 m*M* potassium phosphate buffer at pH 7.0 (*65*). Measuring the bile acids at 250 nm instead of 210 nm improves the sensitivity of the HPLC assay (*66*). Therefore, esterification of the carboxyl group with *p*-nitrobenzyl has been proposed (*67*). In this method, bile acids are separated into glycine- and taurine-conjugated fractions using an Amberlite XAD-s column. The free and glycine-conjugated bile acids are esterified directly, while the taurine-conjugates must be hydrolyzed prior to the esterification. The *p*-nitrobenzyl esters are separated on a $C_{18}$ column using a methanol and $KH_2PO_4$ gradient. Unfortunately, the separation of cholic and chenodeoxycholic acid is incomplete and retention times as long as three hours are observed. Obviously, this method must be improved before it would be generally useful.

The application of HPLC in routine clinical work is sparse and fragmentary, with most assays confined to analysis of bile (*65–68*). Therefore, one cannot recommend HPLC as the method of choice for clinical laboratories until studies have been performed comparing HPLC with the other available methods for bile acid analysis, i.e., gas-liquid chromatography (*69*), capillary gas-liquid chromatography (*70*), mass fragmentography (*71*), enzymatic methods with fluorometric adaptation (*72*), and RIA (*73*).

## VII. Conclusions

At the present time, there are HPLC methods available for the determination of estriol, aldosterone, and cortisol in urine, and for cortisol, cortisone, and exogenous glucocorticoids in plasma. These assays are more specific than conventional methods. However, only the glucocorticoid assays provide more information, in that several steroids may be measured simultaneously using HPLC. Whether or not one may obtain data of clinical importance by HPLC that is qualitatively and/or quantitatively better than that presently available for steroids must still be proven. Once the bile acid and the vitamin D are completely developed and, if possible, simplified, these methods may both give better qualitative and quantitative results than are presently available. However, it is not as yet clear whether the assessment of these levels is important in practical medicine. Therefore, we believe that all the assays discussed here may only be recommended at present as investigational methods, and must await further validation and study before they are appropriate for everyday clinical analyses in laboratories not already possessing HPLC equipment.

## Acknowledgments

Brigitte and Felix Frey were supported during the course of this work by the Swiss National Foundation for the Scientific Research and the Swiss Foundation for Bio-Medical Research. Studies in the authors' laboratory were supported by NIH Center Grant GM26691.

## References

1. Eddy, R. L., Jones, A. L., Gilliand, P. F., Ibarra, J. D., Jr., Thompson, J. Q., and McMurry, J. F., Jr., *Am. J. Med.* **55,** 621 (1973).
2. Finkelstein, M., and Shaefer, J. M., *Physiol. Rev.* **59,** 353 (1979).
3. Smith, B. T., Worthington, D., and Maloney, A. H. A., *Obstet. Gynecol.* **49,** 527 (1977).
4. Sullivan, T. H., Hallmark, M. R., Sakmar, E., Weidler, D. J., Earhart, R. H., and Wagner, J. G., *J. Pharmacokinet. Biopharm.* **4,** 157 (1976).
5. Wilson, C. G., Ssendagire, R., May, C. S., and Paterson, J. W., *Br. J. Clin. Pharmacol.* **2,** 321 (1975).

6. Frey, F. J., Frey, B. M., Amend, W. J. C., Gambertoglio, J. G., and Benet, L. Z., *Clin. Pharmacol. Ther.* **27,** 255 (1980).
7. Frey, B. M., Frey, F. J., Benet, L. Z., and Cochram, K. C., *Int. J. Immunopharmacology* **2,** 1980 (in press).
8. Porter, C. C., and Silver, R. H., *Steroids* **14,** 705 (1969).
9. Rutherford, E. R., and Nelson, D. H., *J. Clin. Endocrinol. Metab.* **23,** 533 (1963).
10. Fieser, L., and Fieser, M., "Bestimmung von Ketosteroiden," in *Steroide,* Verlag Chemie, Weinheim/Bergstrasse, 1961, pp. 570, 571.
11. Goldzieher, J. W., and Besch, P. K., *Anal. Chem.* **30,** 962 (1958).
12. Antoniades, H. M., *Hormones in Human Blood,* Harvard University Press, Cambridge, MA, 1976.
13. Vecsei, P., Penki, B., and Kalty, R., *Experientia* **28,** 1, 104 (1972).
14. Gomez-Sanchez, C., *J. Lab. Clin. Med.* **89,** 105 (1978).
15. Murphy, B. E. P., *J. Clin. Endocrinol. Metab.* **27,** 973 (1967).
16. Brooks, C. J. W., Henderson, W., and Steel, G., *Biochim. Biophys. Acta* **296,** 431 (1973).
17. Loo, J. C. K., Butterfield, A. G., Moffatt, J., and Jordan, M., *J. Chromatogr.* **143,** 275 (1977).
18. Kabra, P. M., Tsai, L. L., and Marton, L. J., *Clin. Chem.* **25,** 1293 (1979).
19. Reardon, G. E., Cardarella, A. M., and Canalis, E., *Clin. Chem.* **25,** 122 (1979).
20. Rose, J. Q., and Jusko, W. J., *J. Chromatogr.* **162,** 273 (1979).
21. Frey, F. J., Frey, B. M., and Benet, L. Z., *Clin. Chem.* **25,** 1944 (1979).
22. Frey, F. J., Frey, B. M., and Benet, L. Z., *Abstr. APhA.,* Acad. Pharmaceut. Sci., November (1979), p. 132.
23. Vetter, W., Vetter, H., and Siegenthaler, W., *Acta Endocrinol.* **74,** 548 (1973).
24. Weidmann, P., Reinhart, R., Maxwell, M. H., Rowe, P., Coburn, J. W., and Nassry, S. G., *J. Clin. Endocrinol. Metab.* **36,** 965 (1973).
25. Bartter, F. C., Pronove, P., Gill, J. R., and MacCardle, R. C., *Am. J. Med.* **33,** 811 (1962).
26. Mitchell, J. R., Taylor, A. A., and Bartter, F. C., *Ann. Intern. Med.* **87,** 596 (1977).
27. Bizollon, C. A., Riviere, J. F., Franchimont, P., Faure, A., and Claustrat, B., *Steroids.* **23,** 809 (1974).
28. Pham-Huu-Trung, Marie, T., and Corrol, P., *Steroids* **24,** 587 (1974).
29. Herkner, K., Nowotny, P., and Waldhausee, W., *J. Chromatogr.* **146,** 273 (1978).
30. DeVries, C. P., Popp-Snijdeos, C., DeKieriet, W., and Akkerman-Faber, A. C., *J. Chromatogr.* **143,** 624 (1977).
31. Craef, V., Furuya, E., and Nishikaze, O., *Clin. Chem.* **25,** 141 (1979).
32. Allen, E. J., Lachelin, G. C. L., *Br. J. Obstet. Gynecol.* **85,** 278 (1978).
33. Kirkish, L. S., Barclay, M. L., Parra, B. P., Compton, A. A., and McCann, D. S., *Clin. Chem.* **24,** 1830 (1978).
34. Jones, G. S., Wentz, A. C., and Rosenwaks, Z., *Am. J. Obstet. Gynecol.* **129,** 760 (1977).

35. Tulchinsky, D., and Abraham, G. L., *J. Clin. Endocrinol.* **33,** 775 (1971).
36. Dolphin, R. J., *J. Chromatogr.* **83,** 421 (1973).
37. Kober, S., *Biochem. Z.* **239,** 209 (1931).
38. Thompson, J., and Haven, G., *Am. J. Clin. Pathol.* **68,** 474 (1977).
39. Adessi, G., Eichenberger, D., Goutte, C., and Jayle, M. F., *Clin. Chim. Acta* **45,** 369 (1973).
40. Gotelli, G. R., Kabra, P. M., and Marton, L. J., *Clin. Chem.* **23,** 165 (1977).
41. Gotelli, G. R., Wall, J. H., Kabra, P. M., and Marton, L. J., *Clin. Chem.* **24,** 2132 (1978).
42. Dolphin, R. L., and Pergande, P. J., *J. Chromatogr. Biomed. Appl.* **143,** 267 (1977).
43. Huber, J. F. K., Hulsman, J. A. R. J., and Meijeos, C. A. M., *J. Chromatogr.* **62,** 79 (1971).
44. Roos, R. W., *J. Pharm. Sci.* **67,** 1735 (1978).
45. DeLuca, H., *Arch. Intern. Med.* **138,** 836 (1978).
46. Jmawari, M., Akanamu, Y., Itakura, H., Muto, Y., Kosaka, K., and Goodman, D. W., *J. Lab. Clin. Med.* **93,** 171 (1979).
47. Chesney, R. W., Mazess, R. B., Hamstra, A. J., and DeLuca, H. F., *Lancet* **ii,** 1, 23 (1978).
48. Davies, M., Taylor, C., and Gill, L., *Lancet* **1,** 55 (1977).
49. Fleischman, A. R., Rosen, H. F., and Nathenson, F., *Arch. Intern. Med.* **138,** 869 (1978).
50. Kooh, S. W., Frazer, D., DeLuca, H. F., and Murray, T. M., *N. Engl. J. Med.* **293,** 840 (1975).
51. Jones, G., *Clin. Chem.* **24,** 287 (1978).
52. Lambert, P. W., Syverson, B. J., Arnaud, C. D., and Spelsberg, T. C., *J. Steroid Biochem.* **8,** 929 (1977).
53. Richard, R. M., Shepard, R. M., Horst, R. L., Hamstra, A. J., and DeLuca, H. F., *Biochem. J.* **182,** 55 (1979).
54. Jones, G., and DeLuca, H. F., *J. Lipid Res.* **16,** 448 (1975).
55. Belsey, R., DeLuca, H. F., and Pows, J. T., Jr., *J. Clin. Endocrinol.* **33,** 554 (1971).
56. Bergstrom, S., and Danielsson, H., "Formation and Metabolism of Bile Acids," in *Handbook of Physiology,* Code, C. F., ed., American Physiological Society, Washington, D.C., 1968, 2391–2407.
57. Hofmann, A. F., *Clin. Gastroenterol.* **6,** 3 (1977).
58. Javitt, N. B., *Clin. Gastroenterol.* **6,** 219 (1977).
59. Matern, S., and Gerok, W., *Acta Hepatogastroenterol.* **26,** 185 (1979).
60. Pennington, C. R., Ross, P. E., and Bouchier, I. A. D., *Gut* **18,** 903 (1977).
61. Lewis, D., Panveliwalla, D., Tabaqchali, S., and Wootton, I. D. P., *Lancet* **i,** 219 (1969).
62. Schwarz, H. P., Paumgartner, G., and Preisig, R., *Schweiz. Med. Wschr.* **105,** 533 (1975).
63. van Berge Henegouwen, G. P., Brandt, K. H., Eyssen, H., and Parmentier, G., *Gut* **17,** 861 (1976).

64. Devers, L. M., Paul, E. M., Buchinsky, V., and Derck, D., *Clin. Chem.* **24,** 1051 (1978).
65. Shaw, R., and Elliorr, W. H., *Lipids* **13,** 971 (1978).
66. Stellaard, F., Hachey, D. L., and Klein, P., *Anal. Biochem.* **87,** 359 (1978).
67. Okuyama, S., Uemura, C., and Hirata, Y., *Bull. Chem. Soc. Japan* **52,** 124 (1979).
68. Bloch, C. A., and Watkins, H. B., *J. Lipid Res.* **19,** 510 (1978).
69. Sandberg, D. H., Sjovall, J., Sjovall, K., and Turner, D. A., *J. Lipid Res.* **6,** 182 (1965).
70. Karlaganis, G., and Paumgartner, G., *J. Lipid Res.* **19,** 771 (1978).
71. Angelin, B., Bjorkhem, I., and Einarsson, K., *J. Lipid Res.* **19,** 527 (1978).
72. Siskos, P. A., Cahill, P. T., and Javitt, N. B., *J. Lipid Res.* **18,** 666 (1977).
73. Beckett, G. J., Hunter, W. M., and Percy-Robb, I. W., *Clin. Chim. Acta* **88,** 257 (1978).
74. Taylor, J. T., Knotts, J. G., and Schmidt, G. J., *Chromatogr. Newsletter.* **7,** 39 (1979).


## Editors' Note

This note is included to bring several points to the attention of our readers that were not discussed in the previous chapter. First, at least within the United States today, improved patient care is frequently adequate justification for increased cost. Second, the concept that LC is more expensive than RIA must be examined more closely. Obviously, if a laboratory does not already own an LC, adequate cost accounting must be considered, just as in purchasing scintillation counting equipment. If both types of equipment are available in the laboratory, LC is frequently the cheaper of the two methods. This is particularly true when a small number of samples are being analyzed; since the large number of controls and standards necessary for an RIA assay need not be run with LC. Lastly, and perhaps most importantly, the quality and quantity of information obtained from LC is significantly better than that obtained from an RIA procedure. In LC, multiple compounds and their metabolites can be assayed in a single run, and a chromatogram allows the chromatographer to consider the possibility of interfering substances. Unless the interfering substances co-elute exactly with the compounds of interest, peaks will be skewed or fused and would thus alert the chromatographer that the result is suspect. With RIA such clues are not available. In addition, the LC peak can be collected for further identification, a process impossible with RIA.

# Chapter 14

# Proteins

Fred E. Regnier and Karen M. Gooding

*Department of Biochemistry, Purdue University, West Lafayette, Indiana*

## I. Introduction

Because of the complexity of cellular material and body fluids, it is seldom possible to analyze a natural product directly. Qualitative and quantitative analyses must often be preceded by some purification step that separates the molecular species being examined from interfering materials. In the case of proteins, column liquid chromatography has been used extensively for these fractionations. With the advent of gel permeation, cation exchange, anion exchange, hydrophobic, and affinity chromatography, it became possible to resolve proteins through their fundamental properties of size, charge, hydrophobicity, and biological affinity. The chromatographic separations used in the early isolation and characterization of many proteins later became analytical tools in their routine analysis. Unfortunately, these inherently simple and versatile column chromatographic techniques introduced in the 50s and 60s have a severe limitation in routine analysis–separation time. It is common to encounter 1–24 h separation times with the classical gel-type supports.

Recent advances in high performance liquid chromatography of proteins have now made it possible to reduce separation times 10–100 fold. Application of this new chromatographic technology to the analysis of isoenzymes, hemoglobins, and proteins associated with bilirubin will be the subject of this review.

## II. Isoenzymes

It is now widely accepted that many enzymes exist in multiple forms known as isoenzymes. In fact, isoenzymes are so ubiquitous in nature that every enzyme seems to be a candidate for this type of heterogeneity. Meister (*1*) and Neilands (*2*) first recognized that beef heart had multiple forms of the enzyme lactate dehydrogenase (LD). With the report by Vessel and Bearn in 1957 (*3*) that the serum profile of LD isoenzymes in humans changed characteristically and dramatically with both myocardial infarcation and acute myelogenous luekemia, a new dimension in clinical diagnosis was introduced. Since that time an enormous quantity of literature has accumulated concerning the structure, function, occurrence, and origin of isoenzymes, in addition to clinical acceptance that isoenzymes are of considerable diagnostic value (*4*).

### A. Lactate Dehydrogenase (LD)

Lactate dehydrogenase is a tetrameric enzyme (mw 150,000) with two structurally distinct subunits, an M type from skeletal muscle and an H type from heart muscle. Five different isoenzymes result from the various combinations of these two subunits. A third type of subunit, C, is found in testicular tissue and spermatoza and usually occurs as the $C_4$ tetramer. However, in the guinea pig, three LD isoenzymes have been found in which the C subunit is hybridized with M or H subunits (*5*).

*1. Separation.* Resolution of the LD isoenzymes in the clinical laboratory is now achieved by both electrophoresis and anion exchange chromatography. Compared to conventional column chromatography, electrophoresis is generally faster, requires less sample, and provides better resolution. When shorter ion exchange columns were used, a smaller number of fractions were collected; sample size and chromatographic separation time were reduced, but so was resolution. Unfortunately, both of these techniques are labor intensive and the separation medium is usually discarded after a single use. Although one can analyze a large number of samples at one time on a single electrophoresis plate, the samples must be accumulated and the precision and accuracy are poor.

High performance liquid chromatography has transformed column liquid chromatography into a rapid, high resolution analytical technique that allows repeated use of one column. The first report of the high performance liquid chromatography of isoenzymes was that of Chang et al. (*6*). The DEAE anion exchange support used in this work was prepared by polymerizing, on the surface of controlled porosity glass (CPG), a thin layer of glycerol that was attached to the support through an organosilane coupling agent. The DEAE stationary phase was bonded to the polymer matrix. A 37–74 μ particle size column of this DEAE Glycophase was used in a 30 min resolution of LD isoenzymes. In a second paper (7), it was reported that by using columns packed with 5–10 μ particles of this support, the resolution of all 5 LD isoenzymes could be achieved in 6 min, as seen in Fig. 1. Essentially the same mobile phases were used to develop these columns that had been used on the classical DEAE cellulose columns.

A second high performance support material that has been applied to the separation of LD isoenzymes is the crosslinked polyethylene imine coating of Alpert (8). This material is prepared by adsorbing a monolayer of polyethylene imine onto the surface of controlled porosity silica and crosslinking adjacent amine molecules with a multifunctional oxirane. This pellicle of anion exchange phase is quite stable and survives hundreds of hours of column operation.

Recoveries of both $LD_1$ and $LD_5$ are greater than 95% on these columns. Elution profiles on these columns are virtually identical to the classical DEAE cellulose columns and are in the order of $LD_5$ to $LD_1$. A careful examination of the elution profiles from these two

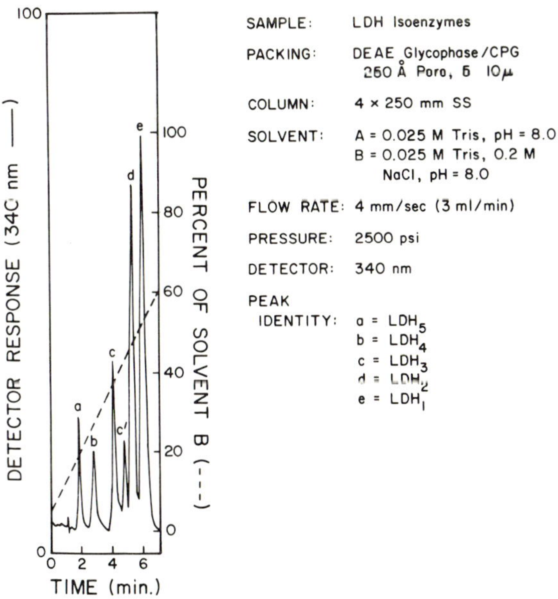


FIG. 1. Separation of LD isoenzymes on an anion exchange column. Reprinted from ref. 7 by courtesy of Elsevier Scientific Publishing Company.

columns indicates that there is a greater resolution between $LD_4$ and $LD_3$ on the DEAE Glycophase than on the crosslinked polyamine column. The only significant difference between these two columns is in their ion exchange capacity. The crosslinked polyamine column has more than twice the hemoglobin ion exchange capacity and five times the small molecule ion exchange capacity. Although this difference has little effect on resolution, the polyamine column has a greater loading capacity.

A third anion exchange support material that we have examined for LD isoenzyme separations is SynChropak AX 300. This anion exchanger has separation and loading characteristics similar to the crosslinked polyamines.

*2. Detection.* Actually, the development of specific detection systems for isoenzymes as they elute from chromatographic columns has been as complex as the development of the columns. These chromatographic systems are capable of generating enough fractions in a few minutes to require several hours of manual enzyme assays. Ideally, one would like an automated post-separation detection system that continuously monitors the column effluent. Several research groups have developed such detection systems. Basically these detectors assay the enzyme with a system that is composed of (1) a post-analytical column substrate or reagent pump, (2) a mixing device, (3) a continuous flow incubation or reaction vessel and, (4) some type of liquid chromatography detector. A block diagram of such a system is seen in Fig. 2. There are several types of problems that must be dealt with in the design of such a detection system. The first is to minimize the distortion of chromatographic peaks during the enzyme–substrate mixing and the passage through the reactor so that chromatographic resolution is not lost. The second problem is to prevent the separation of enzymes, substrates, and products as they move through the detection system. If all of the components for the enzyme assay do not move through the detector at the same velocity, there will be an

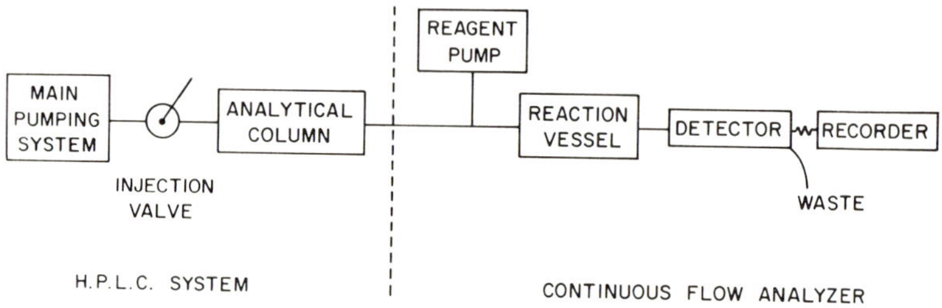


FIG. 2. General diagram of post-column detection systems. Reprinted from ref. *22* by courtesy of the American Association for Clinical Chemistry.

apparent loss in resolution of the enzyme components. All of this must be accomplished in a vessel that can retain the 4–8 mL of column effluent that are generated during the incubation period.

The system described by Chang (*7*) and Schlabach (*8*) has all of the components noted above. Substrate was pumped with a high performance syringe pump while mixing was achieved in a low dead volume mixing tee. The incubation vessel or post-column reactor (PCR) consisted of a chromatography column packed with nonporous glass beads. It was shown in this work that the elimination of pores was essential to prevent the separation by gel permeation of LD from its products in the PCR. A further important conclusion of this work was that the PCR packing material could be 37–100 $\mu$ in diameter instead of the 10 $\mu$ diameter used in the analytical column. These larger particles reduced the pressure drop across the PCR 6–10 fold. These packed bed reactors also gave only 1/3 the chromatographic band spreading that was experienced with a 50 ft × 0.02 in. id open tubular capillary PCR when operated on a very high efficiency chromatographic system. It has since been noted by Schlabach (*9*) that there is little difference in band-spreading between a capillary and packed bed PCR when they are operated in systems with less efficiency.

The advantage of the packed bed PCR is that good mixing of the reactants is easily achieved with minimum band spreading throughout the 3–5 min reactions. There is also adequate surface area on the glass beads to immobilize enzymes for coupled enzyme assays. The disadvantages of the system are that it is hard to obtain uniform temperature regulation and the packed bed may settle after extended use.

An alternative post-column reaction vessel is a simple piece of capillary tubing. Band spreading in open tubular capillaries has been described by Scott and Kucera (*10*) in the equation

$$V_i^2 = d^4 FL/24 D_m \qquad (1)$$

where $d$ is the internal diameter of the tubing, $L$ is the length of the tubing, $F$ is the linear flow-rate, and $D_m$ is the diffusion coefficient of the solute molecule. As can be seen in this equation, band spreading in an open tubular reactor is proportional to the square of the internal diameter. When the reaction time ($t$) necessary to detect the enzyme and the combined volumetric flow-rate ($f$) of the chromatographic column and the substrate pump are known, the necessary post-column reactor length may be calculated by the equation

$$L = 4\ tF/\pi d^2 \qquad (2)$$

For example, at a flow-rate of 1 mL/min, 20 m of 0.01 in tubing is required for each minute of incubation time. Toren's laboratory has used open tubular capillaries successfully for post-column reactors (*11–15*), as have Denton (*16*) and Bostick (*17*). The advantages of the open tubular capillary PCR are its simplicity and the ease of reaction temperature control. The disadvantages are that reactant mixing is slightly harder to achieve, the pressure drop across the reactor is large, and enzyme immobilization in the PCR is not possible.

It should be noted that when an open tubular capillary is used as the PCR, some mixing device other than a simple tee must be used to insure complete mixing of reactants. Schroeder (*11*) used a short packed bed of nonporous particles while Denton (*16*) used a 100-$\mu$/L magnetically stirred mixer.

Post-column detection of LD activity has been accomplished in all cases through the oxidation of lactate to pyruvate in the reaction:

$$\text{L-lactate} + \text{NAD}^+ \xrightarrow{\text{LD}} \text{pyruvate} + \text{NADH} + \text{H}^+ \quad (3)$$

The reduced cofactor, NADH, has been continuously monitored either by its absorbance at 340 nm or its fluorescence. Fluorescence has the obvious advantage of greater sensitivity.

It has been noted in the analysis of serum LD isoenzymes that fluorescent or UV absorbing materials that have spectral characteristics similar to NADH may be present and may coelute with the enzymes. When enzyme activity is being monitored without reference to this interfering material, erroneous results may be obtained. Background interference may be handled in two ways: either by extending reaction time to the point that relative contribution from background material is small or by measuring the absorbance at $t_0$ and subtracting this value from subsequent absorbance measurements on the sample. The use of extended reaction times in a flow-through system is not very practical and, therefore, not really an option. On the other hand, background subtraction has been successfully used (*12–15*) in the determination of LD isoenzymes. The system used to accomplish background subtraction is shown in Fig. 3. This system consists of a mixing tee where sample and substrate enter the detection system followed by a predetector delay coil, a first detector (Detector 1), an open tubular reaction coil, and a second detector (Detector 2). Detector 1 measures the absorbance of the reaction mixture immediately after mixing while Detector 2 measures the absorbance of the sample after some incubation time. A direct subtraction of the signals from the two detectors is not possible because there is a time delay that is equal to the reactor residence time and because the chromatographic band of enzyme detected by Detector 2 is much broader than that detected by Detector 1. Both the time offset and

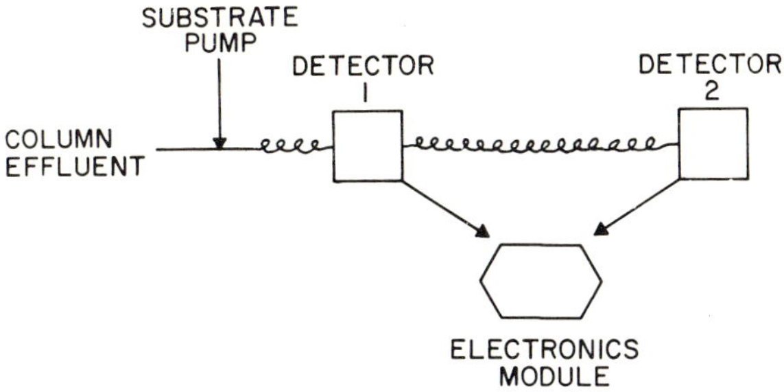


FIG. 3. Diagram of a dual detector post-column reactor system with background subtraction. Described in ref. *12*.

dispersion problems were solved by the development of a computer algorithm based on theoretical models of band dispersion; it predicts the dispersion of the peak detected by Detector 1 by the time it arrives at Detector 2. This allows the point-by-point subtraction of the two signals and the correction for both sample interferences and baseline drift. A typical example of background subtraction from the work of Schlabach is seen in Fig. 4. The utility of this system is obvious.

*3. Applications.* In 1976, Kudirka (*18*) was the first to show that HPLC could be used in the fractionation of serum LD isoenzymes and in the detection of elevated LD-1 that resulted from myocardial infarction. In these studies, manual sample collection and enzyme assays were used to quantitate enzyme activity. This work was followed by simultaneous reports on the use of both packed-bed (*8*) and open-tubular capillary (*11*) post-column reactors for the detection of LD isoenzymes in serum samples.

The recent papers by Schlabach (*12,15*) and Fulton (*13, 14*) describe a computer-based detection system that is capable of background subtraction and has a 1000-fold linear dynamic range. Enzyme activities from 1.5 to 1500 U/L of lactate dehydrogenase were detected with greater than 90% recovery. An example of the use of this system in the confirmation of a myocardial infarction is shown in Fig. 5. The tracing labeled B in this figure is the serum isoenzyme analysis from a patient admitted with precordial chest pain. Although the LD profile in this sample is in the normal range, a slight elevation of LD-1 is seen. After a cardiac episode during the night, the patient was diagnosed as having had a nontransmural, myocardial infarction of the anterior wall. A serum sample obtained 10 h after this episode (18 h after admission) was found to have the LD isoenzyme profile shown in Fig. 5. Comparing these two profiles shows an approximately 4-fold increase in $LD_1$ activity and an elevation of $LD_2$ activity.

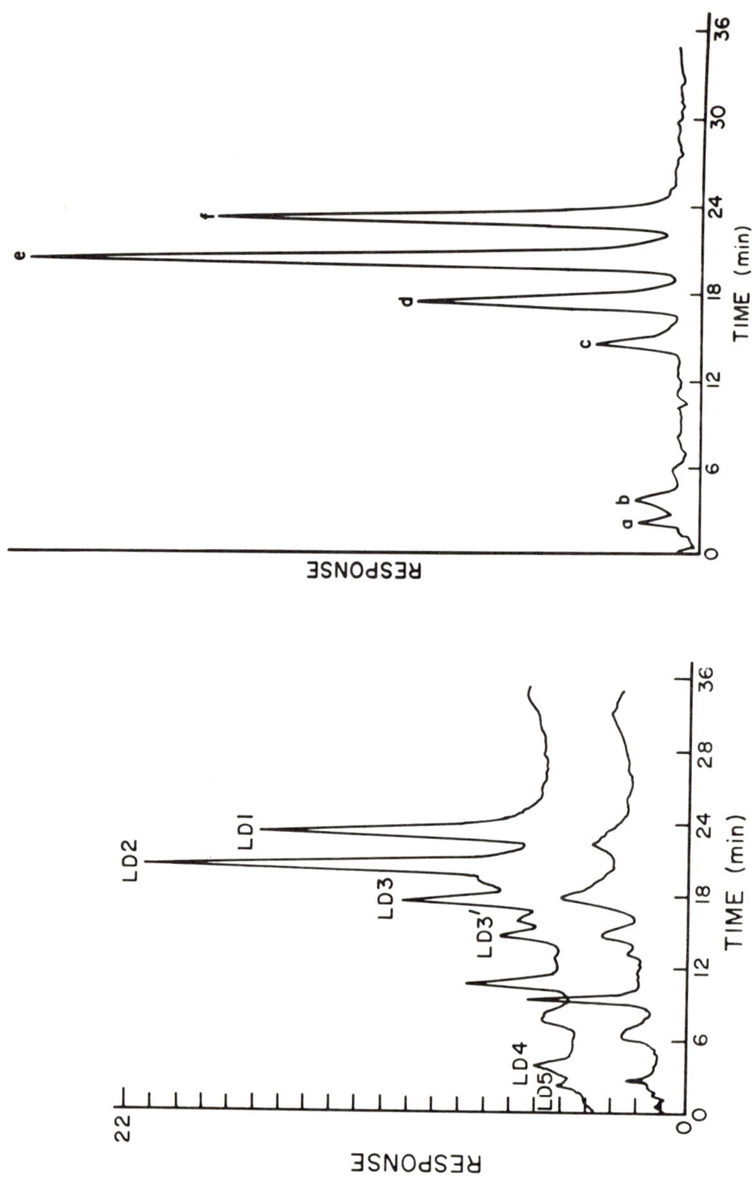


FIG. 4. Profiles of serum LD isoenzymes: (a) Without background subtraction; (b) With background subtraction. Taken from ref. *14* by courtesy of Elsevier Scientific Publishing Company.

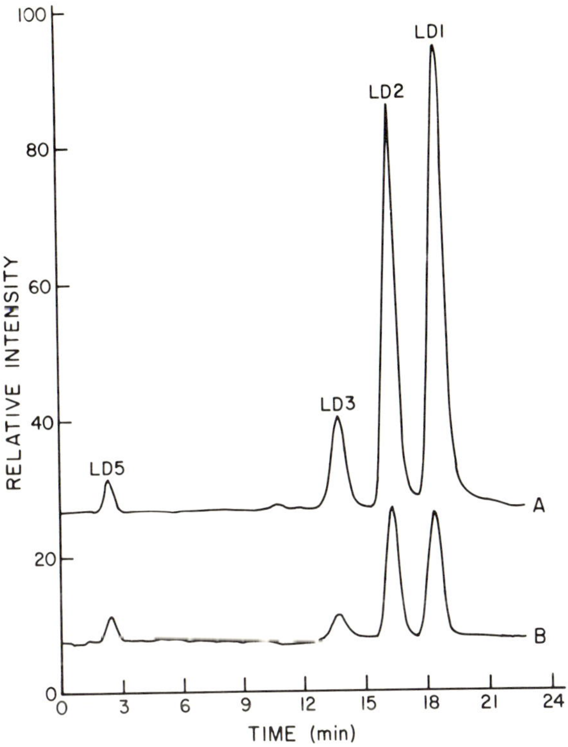


FIG. 5. Comparison of temporarily sequenced LD isoenzyme profiles from the serum of a patient in a cardiac care unit. The profile shown in Trace B was from a serum sample taken at admission. Trace A was from a serum sample drawn 18 h later. Reprinted from ref. *15* by courtesy of the American Chemical Society.

## B. Creatine Kinase (CK)

Creatine kinase is a dimeric molecule (mw 86,000) that is composed of M subunits, B subunits, or a combination of the two. The M subunit and the MM isoenzyme predominate in skeletal muscle, while the B subunit and BB isoenzyme are most abundant in brain. The MB isoenzyme is abundant in the heart. After myocardial infarction, CK–MB increases in blood within 4–6 h and generally peaks within 12–20 h (*19*). CK–MB elevations are virtually specific criteria for myocardial injury. This is particularly significant because it is difficult to differentiate patients with coronary insufficiency from those with myocardial infarctions by electrocardiography and chest pain.

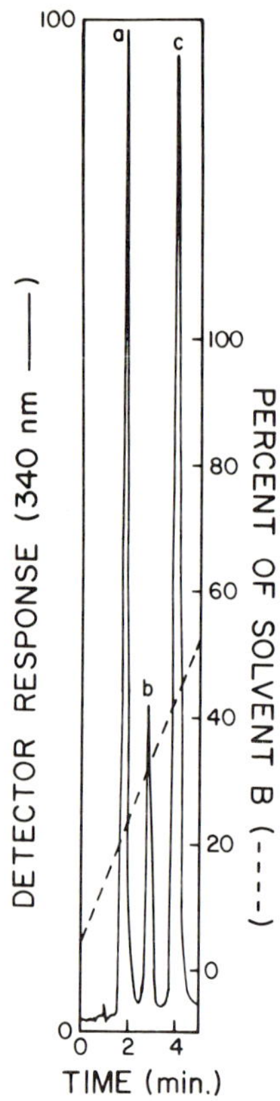


SAMPLE: CPK Isoenzymes

PACKING: DEAE Glycophase/CPG
250 Å pore, 5-10μ

COLUMN: 4 × 250 mm SS

SOLVENT: A = 0.05 M Tris, 0.05 M NaCl, $10^{-3}$ M Mercaptoethanol, pH = 7.5

B = 0.05 M Tris, 0.3 M NaCl, $10^{-3}$ M Mercaptoethanol, pH = 7.5

FLOW RATE: 4 mm/sec (3 ml/min)

PRESSURE: 2500 psi

DETECTOR: 340 nm

PEAK IDENTITY: a = $CPK_3$
b = $CPK_2$
c = $CPK_1$

FIG. 6. Separation of CK isoenzymes on an anion exchange column. Reprinted from ref. *7* by courtesy of Elsevier Scientific Publishing Company.

*1. Separation.* An anion exchange column chromatographic method introduced by Mercer (*20, 21*) for the resolution of CK isoenzymes has been widely used in clinical laboratories. In an effort to decrease the separation time, Chang used the DEAE Glycophase supports in a high performance system as described above for LD. Separation in 4 min and detection of the three CK isoenzymes in a tissue sample is shown in Fig. 6. Subsequently, it has been shown that a variety of hydrophilic anion exchange supports are capable of resolving the CK isoenzymes. Schlabach (*22*) used polymeric coatings with diethylamine, tetraethylpentaamine, and polyethylene imine stationary phases on inorganic supports to resolve the CK isoenzymes. More recently, Schlabach (*12, 15*), Fulton (*13, 14*), and Denton (*16*) have used DEAE Glycophase while Bostick (*23*) has used SynChropak AX 300.

*2. Detection.* The basic design of detection systems used for CK isoenzymes is the same as that used for LD; however, the detection of CK activity is more difficult. Since the products of CK cannot be determined by direct spectrophotometric measurements, the post-column detection of this enzyme must be coupled through other enzymes to the production of a measurable product as indicated below:

$$\text{Creatine phosphate} + \text{ADP} \xrightarrow{\text{CK}} \text{Creatine} + \text{ATP} \quad (4)$$

$$\text{ATP} + \text{D-glucose} \xrightarrow{\text{HK}} \text{glucose-6-phosphate} + \text{ADP} \quad (5)$$

$$\text{Glucose-6-phosphate} + \text{NAD(P)} \xrightarrow{\text{G-6-PDH}} \text{gluconolactone-6-phosphate} + \text{NAD(P)H} + \text{H}^+ \quad (6)$$

HK is hexokinase and G-6-PDH is glucose-6-phosphate dehydrogenase. The use of NAD or NADP in the last reaction depends on whether the G-6-PDH is from *L. mesenteroides* or yeast. In all cases except two to be described below, substrates and coupling enzymes were continuously pumped during the elution and detection of CK isoenzymes.

The continuous monitoring systems used for CK detection are of four types: (1) a simple packed-bed reactor with continuously pumped coupling enzymes and substrates, (2) a simple open tubular capillary reactor with continuously pumped coupling enzymes and substrates and computer based background subtraction, (3) a packed bed reactor with immobilized coupling enzymes and pumped substrate, and (4) an open tubular capillary reactor with pumped substrates and an immobilized coupling enzyme bed used in split-stream background subtraction.

Both the packed bed (*6*) and open tubular capillary (*11*) reactor systems described above for LD detection have also been used for CK detection. HK, G-6-PDH, and all of the substrates for CK and the coupling enzymes were continuously pumped through the detector. The primary disadvantage of this system is that operational costs are high because the coupling enzymes are discarded after a single passage through the system. In an effort to overcome this problem, Schlabach (*24*) immobilized coupling enzymes on the glass beads in a packed bed reactor. This system was capable of detecting CK, but had limited dynamic range and sensitivity. The most recent advancement (*16*) has been to use an open tubular capillary PCR with stream splitting after the PCR and automated background subtraction. This is accomplished by first incubating CK in the PCR and generating only

ATP. At this point, the stream is split and one portion taken directly to the "reference" side of a spectrophotometer while the other part of the stream is carried through a small bed of immobilized coupling enzymes and then to the "sample" side of the detector. This system both eliminates the problem of wasting the coupling enzymes and provides background subtraction of interfering material. A chromatogram of the serum CK profile from a patient with electrophoretically confirmed CK–MB is shown in Fig. 7. At the present time, the ultimate sensitivity, linearity, and dynamic range of this system are unknown.

Denton's CK detection system described above has been noted to provide background subtraction by passing a portion of the sample through the reference side of a detector before the generation of NADH. This system would appear to be both the simplest and cheapest solution to the problem.

The computer based background subtraction system described above for LD (*12–15*) has also been used for CK. The response of this system to applied CK activity was linear from 0.5 to 500 U/L. It may be noted in the work of Schlabach (*15, 25*) that background subtraction is much more important with CK than with LD for two reasons. First, the level of activity is generally lower with CK and second, interfering serum proteins such as albumin elute in the region of CK–MB.

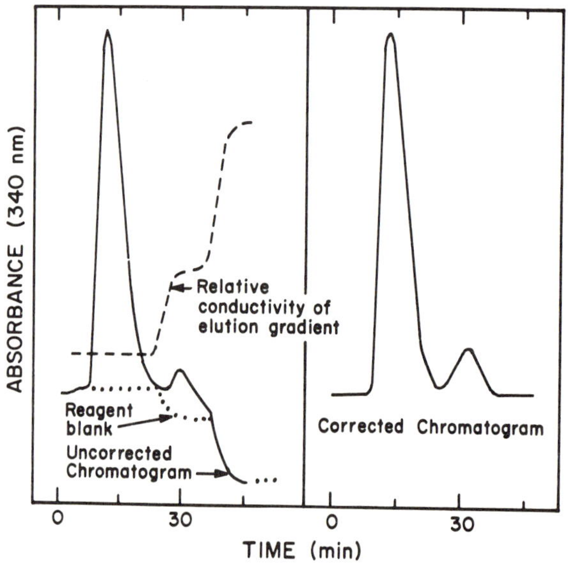


FIG. 7. Chromatographic separation of CK isoenzymes in the serum of a patient with electrophoretically confirmed CK–MB. Reprinted from ref. *16* (by courtesy of the American Association for Clinical Chemistry).

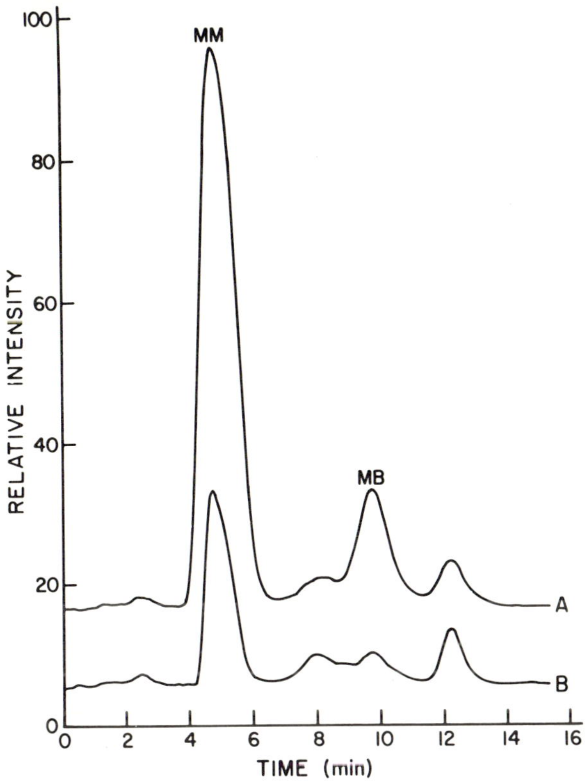


FIG. 8. Comparison of tcmporarily scqucnccd CK isocnzymc profiles from the serum of a patient in a cardiac care unit. The profile in Trace B was obtained from the sample taken at admission. Trace A was from a sample taken 18 h later. Reprinted from ref. *15* by courtesy of the American Chemical Society.

*3. Applications.* The CK profile from a pre-infarction serum sample is shown in Fig. 8A. This is the same sample used to produce the LD profiles in Fig. 5. The presence of CK-MB in the sample leads one to believe that the patient had already suffered cardiac distress before admission. The elevation of CK–MB after myocardial infarction is seen in Fig. 8B.

## C. Arylsulfatase

Two types of arylsulfatases that readily hydrolyze the chromogenic substrates *p*-nitrocatechol sulfate and *p*-nitrophenol sulfate occur in the body fluids of man. Since the natural substrate for arylsulfatase A (ASA) is cerebroside sulfate, a deficiency in this enzyme activity results

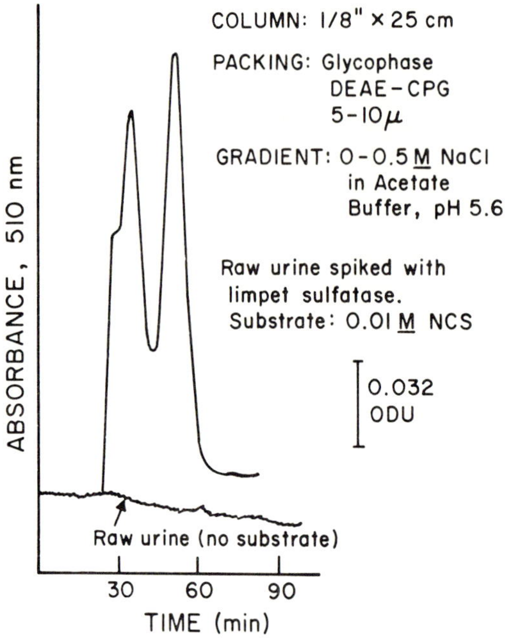


FIG. 9. Continuous separation and assay of limpet arylsulfatase isoenzymes in human urine. Reprinted from ref. *17* by courtesy of the American Association for Clinical Chemistry.

in the accumulation of cerebroside sulfate in the body and causes the metabolic disease metachromatic leukodystrophy. On the other hand, defective arylsulfatase B (ASB) results in Moroteaux-Lamy syndrome.

Bostick (*17*) has accomplished the separation and quantitation of ASA and ASB in an HPLC system equipped with a DEAE Glycophase column and a double-beam photometric detector. In this detector, the nitrocatechol hydrolysis product passes first through the reference cell and then through the sample cell after the addition of NaOH to produce a color specific for arylsulfatase. Figure 9 shows the response of this detector to the arylsulfates in human urine that had been supplemented with purified sulfatase.

## III. Hemoglobins

### A. Background

The analysis of hemoglobins is another clinical problem for which HPLC shows great promise. Hemoglobin is a tetrameric protein consisting of a heme group and two sets of identical globin chains.

Normally, fetal hemoglobin (HbF) is present at birth, but disappears within a few months. For the rest of one's life, adult hemoglobin (HbA) predominates at 96–99% of the total with hemoglobin $A_2$($HbA_2$) comprising the difference. HbF consists of $\alpha$ and $\gamma$ chains; HbA consists of $\alpha$ and $\beta$ chains; and $HbA_2$ has $\alpha$ and $\delta$ chains. In some segments of the population, there are genetically transmitted variants of these hemoglobins, most commonly those involving an amino acid substitution in the chain such as in sickle cell hemoglobin (HbS), HbC, and HbE. Another defect in hemoglobin production is seen in $\alpha$- or $\beta$-thalassemia, where synthesis of either the $\alpha$- or the $\beta$-chain is reduced or absent. Homozygous states of any of the above hemoglobinopathies or combinations of any two abnormalities may result in clinical manifestations especially during pregnancy, surgery, or other times of stress. The importance of screening the population for hemoglobinopathies has been recognized in various sickle cell programs and the Cooley's Anemia Control Act for thalassemia detection.

These hemoglobin variants have been analyzed by electrophoresis and ion exchange chromatography on carbohydrate gel columns (*26, 27*) until recently. Chromatography gives better resolution and quantitation than electrophoresis, but it is slow. The speed was improved with microchromatography on mini-columns (*28*), but such columns are not as versatile as the large ones and they cannot be automated. Chromatography on the large ion exchange columns showed most hemoglobin variants to yield multiple peaks (*26*); the minor peaks have proven to be glycosylated proteins. One of the glycosylated products of HbA, $HbA_{1C}$, appears to be an indicator for diabetes (*29, 30*). $HbA_{1C}$ is normally found in levels of 4–6%, but is doubled in the blood of uncontrolled diabetics. The quantity of $HbA_{1C}$ reflects the glucose tolerance over the life of the erythrocyte and, as such, is independent of factors such as recent food intake. Preliminary clinical studies have shown that $HbA_{1C}$ monitoring may replace more inaccurate tests such as glucose tolerance tests.

## B. Applications

*1. Hemoglobinopathies.* Hemoglobin variants have been analyzed by HPLC of the intact proteins (*7, 31*), the globin chains (*32*), and the tryptic digests (*33*). No clinical trials have been performed as yet, and some of the analyses were of commercial hemoglobin standards. Chang et al. (*7*) separated a control sample of HbA, F, S, and $A_2$ on DEAE Glycophase G. Figure 10 shows the analysis of $HbA_2$, S, A and F standards on a SynChropak anion exchange column. Hearn and Hancock (*32*) analyzed a $HbA_1A_2$ hemolysate by reversed-phase chromatography.

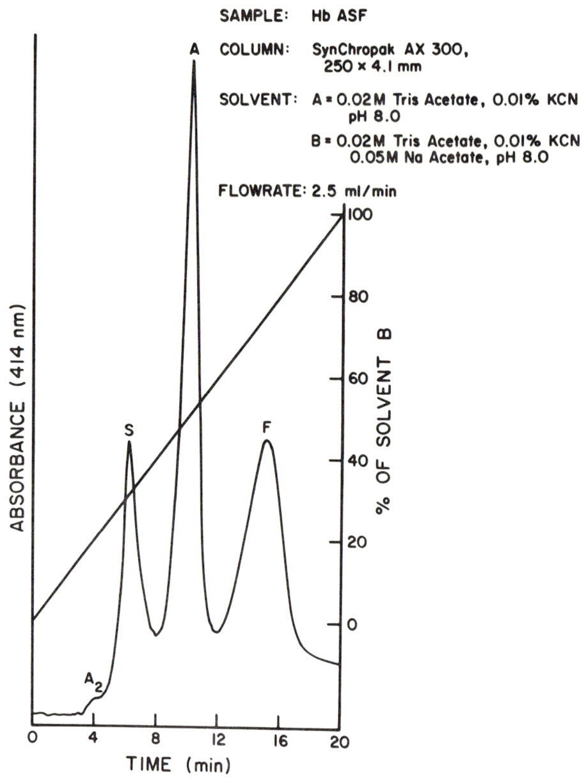


FIG. 10. Analyses of $HbA_2$, S, A, and F on a SynChropak AX 300 anion-exchange column.

HbE was identified by Schroeder et al. (*33*) by the resolution on a reversed-phase column of the peptide fragments after tryptic digestion (Fig. 11). They also examined the tryptic peptides of the $\alpha$- and $\beta$-chains of HbC (*33*). Hancock and Hearn analyzed the tryptic peptides of HbJ Cambridge by reversed-phase (*32*). Quantitation of HbA and $A_2$ can serve to screen for $\beta$-thalassemia. The difference between profiles from normal blood and from blood from a person with $\beta$-thalassemia trait is seen in Figs. 12 and 13 (*31*). Since $\beta$-chain synthesis is reduced, $HbA_2$ comprises a larger than normal percentage of the total hemoglobin. HbA and $A_2$ in blood were also analyzed by Chang et al. (*7*).

*2. Hemoglobin $A_{1C}$.* Hemoglobin $A_{1C}$ has been analyzed by HPLC on BioRex 70, a cation exchange support based on a semirigid gel (*34–36*). Since the procedures for this analysis have been developed for several years, they are more established than those for

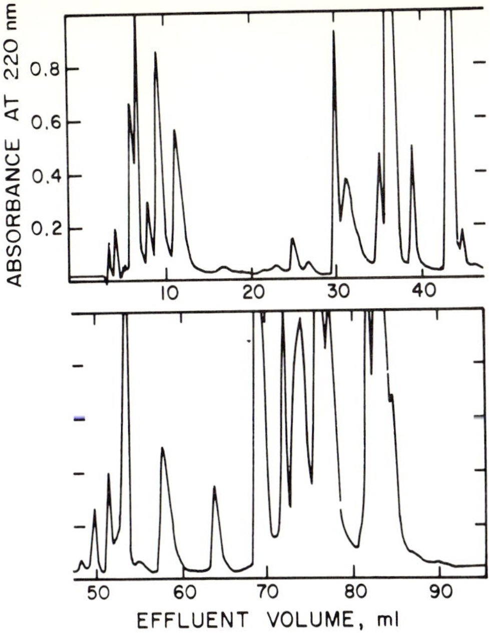


FIG. 11. Separation of the tryptic peptides of the $\alpha$- and $\beta$-chains of HbE on a reversed-phase column. For identification of components in the chromatogram see ref. *33*. Reprinted by courtesy of Elsevier Scientific Publishing Company.

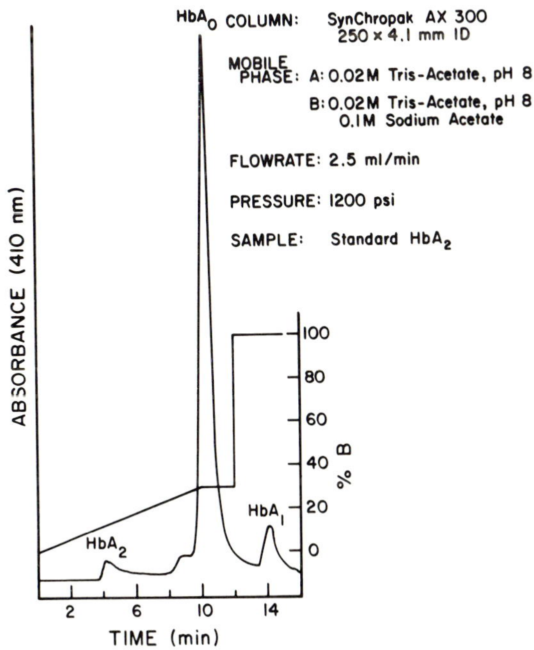


FIG. 12. Analysis of a standard $HbAA_2$ sample on a SynChropak AX 300 anion-exchange column. Reprinted from ref. *31* by courtesy of Elsevier Scientific Publishing Company.

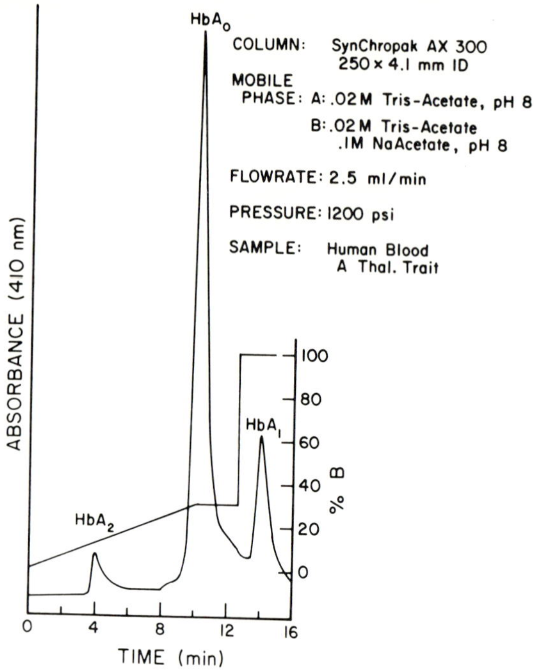


FIG. 13. Analysis of the hemoglobin components in the blood of a person with β-thalassemia trait. Reprinted from ref. *31* by courtesy of Elsevier Scientific Publishing Company.

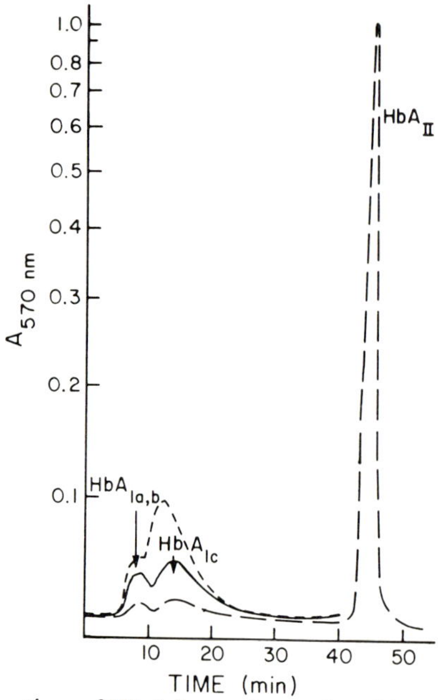


FIG. 14. Separation of HbA from glycosylated hemoglobins. Reprinted from ref. *36* by courtesy of Elsevier Scientific Publishing Company.

hemoglobinopathies and clinical trials are being conducted. Figure 14 shows an example of the separations achieved by these methods (*36*).

## IV. Protein-Associated Bilirubin in Neonatal Serum

### A. Background

Jaundice is a relatively common condition in neonates, especially those who are premature or ill. Such jaundice may be mild and disappear naturally or through phototherapy; or the jaundice may be severe, and result in bilirubin encephalopathy (kernicterus) and eventually death. The line between the two conditions is not distinct. When Rh disease was common, many infants with nonhemolytic jaundice were ignored since they had the "lesser" problem. Now that a major cause of jaundice is prematurity and related conditions, this condition takes on greater significance and is seen to be responsible for neurological abnormalities and death. The physician would like criteria (tests) to indicate the severity of the jaundice and which kind of therapy—phototherapy or exchange transfusion—is necessary. Measurement of unbound, unconjugated bilirubin in serum has proven a difficult task and the interpretation of previous data is uncertain. Several reviews have recently addressed themselves to these problems (*37–41*).

It has been postulated that the nonpolar character of bilirubin is responsible for its toxicity. Bilirubin is a degradation product of hemoglobin and in its acid form has the structure indicated in Fig. 15. The intramolecular hydrogen bonding indicated in the figure leaves bilirubin only sparingly soluble at physiological pH (7.4). At this pH the solubility of bilirubin is 0.1 μmol/L or 5.85 μg/dL (*42*). Normally, this insoluble molecule is carried by albumin and other proteins to the liver where the enzyme system glucuronyl transferase converts bilirubin to the glucuronides (*43*). These bilirubin glucuronides are

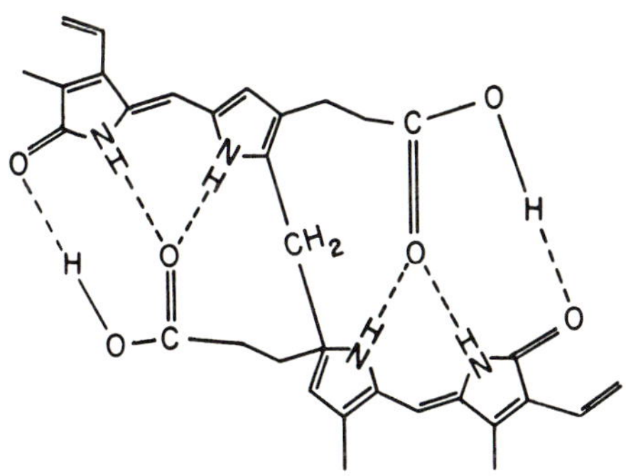

FIG. 15. Structure of bilirubin IX-α(Z).

soluble and are excreted in bile. These enzymes are not operative until birth, which accounts for the high incidence of jaundice in premature infants. When this normal mechanism for bilirubin disposal is inoperative, the hydrophobic species shown in Fig. 15 binds to proteins and other hydrophobic materials. The lipophilic nature of bilirubin accounts for its accumulation in the skin and brain.

Studies have shown that there is one high affinity and several lower affinity bilirubin binding sites on human serum albumin (HSA). The first mole of bilirubin is bound with a $K_a \approx 10^8 M^{-1}$, and for all practical purposes reduces the concentration of free bilirubin to zero. The secondary binding site on albumin has a binding constant at least 100-fold smaller. It has been shown that lipoproteins (*44*) and globulins ($\alpha_1$-, $\beta_2$-, and $\alpha_2$-) (*45*) also bind bilirubin; however, globin chains do not (*46*).

A series of problems and facts must be recognized when dealing with the jaundice problem clinically. The first is that there is often not a good correlation between free and/or total bilirubin and the incidence or severity of brain damage and kernicterus. This probably owes to the vast differences in birth weights among neonates. Further problems arise from the lack of knowledge concerning the binding affinity of neurological tissue for bilirubin and the fact that multiple proteins of variable bilirubin binding affinities may be involved in partitioning bilirubin in serum.

The analytical problem becomes one of trying to define the system sufficiently that part of the uncertainty noted above can be eliminated. By being able to determine the free bilirubin, protein bound bilirubin, total bilirubin, and the percentage of the high affinity binding sites that are loaded, one has a much better picture of the physiological status of the patient. The following is an initial attempt to do this with HPLC.

## B. Separation of Components

On the basis of prior separations of protein-associated and free bilirubin on Sephadex gel permeation supports, it was reasoned that one of the new high performance gel permeation supports might function for the same separation (*47*). SynChropak GPC 100, a bonded phase silica support designed for protein separations, was used. The 100 Å pore diameter support was chosen since it provides a good separation of proteins and small molecules. Bilirubin concentration was monitored at 453 nm.

When an adult human serum sample that was enriched with bilirubin was applied to a column and eluted with 0.1 $M$ $K_2HPO_4$ (pH 7.0), the protein-associated bilirubin was eluted from the column and

Table 1
Bilirubin Released by BSA Injection

| | Bilirubin control, mg/L | | Bilirubin addition to adult serum, mg/L | | | | |
|---|---|---|---|---|---|---|---|
| | 92 | 200 | 0 | 100 | 150 | 200 | 250 |
| Protein bound, % | 99.05 | 98.96 ± 0.065 | 100 | 97.12 | 97.71 | 96.06 | 87.66 |
| After 1st injection, % | 0.77 | 0.87 ± 0.070 | 0 | 2.53 | 2.02 | 3.65 | 11.07 |
| After 2nd injection, % | 0.18 | 0.16 ± 0.038 | 0 | 0.35 | 0.27 | 0.29 | 1.26 |
| After 3rd injection, % | | | | | | | 0.005 |

free bilirubin was adsorbed at the head of the column as was reported with Sephadex (*48*). In the case of the Sephadex columns, bilirubin adsorbed at the head of the column was eluted with 0.1 *N* NaOH. Since that is not possible with a silica based HPLC column, the adsorbed bilirubin was desorbed and eluted by an injection of bovine serum albumin (BSA). A single injection of BSA was sufficient to desorb free bilirubin in most cases as is shown in Table 1. It was noted, however, that two BSA injections insured that there would be no carryover between samples.

A separation of protein-associated and free bilirubin according to the conditions specified above is shown in Fig. 16. The major peak eluting at 4 min is bilirubin associated primarily with albumin and the small peak eluting at 3 min is bilirubin associated with a higher molecular weight species, possibly lipoprotein (*44*).

The effect of pH and ionic strength on bilirubin recovery is shown in Table 2. The effect of ionic strength is obviously more important than pH. A pH of 7.4 was chosen because it is physiological and is less destructive to the column than more basic pH values. When the 0.01 *M* $K_2HPO_4$ (pH 7.4) buffer was used, the higher molecular weight bilirubin associated peak noted in Fig. 16 disappeared, as is seen in Fig. 17. It is likely that the higher ionic strength buffer forces a solvophobic association of bilirubin with some hydrophobic protein(s).

During the course of these studies it was noted that all albumin samples (both bovine and human) showed a weak absorbance at 453 nm regardless of their purity. This necessitated that this small background absorbance be subtracted from all free bilirubin peaks. Failure to do this results in a significant error in samples that have small amounts of free bilirubin.

It was found that chromatographic analyses could be carried out on 10 μL of serum as opposed to 250 μL by the thin-layer Sephadex method (*49*).

Table 2
Effect of Mobile Phase on Bound Bilirubin Recovery

| Mobile phase | % Recovery of bound bilirubin[a] 92 mg/L | 200 mg/L |
|---|---|---|
| 0.1 mol/L $K_2HPO_4$, pH 7.4 | 73.89 | 78.55 |
| 0.1 mol/L $K_2HPO_4$, pH 8.0 | 80.42 | 86.20 |
| 0.1 mol/L $K_2HPO_4$, pH 7.4 (5% methanol) | 67.44 | 76.14 |
| 0.02 mol/L $K_2HPO_4$, pH 8.0 | 90.42 | 92.93 |
| 0.01 mol/L $K_2HPO_4$, pH 8.0 | 92.93 | 95.73 |
| 0.01 mol/L $K_2HPO_4$, pH 7.4 | 93.76[a] | 95.22[b] |
| | 92.99[b] | 95.69[b] |
| | 94.58[c] | 96.57[c] |
| | 94.55[d] | 96.23[d] |
| | | 96.77[e] |

[a]Background was not subtracted.
[b–e]Denote different columns.

## C. Linearity and Precision

The relationship between total absorbance and total bilirubin concentration was examined for linearity in a human serum albumin solution, adult serum, and neonatal serum. Total absorbance was determined in all cases by summing the protein-associated and free bilirubin absorbance, while total bilirubin concentration was obtained from the amount of bilirubin added to the sample. The sum of the total absorbance at 453 nm was directly proportional to total bilirubin in all cases as seen in Figs. 18–20. The line for the neonatal sample in Fig. 20 is offset because all neonatal samples initially contained some bilirubin.

The importance of a clean system should also be emphasized. Owing to the small quantity of sample used, any background absorbance is intolerable. The detector and injector must be cleaned periodically.

## D. Bilirubin Binding Curves

The association of bilirubin with proteins was examined by adding increasing amounts of bilirubin to a fixed concentration of protein and determining the amount of protein-associated and free bilirubin as

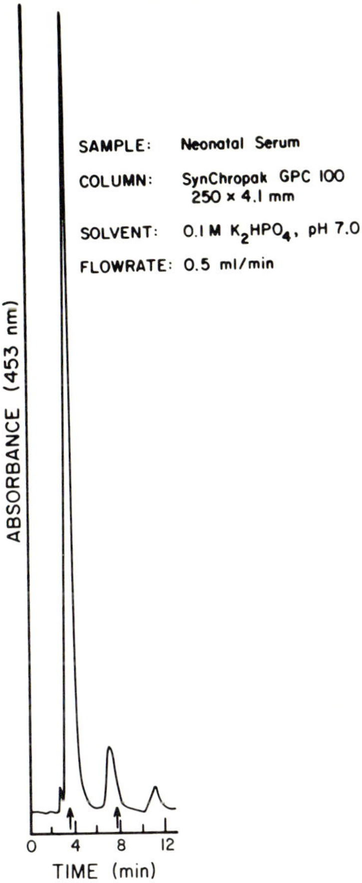


FIG. 16. Analysis of protein-bound bilirubin in neonatal serum. Arrows indicate injection of 20 $\mu$L of 0.1 mmol/L bovine serum albumin reprinted from ref. *47* by courtesy of the American Association for Clinical Chemistry.

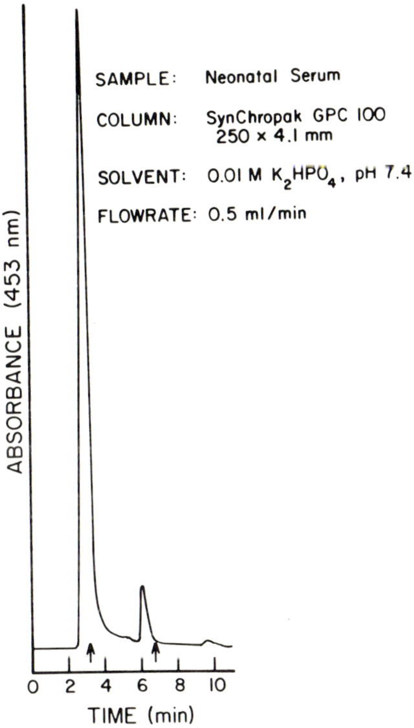


FIG. 17. Analysis of protein-bound bilirubin in neonatal serum. Arrows indicate injection of 20 $\mu$L of 0.1 mmol/L bovine serum albumin reprinted from ref. *47* by courtesy of the American Association for Clinical Chemistry.

described above. The binding curve for 0.1 mmol/L of HSA is seen in Fig. 21. It will be seen that, in this system, albumin can be saturated with bilirubin. As albumin approaches saturation, the concentration of free bilirubin increases rapidly. The inflection point for the bilirubin both in the case of HSA and in that of adult human serum (Fig. 22) occurs at approximately 80% saturation of albumin and/or the high affinity binding sites in serum. As the concentration of bilirubin in

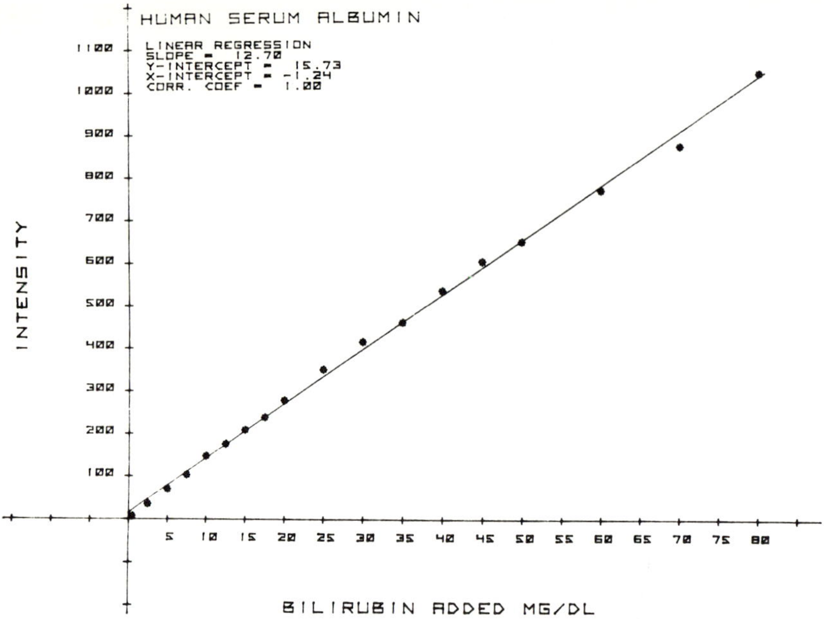


FIG. 18. Relationship of detector response to the amount of bilirubin added to a human serum albumin solution. Conditions as in Fig. 17 reprinted from ref. *47* by courtesy of the American Association for Clinical Chemistry.

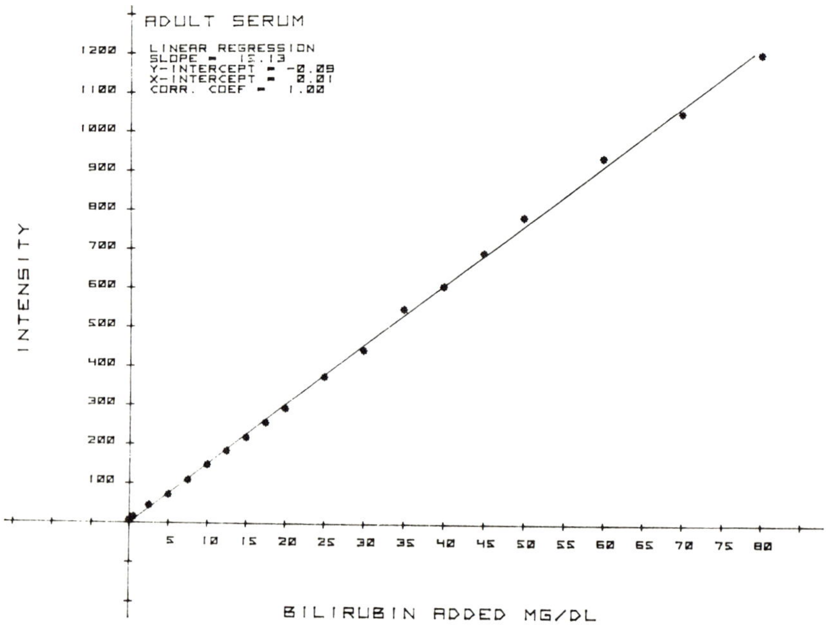


FIG. 19. Relationship of detector response to the amount of bilirubin added to adult serum. Conditions as in Fig. 17 reprinted from ref. *47* by courtesy of the American Association for Clinical Chemistry.

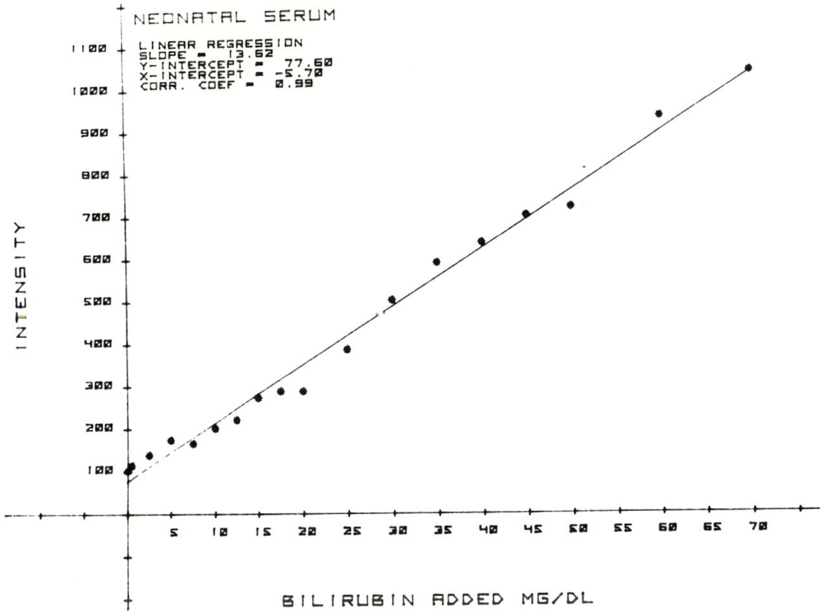


FIG. 20. Relationship of detector response to the amount of bilirubin added to neonatal serum. Conditions as in Fig. 17 reprinted from ref. *47* by courtesy of the American Association for Clinical Chemistry.

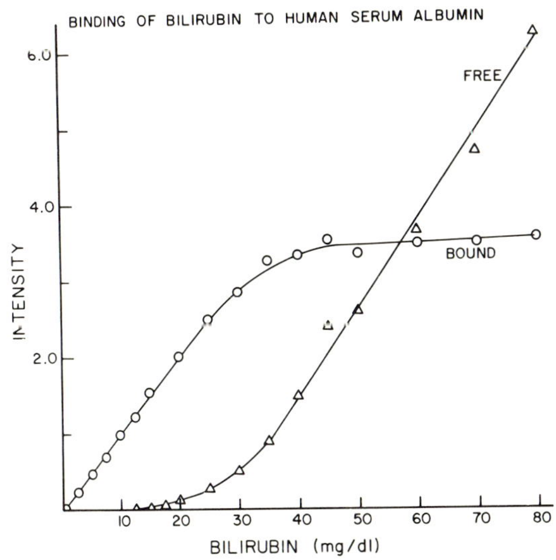


FIG. 21. Binding of bilirubin to purified human serum albumin. Conditions as in Fig. 17 reprinted from ref. *47* by courtesy of the American Association for Clinical Chemistry.

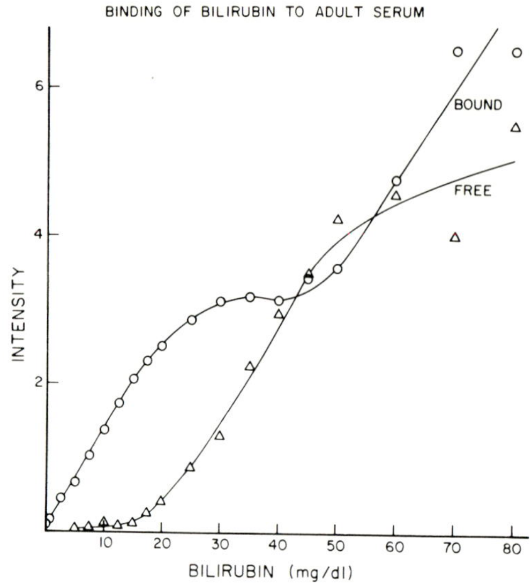


FIG. 22. Binding of bilirubin to proteins in adult serum. Conditions as in Fig. 17 reprinted from ref. *47* by courtesy of the American Association for Clinical Chemistry.

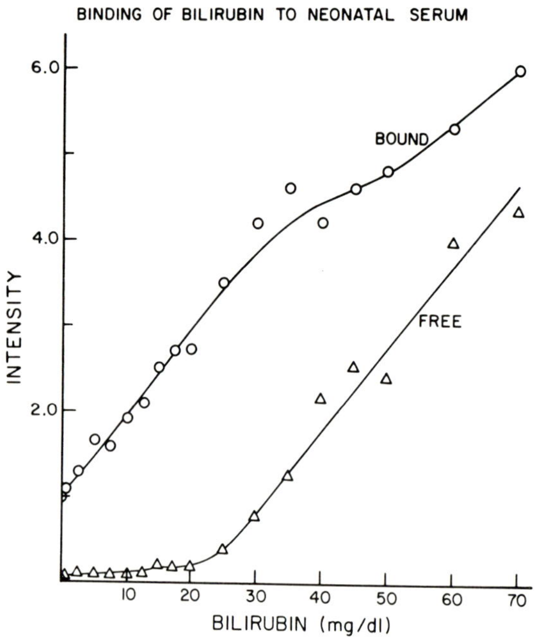


FIG. 23. Binding of bilirubin to proteins in neonatal serum. Conditions as in Fig. 17 reprinted from ref. *47* by courtesy of the American Association for Clinical Chemistry.

adult serum increases, it is seen in Fig. 22 that a second family of lower affinity binding sites begin to load. The binding curve for neonatal serum shown in Fig. 23 is quite different. The free bilirubin curve indicates that there are sets of high and low affinity sites, but the protein-associated bilirubin curve is almost continuous. These results suggest that both the proteins binding bilirubin and the binding affinities are different in neonatal than in adult serum.

## E. Protein Profiles

When the adult serum and neonatal serum samples were examined by high performance anion exchange chromatography (Figs. 24 and 25, respectively), there were obvious differences in protein composition that could account for the differences in binding characteristics. No attempts were made to isolate and determine the bilirubin binding characteristics of these various components.

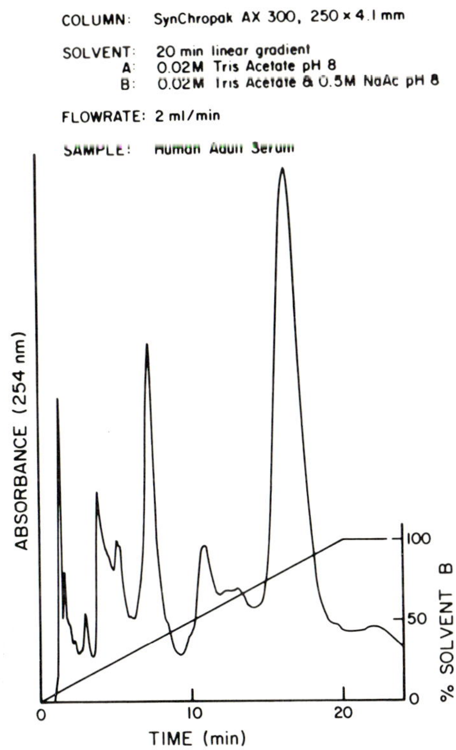


FIG. 24. Analysis of adult serum proteins reprinted from ref. *47* by courtesy of the American Association for Clinical Chemistry.

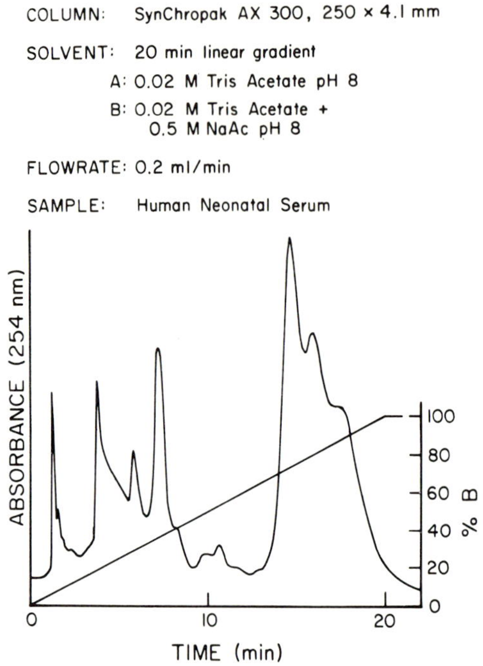


FIG. 25. Analysis of neonatal serum proteins reprinted from ref. *47* by courtesy of the American Association for Clinical Chemistry.

## F. Relevance

These studies raise a question whether model system studies carried out with adult serum proteins may be applied to neonatal serum as has been done in the past. This observation is further substantiated by the report of Kapitulnik et al. (*50*) that bilirubin binding capacity of albumin increases during the first few weeks of life and continues to increase until about 5 months of age. Since albumin concentration does not change during this time (*51*), it is possible that the binding affinity of albumin or other proteins is changing.

These studies further show that the binding of bilirubin to albumin and other proteins is a function of pH, ionic strength, and the presence of compounds that compete for the same binding sites. This must be taken into account when developing an assay. Failure to control these variables is probably responsible for some of the conflicting data in the literature.

Clinically, the most widely used test for the determination of free bilirubin and bilirubin binding capacity of serum is the Sephadex gel method. On the basis of the limited data presented here, we would conclude that the HPLC technique is comparable to the Sephadex gel method, but requires much less sample, is much quicker, and is more quantitative. However, both of these methods may be criticized because the adsorption of bilirubin to the column disturbs the equilibrium of bound and free bilirubin and the columns probably strip part of the loosely bound bilirubin from the protein.

## V. Future Trends

We would predict that protein separations in research laboratories will change dramatically in the next decade. At least 80% of all column fractionations of proteins will be carried out on high performance supports because of their greater resolving power and shorter separation times. As column technology for proteins continues to develop, we can expect still higher resolution and the introduction of new separation modes not being used today.

From the work presented in this review, we would conclude that HPLC is fast and more quantitative than other techniques for protein separation. With the 1000-fold dynamic range and the excellent sensitivity of the detection systems that were described above, HPLC systems are capable of handling the wide variations encountered in biological samples. Additionally, the ease of automating HPLC systems and the absence of lengthy sample preparations before analysis are major reasons why protein separations by HPLC have the potential to gain wide acceptance.

Although the 10–20 min analyses described in this review are quite short, there will be a need to further increase sample throughput if the techniques are to become widely used in routine analyses, such as that for $HbA_{1C}$. It should also be noted that during the course of separating one series of proteins on a column, other proteins are also resolved. At the present time, these additional materials are discarded or ignored. Multiple detectors could be used for the simultaneous assay of several classes of compounds.

## Acknowledgments

This research was supported by Grant No. GM 25431 from NIH-USPHS. Journal paper #7915 from Purdue Agricultural Experiment Station.

## References


1. Meister, A., *J. Biol. Chem.* **184,** 117 (1950).
2. Neilands, J. B., *Science* **115,** 143 (1952).
3. Vesell, E. S., and Bearn, A. G., *Proc. Soc. Exptl. Biol. Med.* **94,** 96 (1957).
4. Galen, R. S., Reiffel, J. A., and Gambino, S. R., *J. Am. Med. Assoc.* **232,** 145 (1975).
5. Evrev, T. I., Autoantigenicity of LDH-X isoenzymes, in *Isozymes,* Vol. II, Markert C., ed., Academic Press, New York, 1975, p. 129.
6. Chang, S. H., Noel, R. N., and Regnier, F. E., *Anal. Chem.* **48,** 1839 (1976).
7. Chang, S. H., Gooding, K. M., and Regnier, F. E., *J. Chromatogr.* **125,** 103 (1976).
8. Schlabach, T. D., Chang, S. H., Gooding, K. M., and Regnier, F. E., *J. Chromatogr.* **134,** 91 (1977).
9. Schlabach, T. D., personal communication.
10. Scott, R. P. W., and Kucera, P., *J. Chromatogr. Sci.* **9,** 641 (1971).
11. Schroeder, R. R., Kudirka, P. J., and Toren, E. C., Jr., *J. Chromatogr.* **134,** 83 (1977).
12. Schlabach, T. D., Fulton, J. A., Mockridge, P. B., and Toren, E. C., Jr., *Clin. Chem.* **25,** 1600 (1979).
13. Fulton, J. A., Schlabach, T. D., Kerl, J. E., and Toren, E. C., Jr., *J. Chromatogr.* **175,** 269 (1979).
14. Fulton, J. A., Schlabach, T. D., Kerl, J. E., and Toren, E. C., Jr., *J. Chromatogr.* **175,** 269 (1979).
15. Schlabach, T. D., Fulton, J. A., Mockridge, P. B., and Toren, E. C., Jr., *Anal. Chem.* **52,** 729 (1980).
16. Denton, M. S., Bostick, W. D., Dinsmore, S. R., and Mrochek, J. E., *Clin. Chem.* **24,** 1408 (1978).
17. Bostick, W. D., Dinsmore, S. R., Mrochek, J. R., and Waalkes, T. P., *Clin. Chem.* **24,** 1305 (1978).
18. Kudirka, R. J., Schroeder, R. R., Hewitt, T. E., and Toren, E. C., Jr., *Clin. Chem.* **22,** 471 (1976).
19. Roberts, R., and Sobel, B. E., *Am. Heart J.* **95,** 521 (1978).
20. Mercer, D. W., *Clin. Chem.* **20,** 36 (1974).
21. Mercer, D. W., *Clin. Chem.* **21,** 1102 (1975).
22. Schlabach, T. D., Alpert, A. J., and Regnier, F. E., *Clin. Chem.* **24,** 1351 (1978).
23. Bostick, W. D., Denton, M. S., and Dinsmore, S. R., *Clin. Chem.* **26,** 712 (1980).
24. Schlabach, T. D., and Regnier, F. E., *J. Chromatogr.* **158,** 349 (1978).
25. Schlabach, T. D., Fulton, J. A., Mockridge, D. B., and Toren, E. C., *Clin. Chem.* **25,** 707 (1980).
26. Schroeder, W. A., Pace, L. A., and Huisman, T. H. J., *J. Chromatogr.* **118,** 295 (1976).
27. Abraham, E. C., Reese, A., Stallings, M., and Husiman, T. H. J., *Hemoglobin* **1,** 27 (1976).

28. Abraham, E. C., Huisman, T. H. J., Schroeder, W. A., Pace, L. A., and Grussing, L., *J. Chromatogr.* **143,** 57 (1977).
29. Bunn, H. F., Gabbay, K. H., and Gallop, P. M., *Science* **200,** 21, (1978).
30. Gonen, B., and Rubenstein, A. H., *Diabetologia* **15,** 1 (1978).
31. Gooding, K. M., Lu, K. -C., and Regnier, F. E., *J. Chromatogr.* **164,** 506 (1979).
32. Hearn, M. T. W., and Hancock, W. S., *TIBS* **4,** N58-62 (1979).
33. Schroeder, W. A., Shelton, J. B., Shelton, J. R., and Powars, D., *J. Chromatogr.* **174,** 385 (1979).
34. Davis, J. E., McDonald, J. M., and Jarett, L., *Diabetes* **27,** 102 (1978).
35. Cole, R. A., Soeldener, J. S., Dunn, T. J., and Bunn, H. F., *Metabolism* **27,** 289 (1978).
36. Wajcman, H., Dastugue, B., and Labie, D., *Clin. Chem. Acta* **92,** 33 (1979).
37. Karp, W. B., *Pediatrics* **64,** 361 (1979).
38. Gitzelmann-Cumarasamy, N., and Kuenzle, C. C., *Pediatrics* **64,** 375 (1979).
39. Levine, R. L., *Pediatrics* **64,** 380 (1979).
40. Lee, K. -S., and Gartner, L. M., *Rev. Perinatal Med.* **2,** 319 (1978).
41. Cashore, W. J., Gartner, L. M., Oh, W., and Stern, L., *J. Pediatrics* **93,** 827 (1978).
42. Brodersen, R., *Acta Pediatr. Scand.* **66,** 625 (1977).
43. Lightner, D. A., *In vitro* photooxidation products of bilirubin, in *Phototherapy in the Newborn: An Overview,* Odell, G. B., Schaffer, R., and Simopoulos, A. P., National Academy of Sciences, Washington, D. C., 1974, p. 34.
44. Cooke, J. R., and Roberts, L. B., *Clin. Chim. Acta* **26,** 425 (1969).
45. Athanassiadis, S., Chopra, D. R., Fisher, M. A., and McKenna, J., *J. Lab. Clin. Med.* **83,** 968 (1974).
46. Klatskin, G., and Bungards, L., *J. Clin. Invest.* **35,** 537 (1956).
47. Lu, K., Gooding, K. M., and Regnier, F. E., *Clin. Chem.* **25,** 1608 (1979).
48. Jersova, V., Jirsa, M., Herengova, A., Koldovsky, O., and Weirichova, J., *Biol. Neonat.* **11,** 204 (1967).
49. Irivin, R., Odievre, M., and Lemonnier, A., *Clin. Chem.* **23,** 541 (1977).
50. Kapitulnik, J., Horner-Metashau, R., Blondheim, S. H., Kaufmann, N. A., and Russell, A., *Pediatrics* **86,** 442 (1975).
51. Cashore, W. J., Horwick, A., Laterra, J., and Oh, W., *Biol. Neonate* **32,** 304 (1977).

# Chapter 15

## Bilirubin and Its Carbohydrate Conjugates

Norbert J. C. Blanckaert†

*Department of Laboratory Medicine and Liver Center*
*University of California Medical Center*
*San Francisco, California*

### I. Introduction

In mammals, the open tetrapyrrole bilirubin (structure 2, Fig. 1) is the principal degradation product of iron-protoporphyrin-IX (heme). The latter molecule is a tetrapyrrolic macrocycle and plays a critical role in aerobic metabolism by reversibly binding oxygen in hemoglobin and myoglobin, and by serving as the active site in oxidation reactions catalyzed by hemoprotein enzymes. Important cyclic tetrapyrroles in nature related to heme are chlorophylls, which contain magnesium and are derived from protoporphyrin-IX, and vitamin $B_{12}$, a corrinoid derived from uroporphyrinogen-III.

Whereas open tetrapyrroles have an important physiological role in algae, serving as the prosthetic group of the photosynthetic biliproteins, bilirubin in mammals merely is a waste product without any obvious function. Yet bilirubin metabolism has piqued the curiosity of many generations of clinicians and investigators, largely

†Current address: Department of Medical Research, Campus Gasthuisberg, University of Leuven, Leuven, Belgium.

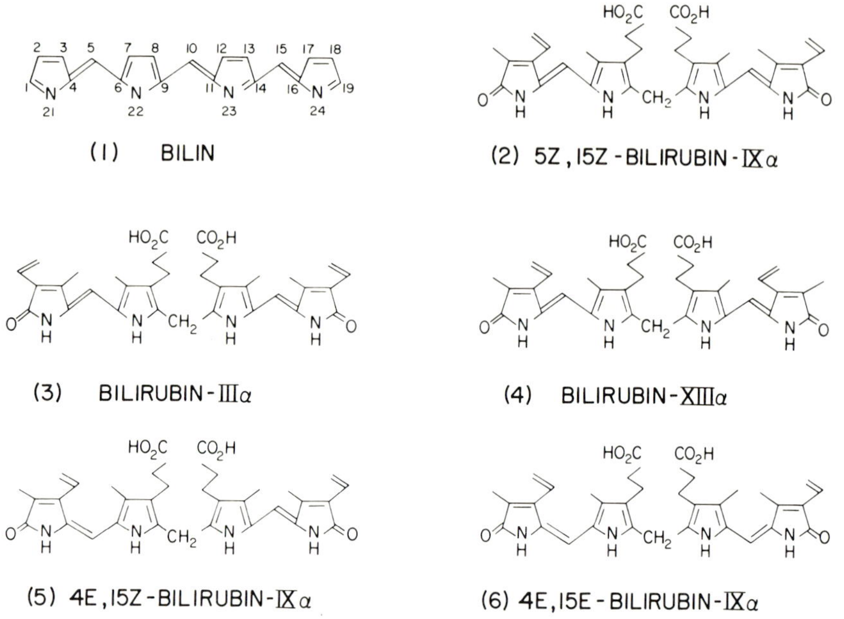


FIG. 1. Structures of bilin and bilirubins.

because accumulation of this yellow pigment in tissues is such an obvious and frequent sign of liver dysfunction or hemolytic disease. Jaundice also frequently develops in neonates, and it is now well realized by pediatricians that bilirubin is a potentially neurotoxic compound, which in neonates with severe hyperbilirubinemia can cause encephalopathy and psychomotor retardation.

Protection against the cytotoxic effects of bilirubin depends on its binding to intra- and extracellular "carrier proteins," and conversion into polar glycosides. This conjugation is catalyzed by microsomal UDP-glucuronosyltransferase, an enzyme system that also plays a key role in detoxification and disposition of many other endogenous compounds and xenobiotics. Bilirubin can be regarded, therefore, as a probe for studying transport and biotransformation of many other nonpolar substances, and thus has attracted the attention of numerous biochemists and pharmacologists. Bilirubin is also assured of continued physiological interest because it is such an obvious solute in bile and shares the same or similar "carriers" for hepatic uptake and biliary transport with many other organic anions.

In recent years, the most significant development in bilirubin research undoubtedly has been the awakening to the complexities of the chemistry of bilirubin and its congeners. These aspects currently are being investigated by analytical, organic, physical, and photochemists and further progress in this area unquestionably will lead to significant improvement in our understanding of the molecular biology of these pigments.

This article is focused primarily on those aspects of bilirubin metabolism and methodology for bile pigment analysis that are of interest to the clinical scientist, and on which new information recently has become available. For a more detailed discussion of bilirubin metabolism and its disorders, the reader is referred to several recent review articles (*1–5*).

## II. Nomenclature

The appropriate commissions have not, as yet, pronounced on nomenclature rules for bile pigments and their azoderivatives. The naming of bile pigments and related compounds that is outlined below and used throughout this chapter is not universally accepted, but generally reflects the most appropriate names presently used amongst those who are actively involved with bile pigment research. The nomenclature described is largely based on the recently published recommendations of Bonnett (*6*), who is a member of a working party that was set up jointly by the Commission for Nomenclature in Organic Chemistry and the Commission for Biochemical Nomenclature.

### A. Tetrapyrroles

The term bilinoids, or bile pigments, is used to describe open-chain tetrapyrroles with the skeletal structure 1 in Fig. 1. This basic structure is called bilin. By convention, (i) in the absence of specific information on the imino hydrogen location the 22-H tautomer is drawn, and (ii) number 20 is omitted; C-20 is a phantom atom that corresponds to the extra carbon atom that would be required to transform the bilinoid into a porphyrin ring. Naturally occurring bilinoids have oxygen at the terminal positions, and are formally 1,19-dihydroxy derivatives of bilin or 10,23-dihydrobilin.

Many trivial names, including bilirubin (structure 2, Fig. 1), biliverdin, and mesobilirubin are used to denote specific bile pigments. 1,19-Dihydroxy derivatives of 10,23-dihydrobilin are commonly called "(bili)rubins," and 1,19-dihydroxy derivatives of bilin called "(bili)verdins." The configuration of the $\beta$-substituents in rubins and verdins is conveniently denoted by reference to the corresponding protoporphyrin isomer that has the same sequence of substituents.

Thus, the trivial name is followed by a Roman numeral (e.g., IX), corresponding to that used to designate the isomeric type of the precursor porphyrin, and by a Greek letter (e.g., $\alpha$) indicating which one of the porphyrin meso-bridges ($\alpha$, $\beta$, $\gamma$, or $\delta$) corresponds to the phantom C-20 carbon atom of the bilinoid. For example, bilirubin-IX$\alpha$ (structure 2, Fig. 1), corresponds to the rubin that is formed by cleavage of protoporphyrin-IX at the $\alpha$ meso bridge. By convention, and for convenience, the term "bilirubin" can be used to specifically denote bilirubin-IX$\alpha$. Bilirubin-III$\alpha$ (structure 3, Fig. 1) and bilirubin-XIII$\alpha$ (structure 4, Fig. 1) normally do not occur in bodily fluids in significant amounts, but can be formed in vitro by dipyrrole exchange of bilirubin-IX$\alpha$ (see below).

Depending on the configurations (Z or E) at the C-5 and C-15 methine bridges, several geometrical isomers of bilirubins exist. Thus, bilirubin can occur in the form of four geometrical isomers (Fig. 1): 4Z,15Z-bilirubin (structure 2), 4E,15Z-bilirubin (structure 5), 4Z,15E-bilirubin (not shown), and 4E,15E-bilirubin (structure 6).

### B. Azoderivatives

Bilirubin pigments are routinely determined by diazo methods. These procedures involve cleavage of the tetrapyrrolic bilirubin molecule by reaction with a diazotized aromatic amine (e.g., diazotized sulfanilic acid), with formation of two dipyrrolic azoderivatives and formaldehyde. It should be noted that diazotation refers to conversion of the aromatic amine to diazo reagent, and not to conversion of bilirubins to azoderivatives. The term diazo cleavage or diazo coupling is used to denote the reaction of the bilirubin pigment with the diazo reagent. Usage of the term azobilirubin to denote the dipyrrolic azoderivatives is misleading since azobilirubin refers to a tetrapyrrole, and the term azodipyrrole has been proposed instead (*7*).

## III. Bilirubin Chemistry and Metabolism

### A. Bilirubin Chemistry

Bilirubin is only sparingly soluble in aqueous solution at physiologic pH. Although accurate data are not available, its estimated solubility in 0.1 *M* phosphate buffer, pH 7.4, at 25°C is $10^{-7}$ *M* (0.006 mg%) (*8*). A biological implication of this appears to be that conjugation of bilirubin, to form polar conjugates, is required for efficient excretion. However, the nonpolar character of bilirubin is totally unexpected considering the presence of two carboxyl groups in the molecule that, at physiological pH, would be expected to be present in the carboxylate form. A possible explanation for the nonpolar properties of bilirubin was first suggested by Fog and Jellum, who postulated that both propionic acid groups of the pigment are involved in intramolecular hydrogen bonds that make the carboxyl groups unavailable for ionization (*9*). Such a hydrogen-bonded structure recently has been demonstrated by X-ray diffraction studies for crystalline bilirubin in which each carboxyl group is involved in three intramolecular hydrogen bonds (*10*). These hydrogen bonds effectively shield all polar functions and confer to the molecule a rigidly fixed, chiral ridge-tile conformation in which the two dipyrrylmethene parts of the molecule form two planes with an interplanar angle of approximately 97°. A similar angular structure with four intramolecular hydrogen bonds has been demonstrated for the crystal structure of the diisopropylammonium salt of bilirubin (*11*), and it is now postulated that bilirubin present in the body also has a rigidly fixed, angular structure that apparently makes it impossible for the unconjugated pigment to become secreted in the bile canaliculus. Disruption of the bilirubin conformation that is imposed by the multiple intramolecular hydrogen bonds and/or exposure of polar functions of the molecule may be required for efficient biliary excretion, and the principal mechanism that has evolved in nature to ensure excretion of bilirubin, is esterfication of one or both propionic acid side chains with a carbohydrate. Experimental evidence in support of this concept has come from studies of model compounds, including mesobilirubinogen (*12*) and bilirubin-IX$\beta$, $\gamma$, and $\delta$ isomers (*13*), in which intramolecular hydrogen bonds are absent and conjugation is not required for efficient biliary secretion. The particular conformation and/or poor solubility of bilirubin may also explain the "indirect" diazo reaction of unconjugated bilirubin since coupling of diazo reagent with pigment may require a conformational change in bilirubin that possibly can be achieved by addition of an "accelerator" substance.

## B. Bilirubin Metabolism

Bilirubin formation reflects the continuous turnover of heme and essential hemoproteins such as hemoglobin, myoglobin, cytochromes, and other hemoprotein enzymes. In normal adults, the daily production of bilirubin averages 250–350 mg. Its major source is hemoglobin of senescent erythrocytes that are being destroyed in the mononuclear phagocytic cells (reticuloendothelial cells) of the spleen and bone marrow. Bilirubin formed in these organs is released into the circulation and transported to the liver. Turnover of hemoglobin-heme normally accounts for approximately 70% of the bilirubin formed in humans. Another significant source of bilirubin is degradation of nonhemoglobin heme in the liver, which contains relatively large amounts of hemoprotein enzymes with high turnover rates. Recent observations indicate that the hepatic contribution to total bilirubin formation in normal humans ranges from 23 to 37% (*14*).

Heme catabolism in man and other mammals normally involves cleavage of the porphyrin macrocycle and results in nearly stoichiometric formation of CO and bilirubin. This cleavage reaction is remarkably regioselective since virtually all natural bile pigments have the IX$\alpha$ configuration, indicating that opening of heme occurs almost exclusively at the $\alpha$-methene bridge. Physiological heme degradation appears to be catalyzed by two enzyme systems, one microsomal, the other cytosolic (*5*). Oxidative attack on the $\alpha$-methene bridge of heme, resulting in cleavage of the ring tetrapyrrole and formation of bilirubin and CO, is catalyzed by heme oxygenase, a microsomal enzyme system whose highest activity is in the spleen, but which is also present in liver, macrophages, and other tissues that convert heme to bile pigment. Biliverdin is reduced to bilirubin by biliverdin reductase, a cytosolic enzyme abundantly present in most mammalian tissues.

Unconjugated bilirubin, which is virtually insoluble in water at neutral pH, is maintained in solution in body fluids by reversible binding to proteins, albumin in plasma, and predominantly ligandin in the cytoplasm. Both proteins contain one high-affinity binding site, with an estimated affinity constant ($K_a$) of the order of $10^8$–$10^9$ $M^{-1}$ and are present in abundance. Therefore, they provide a large binding reservoir and the concentration of unbound bilirubin in plasma and tissues is normally vanishingly small, probably around $10^{-8}$–$10^{-9}$ $M$. Since it is postulated that the cytotoxicity of bilirubin is directly related to the concentration of unbound bilirubin, binding proteins might have a detoxification function, in addition to their solubilization role.

Bilirubin in the circulation normally is rapidly cleared and excreted in bile by the liver, which under physiological conditions is the only organ that removes bilirubin from the plasma. Hepatic uptake of the pigment probably is mediated by carrier proteins in the sinusoidal membrane of the liver cell. In the hepatocyte, bilirubin is tightly bound to so-called carrier proteins, including ligandin and fatty acid binding, or Z-, protein. Similar to albumin in the plasma, these proteins serve such functions as solubilization, storage, and possible detoxification of bilirubin.

Detoxification of bilirubin occurs by esterfication with carbohydrates. The formed conjugates are presumably nontoxic and

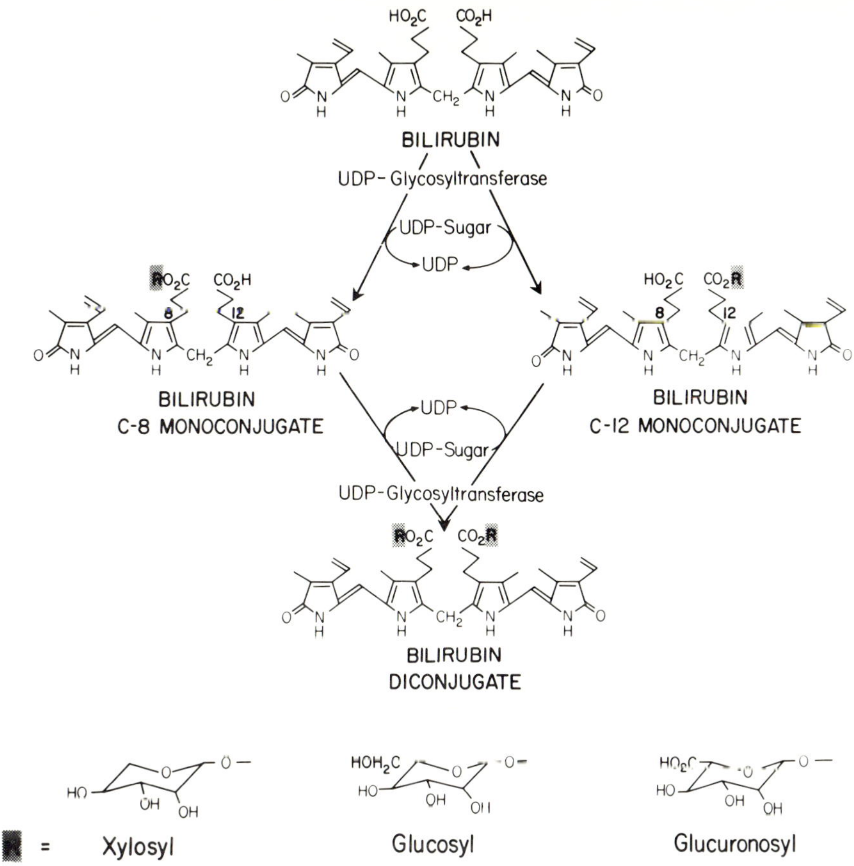


FIG. 2. Hepatic bilirubin conjugation.

readily excretable in bile. Conjugation occurs by esterfication of one or both propionic acid side-chains of the pigment with a sugar residue to form a mono- or di-ester. Conjugates with glucuronic acid, glucose, and xylose (Fig. 2), have been demonstrated and the relative abundance of these glycosides excreted in bile has been found to be species-dependent (*15*). In most species thus far examined, including humans, bilirubin glucuronides constitute the major fraction of pigments in bile. Whereas the conjugation process results in formation of glycosides (1-*O*-acyl esters), rapid, presumably nonenzymatic isomerization in the bilirubin conjugates may occur with stored bile in vitro and with cholestasis in the body by sequential migration of the bilirubin acyl group from position C-1 to positions C-2, C-3, and C-4 of the carbohydrate residue (Fig. 3) (*16, 17*). This positional

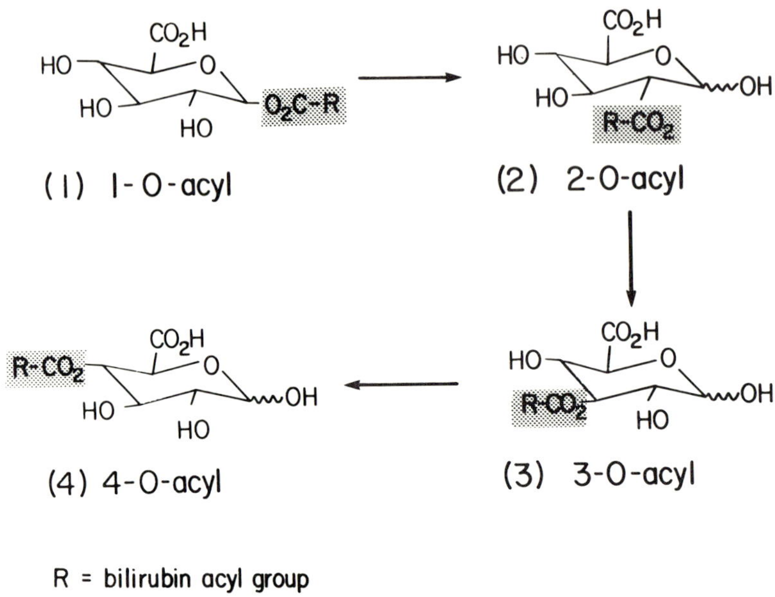


FIG. 3. Structures of the four positional isomers of a bilirubin glucuronide.

isomerization is responsible for the increased complexity of the azopigments derived from body fluids of patients with hepatobiliary disease (*3*).

The existence of bilirubin sugar conjugates with a disaccharidic conjugating residue as reported by Kuenzle (*18*) has not been confirmed in recent studies (*19*). There is also some indirect evidence that bilirubin derivatives such as sulfates, phosphates, and polypeptide conjugates may occur in mammalian bile (*20–23*). Their existence is at best unproven (*24*), and even if these pigments occur naturally, their amounts in bile must be minute compared to those of the sugar conjugates.

The formation of bilirubin monoglycosides is catalyzed by an enzyme system in the endoplasmatic reticulum of the liver cell, the glycosyl moieties being transferred to bilirubin from their respective UDP-sugars (*15*). The exact relationship between xylosyl-, glucosyl-, and glucuronosyltransferase is unknown.

It recently has been postulated that diglucuronide is formed in the liver by transglucuronidation of monoglucuronide, with conversion of two moles monoglucuronide to one mole of bilirubin and one mole of diglucuronide (*25*). Recent studies have demonstrated, however, that formation of diglucuronide from monoglucuronide in rat liver microsomal fraction is catalyzed by a UDP-glucuronic acid-dependent glucuronosyltransferase system (*26, 27*). Moreover, studies in intact rats indicate that transglucuronidation of monoglucuronide does not occur in vivo, whereas the results are compatible with diglucuronide formation by a UDP-glucuronosyltransferase system (*28*).

Once formed in the endoplasmatic reticulum of the liver cell, bilirubin conjugates are rapidly excreted in the bile canaliculus and drained to the gut lumen by the biliary tree. Whereas it has been reported that some monoconjugated pigment may normally accumulate in the hepatocyte, biliary excretion of bilirubin glucuronides seems to be a very efficient process. Reflux of conjugates into plasma is probably minute or nonexistant because bilirubin conjugates are undetectable in normal serum even when using a sensitive and specific method (see below). The available experimental data suggest that pigment secretion is a carrier-mediated, probably active, transport process that is normally the rate-limiting step in the overall transport of bilirubin from plasma to bile.

Bilirubin conjugates are not appreciably absorbed by the gall bladder or intestinal mucosa, so that there is no significant enterohepatic circulation of conjugated pigment. Catabolism of the excreted bilirubin conjugates largely occurs in the terminal ileum and large intestine and involves hydrolysis to unconjugated pigment and reduction by the action of intestinal bacteria to a group of colorless tetrapyrroles, collectively termed the urobilinogens (*29*). These pigments and their oxidized, orange derivatives, the urobilins, are excreted in the stool. Urobilinogens are partially absorbed and undergo enterohepatic circulation. On the other hand, a substantial fraction (approximately 50%) of bilirubin conjugates secreted in bile appear to be converted to unknown derivatives that differ from urobilinogens and might correspond to dipyrrolic oxidation products of bilirubin (*30*).

Hyperbilirubinemia corresponds to accumulation in serum of unconjugated bilirubin and/or conjugated bilirubin mono- and diglucuronides, each as a mixture of the four positional isomers, and can be caused by a variety of disorders in bilirubin metabolism including bilirubin overproduction (e.g., hemolysis), defective hepatic uptake and/or conjugation (e.g., neonatal jaundice, Gilbert syndrome), and impaired biliary secretion and/or drainage of the conjugated bilirubins. Differential diagnosis of hyperbilirubinemia can be considerably aided by differential determination of the various bilirubins in serum. Thus unconjugated hyperbilirubinemia occurs in hemolytic disorders or when hepatic uptake and/or conjugation are deficient, whereas predominantly conjugated hyperbilirubinemia is caused by hepatobiliary disease. For further details on differential diagnosis of jaundice by analysis of bilirubins, the reader is referred to reference *3*.

## IV. Analysis of Serum Bilirubins

Apart from its importance in studies on bilirubin metabolism, measurement of unconjugated bilirubin and its mono- and diconjugates in bodily fluids is important in clinical diagnosis. In fact, determination of the serum bilirubin concentration is one of the oldest "liver function tests" used in clinical laboratories. The presence of unconjugated bilirubin and its mono and di-ester conjugates in serum of jaundiced patients with hepatobiliary disease and bile was recognized as long ago as 1957 (*31–33*). Yet an accurate and precise method for determination of these pigment fractions has not been developed until recently (*34*). This is not for want of trying, since numerous methods and many more modifications have been devised. Problems that have impeded progress in this area have been the unavailability of pure bilirubin mono- and diconjugates, the notorious instability of the native tetrapyrroles, the adherence of the pigments, particularly of the conjugated ones, to denatured protein, and the difficulty of extracting bilirubin diconjugates into an organic phase.

In general, two basically different approaches to the measurement of bilirubin and its conjugates can be discerned. The first is to analyze the pigments in their native forms, as tetrapyrroles. In the second, the pigments are converted into their more stable dipyrrolic azoderivatives (Fig. 4) prior to measurement and further qualitative analysis (diazo methods).

### A. Conventional Methods

*1. Diazo Procedures.* Diazo methods are most frequently used for the measurement of bilirubins in bodily fluids. This approach has been chosen because it minimizes interference by dietary lipochromes, yields fairly stable reaction products, and is believed to circumvent the calibration problems for conjugated bilirubin pigments. The diazo reaction of diazotized sulfanilic acid with bilirubin was discovered by Ehrlich in 1883 (*35*), and almost all of the bilirubin assays presently used in clinical laboratories are still based on this diazo reagent. A milestone in bilirubin research was the discovery by Van den Bergh and Muller in 1916 that there were at least two different types of bilirubin in serum, one that requires addition of an accelerator, such as an alcohol, to the sample to rapidly react with

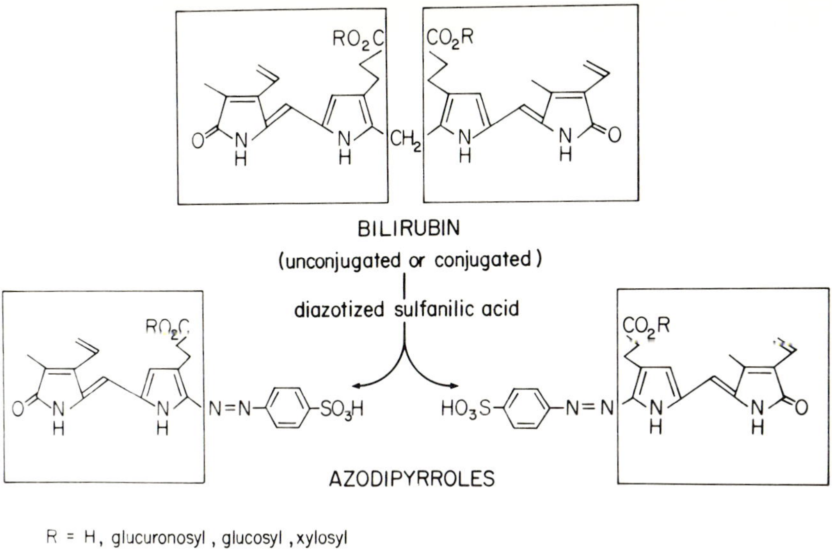


FIG. 4. Conversion of bilirubins to azoderivatives.

diazotized sulfanilic acid (indirect-reacting bilirubin), and another one that readily reacts with the diazo reagent even in the absence of an accelerator substance (direct-reacting bilirubin) (*36*). In 1957, three groups of investigators independently postulated that "direct-reacting bilirubin" is an alkali-labile sugar conjugate, and that "indirect-reacting bilirubin" corresponds to unconjugated bilirubin (*31–33*). Measurement of direct- and indirect-reacting bilirubin has proven to be very valuable for clinical diagnosis, since it generally permits one to ascertain the relative degree to which conjugated or unconjugated bilirubin predominates in the sample. These diazo procedures are inadequate, however, for accurate differential determination of unconjugated and conjugated pigment (*7, 37*). Countless modifications have been devised to achieve a more specific quantification of the various pigment fractions. Much of this work is repetitious and irrelevant, and the huge volume of literature on this subject merely reflects the inadequacy of this forest of diazo methods.

Significant progress in methods of analysis of bilirubins was achieved in the late sixties, with the development of diazo reagents that, in contrast to diazotized sulfanilic acid, yield azoderivatives that are readily extractable into an organic solvent (*7, 18*). Thus, extraction of the azopigments results in an increased sensitivity of the assay, renders the method suitable for measurement of bilirubins in turbid aqueous samples or tissue preparations, and makes it possible to apply the extracted pigments directly to thin-layer chromatographic plates for analysis. The ethyl anthranilate method offers the additional advantage that selective coupling of the diazo reagent with conjugated bilirubin can be obtained under well-defined reaction conditions. These features, namely the possibility to easily extract the azopigments in an organic solvent, and the selective reaction with conjugated bilirubins even in the presence of an excess of unconjugated bilirubin, have been the basis for the determination of mono- and diconjugates in bodily fluids (*7*) and for the development of the first reliable assay for measurement of bilirubin UDP-glucuronosyltransferase activity by Van Roy and Heirwegh in 1968 (*38*).

The aniline and, particularly, the ethyl anthranilate azoderivatives have also proven to be of great value for structural analysis of bilirubins. Thus, these azopigments were found to be suitable for chromatographic analysis and NMR studies, stable enough for derivatization, and sufficiently volatile for electron impact mass spectrometric analysis.

*2. Analysis of Tetrapyrroles.* Unconjugated bilirubin in organic solvents or in artificial protein-containing aqueous solutions can be readily measured by direct spectrophotometry, using appropriate absorption coefficients and/or calibration curves. It is important, however, to ascertain that Beer's law is obeyed and that the absorption spectrum matches that of bilirubin in the pure solvent. For bodily fluids, however, direct spectrophotometric determinations generally are unreliable because turbidity and natural pigments, such as carotenoids, flavins, and hemoproteins, may significantly interfere. Moreover, this approach cannot be used for samples that contain bilirubin conjugates as these pigments are not available in pure form for calibration purposes, and because the proportion of the various pigment fractions in an individual sample would be unknown. Direct spectrophotometric measurement probably allows a fairly reliable estimation of bilirubin in serum from neonates, which generally does not contain significant amounts of carotenoids or bilirubin conjugates. Moreover, it is possible to correct for the presence of hemoproteins by measurement of the absorbance of the sample at two different wavelengths (*39, 40*). Although the technique does not allow an accurate determination of bilirubin, direct spectrophotometry on amniotic fluid is widely used, and has been shown to be valuable for assessment of the degree of hemolytic disease in a fetus of a blood group-incompatible pregnancy (*41*).

In general, two approaches have been used for separation and individual quantification of unconjugated bilirubin and its mono- and diconjugates. One approach attempts to separate the various pigment fractions by solvent-partitioning, using solvent systems consisting of two (*42–44*) or three (*45*) phases. Critical evaluation of these methods has shown that none of them is truly analytic and accurate (*7, 46–48*).

In a second approach, numerous chromatographic procedures have been devised. Based on the solvent-partitioning principle, satisfactory separation of bilirubin and its mono- and diglucuronides has been achieved by reversed-phase chromatography on siliconized kieselguhr (*31, 49, 50*). Column chromatographic separation based on adsorption chromatography has also been reported (*42, 43, 51*). None of these column chromatographic techniques, however, has been developed or validated as an analytical method, and most of them are elaborate and/or cumbersome. Billing's refined version of her original reversed-phase chromatographic procedure (*52*) comes closest to being a valid quantitative method, but is not sensitive and is inaccurate, since predominantly bilirubin conjugates are lost by their adherence to the protein precipitate formed during preparation of the sample.

Over the last decade, solvent-partitioning methods and column techniques have been largely superseded by thin-layer chromatography, which offers far greater versatility (*53–58*). Although extremely useful for structural analysis of the pigments, these TLC methods are only semiquantitative. A thin-layer chromatographic procedure for separation of bilirubin and its mono- and dicarboxyl amide derivatives has been reported (*59*), and is useful for qualitative analysis of the various pigment fractions. All of these reported methods are at best semiquantitative, and generally not truly "direct" methods, as the separated pigment fractions are usually measured by diazo methods.

## B. High-Performance Liquid Chromatography

*1. Separation of Bilirubin and Its Mono- and Diglucuronides by HPLC.* Preliminary work on separation of bilirubin and its mono- and diglucuronides by HPLC recently has been reported (*60–62*). Lim described procedures for separation of unconjugated bilirubin from mono- and diglucuronides in bile by $C_{18}$ reversed-phase liquid chromatography and also for separation of bilirubin monoglucuronide from diglucuronide using a $\mu$-Bondapak carbohydrate column, with direct injection of bile (*60*). Onishi and coworkers recently reported separation of various bilirubin conjugates from bile or enzymic incubation mixtures by ion-pair reversed-phase HPLC (*61*). None of these methods, however, has yet been validated as an analytical tool or tested for analysis of serum samples. On the other hand, reports from another group of Japanese investigators on HPLC analysis of bilirubins cannot yet be evaluated, since details on the actual chromatographic procedures and analytical variables have not been published (*62*).

*2. Determination of Unconjugated Bilirubin and Its Mono- and Di-Carbohydrate Conjugates in Serum by Alkaline Methanolysis and High Performance Liquid Chromatography.* Major problems with direct chromatographic analysis of bilirubin and its conjugated derivatives include the large difference in polarity of unconjugated and conjugated pigment, the unavailability of pure, well-characterized reference bilirubin conjugates, preferential adsorption of bilirubin conjugates to precipitated serum proteins and difficulty in extracting the polar bilirubin conjugates in organic solvent. These problems can be largely circumvented by first converting the bilirubin mono- and di-conjugates to the corresponding mono- and dimethyl esters by alkaline methanolysis (Fig. 5) (*63*). Similar to bilirubin, these methyl ester derivatives are nonpolar, and therefore easily extractable into chloroform, and pure, well-characterized reference pigments are available. Virtually quantitative conversion (approx. 97%) of the sugar conjugates to methyl ester derivatives can be obtained, while unconjugated bilirubin remains

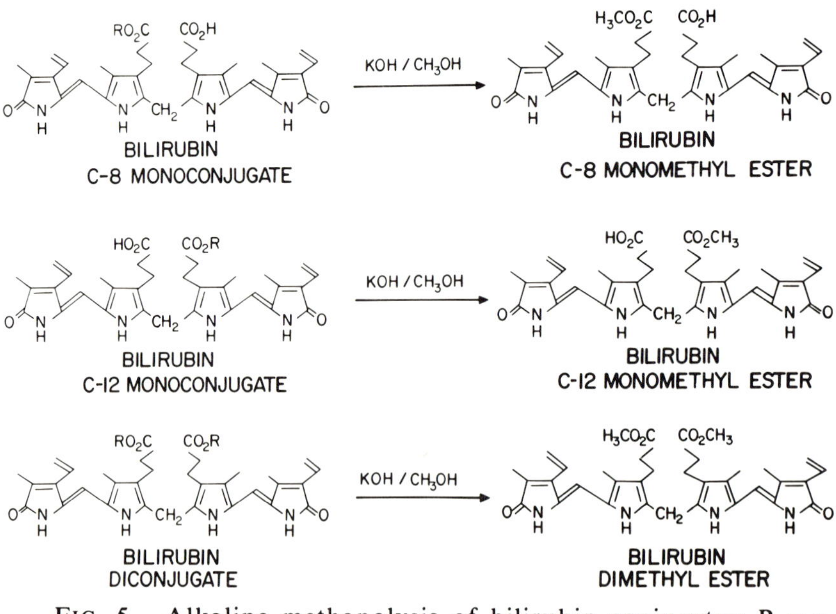


FIG. 5. Alkaline methanolysis of bilirubin conjugates; R = a carbohydrate residue.

intact. Bilirubin and its mono- and dimethyl ester can be separated by thin-layer chromatography (*63*) or HPLC (*34*), and then individually quantitated by photometry. In comparison with thin-layer chromatography, HPLC analysis offers the advantage of higher sensitivity and better precision and resolution. Moreover, an internal standard can be employed to measure the various pigment fractions directly.

Bilirubin and its mono- and dimethyl esters are separated by normal-phase chromatography on a silica-gel (LiChrosorb Si 60) 5$\mu$ particle-size column at 45°C, with detection of the eluted pigments at 430 nm and determination of peak areas in the chromatogram by an electronic integrator. With chloroform (containing 1% ethanol)/acetic acid (199/1; v/v) as mobile phase, excellent separation of unconjugated bilirubin and the various isomeric forms of bilirubin monomethyl ester (III$\alpha$, IX$\alpha$ C-8, IX$\alpha$ C-12, XIII$\alpha$; Fig. 6) is obtained, but bilirubin dimethyl esters (III$\alpha$, IX$\alpha$, XIII$\alpha$; Fig. 6) appear in the chromatogram as flat peaks, with pronounced tailing and long retention times (up to 35 min for the XIII$\alpha$ isomer). To maintain good separation of the early eluting compounds and improve chromatography of the dimethyl esters, gradient elution is used, with a slightly convex gradient, starting with chloroform/acetic acid (199/1; v/v) and ending after 8 min with chloroform/methanol/acetic acid (197/2/1 by vol). Elution with the latter solvent is continued for 6 min and the column is then re-equilibrated again for 10 min with the former

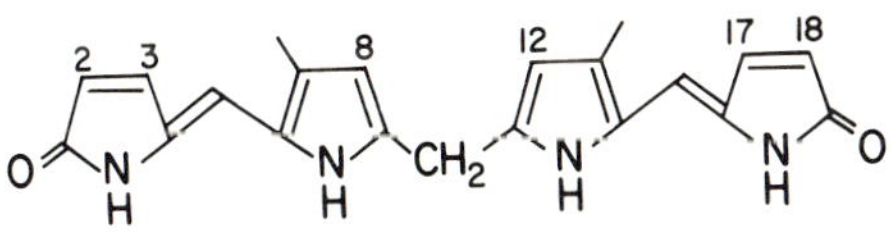


| Compound | Substituents in position | | | | | |
|---|---|---|---|---|---|---|
| | 2 | 3 | 8 | 12 | 17 | 18 |
| Bilirubin-III$\alpha$ | V | Me | P | P | Me | V |
| Bilirubin-IX$\alpha$ | Me | V | P | P | Me | V |
| Bilirubin-XIII$\alpha$ | Me | V | P | P | V | Me |
| Bilirubin-III$\alpha$ monomethyl ester | V | Me | P | Pme | Me | V |
| Bilirubin-IX$\alpha$ C-8 monomethyl ester | Me | V | Pme | P | Me | V |
| Bilirubin-IX$\alpha$ C-12 monomethyl ester | Me | V | P | Pme | Me | V |
| Bilirubin-XIII$\alpha$ monomethyl ester | Me | V | P | Pme | V | Me |
| Bilirubin-III$\alpha$ dimethyl ester | V | Me | Pme | Pme | Me | V |
| Bilirubin-IX$\alpha$ dimethyl ester | Me | V | Pme | Pme | Me | V |
| Bilirubin-XIII$\alpha$ dimethyl ester | Me | V | Pme | Pme | V | Me |

FIG. 6. Structures of unconjugated bilirubins and bilirubin methyl esters.

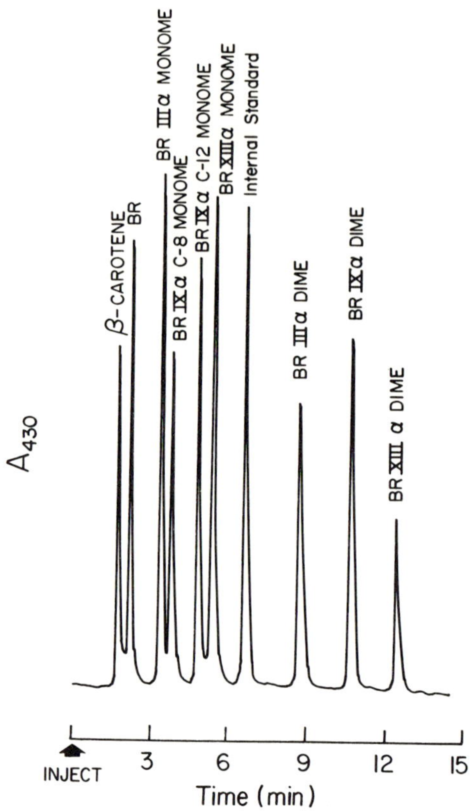


FIG. 7. Separation of reference compounds by high-performance liquid chromatography; BR = bilirubin.

solvent before injection of the next sample. Under these chromatographic conditions, β-carotene is separated from unconjugated bilirubin, and all isomeric forms of bilirubin mono- and dimethyl esters are resolved (Fig. 7). The method is calibrated with pure crystalline reference bilirubin, bilirubin monomethyl esters, bilirubin dimethyl ester, and xanthobilirubic acid methyl ester, which is used as internal standard.

When applied to human serum samples, the assay shows good precision (less than 5% for within-day analysis of bilirubin in normal or hyperbilirubinemic serum, and less than 8% for day-to-day analysis of hyperbilirubinemic serum). The sensitivity is excellent since 0.2 $\mu M$ bilirubin in 0.6 mL serum can be satisfactorily determined at a signal-to-noise ratio of 5/1. Neither β-carotene nor hemoglobin interferes with the determination of any of the bilirubin fractions. There is, however, an unknown carotenoid in normal serum whose retention time is almost identical to that of bilirubin C-8 monomethyl ester (Fig. 8); interference by this pigment may therefore falsely increase the observed concentration of the C-8 mono-conjugate in hyperbilirubinemic serum by approximately 1–2 μM.

Preliminary studies of the composition of serum bilirubins (*64, 65*) have shown that only unconjugated bilirubin is detectable in serum of normal adults and cord blood (Fig. 8). Small amounts of mono- and diconjugates appear in serum of newborn infants in the course of the first postpartum week. In serum of jaundiced adults with hepatobiliary disease, both mono- and diconjugates are present, and the unconjugated bilirubin concentration is increased (Fig. 8).

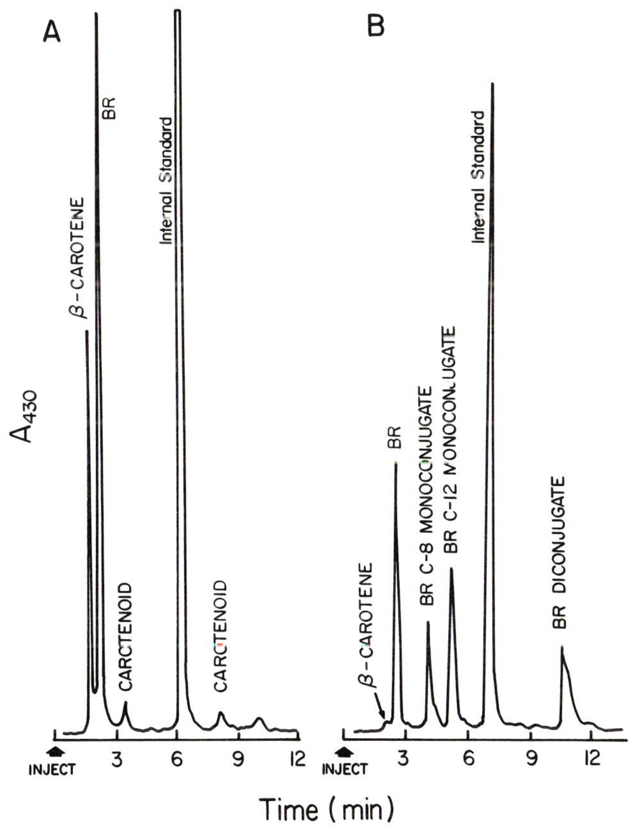


FIG. 8. Chromatogram of bilirubins from serum of a normal adult (total bilirubin, 10 μM; panel A) and a patient with obstructive jaundice (total bilirubin concentration, 166 μM; panel B); BR = bilirubin.

Monoconjugates comprise most of the conjugated pigments, and the C-12 isomer generally predominates in the monoconjugated fraction. No differences between the patterns of serum bilirubins in different etiologic groups (hepatocellular dysfunction, intrahepatic or extrahepatic cholestasis) became apparent in these preliminary studies. It is anticipated, therefore, that it is unlikely that individual determination of mono- and diconjugated bilirubins in serum will be useful in differential diagnosis of patients with hepatobiliary disease. However, based on its high sensitivity and ability to specifically measure conjugated bilirubins that are not detectable in normal serum, the alkaline methanolysis HPLC method is expected to prove of clinical value for early detection of disordered liver secretory function in anicteric patients.

Values for total bilirubin concentration for serum samples of jaundiced patients with hepatobiliary disease generally were considerably lower with the HPLC method than with the conventional diazo procedures including the ACA method (DuPont Instruments, Wilmington, DE) and Michaelsson procedure (*41*) (Fig. 9). The

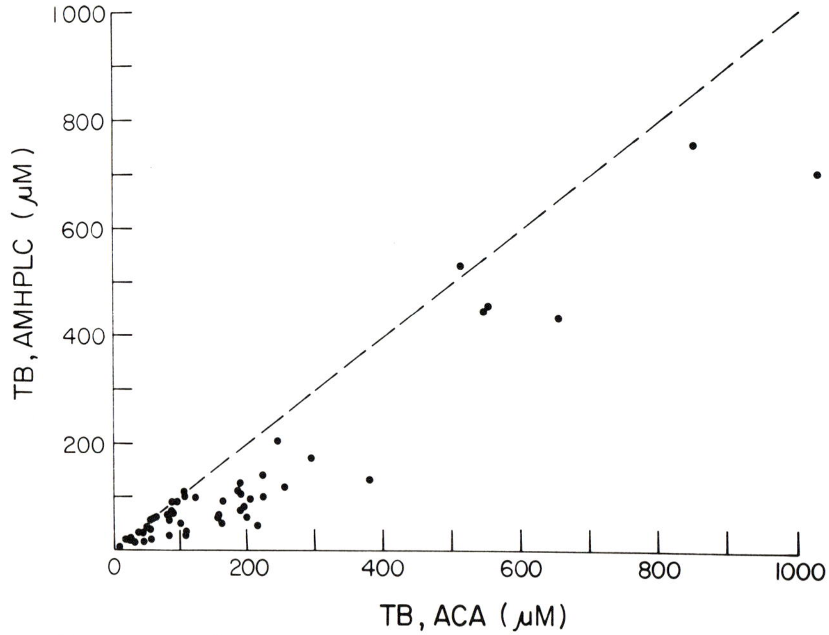


FIG. 9. Relationship between results of alkaline methanolysis HPLC method (AMHPLC) and conventional diazo method (ACA) for total serum bilirubins (TB) in human serum. The dashed line corresponds with equal values in abcissa and ordinate.

accuracy of the HPLC assay was verified in several manners, including by a radioisotope dilution procedure using $^{14}$C-bilirubin, and the findings collectively indicate that unidentified compounds, that are diazopositive, but distinct from bilirubin and its ester conjugates, are present in the pathological serum samples (*34*). Interestingly, Bratlid and Winsnes found considerably higher values for total bilirubin concentration in patients with direct-reacting hyperbilirubinemia with a conventional diazo method (Jendrassik-Grof) than with the *p*-iodoaniline diazo procedure (*66*). A possible explanation for this discrepancy might be that the newer *p*-iodoaniline diazo reagent is more selective towards bilirubin and its conjugates than diazotized sulfanilic acid.

Not unexpectedly, a poor correlation was found between the serum concentration of bilirubin ester conjugates, as measured by the HPLC-method, and that of "direct-reacting" compounds (Fig. 10). This finding reemphasizes the limited value of measurement of the "direct-reacting" fraction as a measure of conjugated bilirubins.

With its greater selectivity and potential to specifically measure

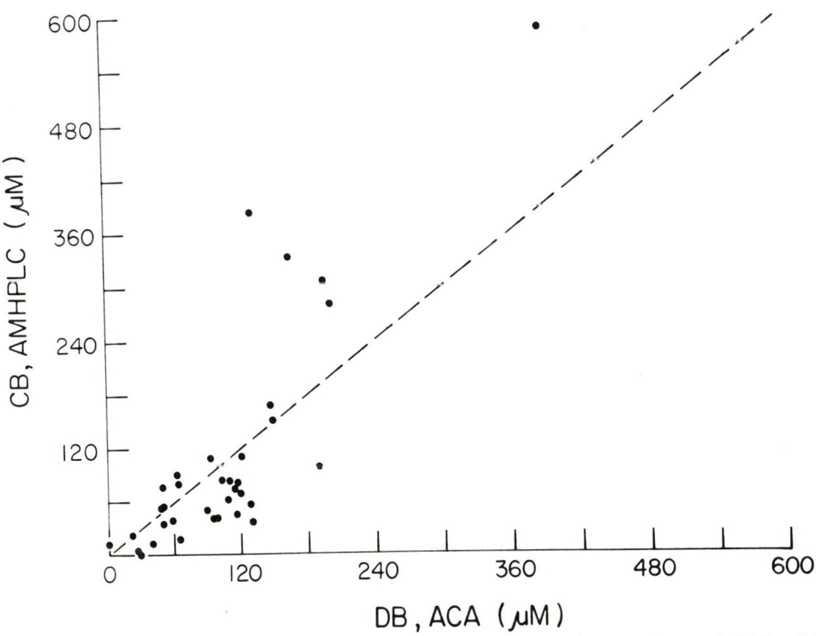


FIG. 10. Relationship between concentrations of total bilirubin conjugates, determined by alkaline methanolysis HPLC (CB, AMHPLC), and concentrations of "direct-reacting" bilirubins (DB, ACA) in human serum. The dashed line corresponds with equal values in absicca and ordinate.

bilirubins and its mono and diconjugates, we suggest that this HPLC assay should serve as a reference method for the determination of bilirubins in biological samples. Whereas the usefulness of the present assay in routine clinical laboratories remains to be established, it undoubtedly will be very valuable to many investigators engaged in the study of bile pigments and bilirubin metabolism.

*3. Analysis of Bilirubin Isomers and Biliverdins.* Whereas heme degradation predominantly involves cleavage of the macrocycle at the α-methine bridge, a minor fraction of heme is cleaved at the non-α-meso positions, resulting in formation of small amounts of bilirubin-IXα, β, γ, and δ (*67*). The concentration of these non-α isomers in bodily fluids normally is minute, but these pigments possibly are increased in certain conditions (*68*) and they probably accumulate in serum of patients with hepatobiliary disease. Separation of bilirubin-IXα, β, γ, and δ is possible by thin-layer chromatography (*69*), and an HPLC procedure recently has been reported, though experimental details of this technique are not yet available (*70*).

Although not present in biological fluids in significant amounts, bilirubin-IIIα and -XIIIα can arise from bilirubin-IXα by dipyrrole exchange.* For example, these nonphysiological isomers may constitute a significant fraction of commercial bilirubin preparations (*71*). Such a disproportionation reaction probably also occurs in bilirubin conjugates, resulting again in an increased fraction of bilirubins of the IIIα and XIIIα isomeric type. HPLC methods for separation of the IIIα, IXα, and XIIIα isomers of bilirubin, bilirubin monomethyl esters, and bilirubin dimethylesters are now available (*34, 72*).

It recently has been postulated that accelerated clearance of bilirubin in the course of phototherapy in jaundiced newborns is related primarily to isomerization of 4Z,15Z-bilirubin (structure 2, Fig. 1) to the geometrical isomers 4E,15Z-bilirubin (structure 5), 4Z,15E-bilirubin, and possibly also 4E,15E-bilirubin (*73*). These isomers are thought to be generated in the skin and then transported in the circulation to the liver and rapidly excreted in bile. Recently, HPLC procedures have been reported for separation of photoproducts of bilirubins, believed to correspond with geometrical isomers of unconjugated bilirubins (*73, 74*). These studies need further confirmation, and it presently is unclear whether determination of

*Dipyrrole exchange is used to denote disproportionation of bilirubins. This reaction involves cleavage of each molecule at either side of the central C-10 bridge into two dipyrrolic fragments, followed by random recombination of dipyrrole moieties from different molecules (*71*).

bilirubin metabolites in serum during phototherapy of jaundiced newborn babies will be clinically useful.

It generally is assumed that the risk for development of bilirubin encephalopathy in patients with unconjugated hyperbilirubinemia is directly related to the concentration of the unbound bilirubin in serum. Unfortunately, an accurate method presently is not available for measurement of this pigment fraction (*75*). It should be noted that a recently published HPLC method for the assessment of free bilirubin (*76*) is no exception, since "free" bilirubin merely corresponds to pigment adsorbed to the stationary phase and therefore largely overestimates the truly unbound fraction (for a discussion, see ref. *75*).

Biliverdinemia rarely occurs in humans, and in those cases, it probably is related to degradation of bilirubins accumulated in the body. There is no need, therefore, for measurement of serum biliverdins in clinical diagnosis. For basic research purposes, HPLC procedures for separation of biliverdin-IX isomers ($\alpha$, $\beta$, $\gamma$, and $\delta$) and algal bile pigments (phycoerythrobilin, phycocyanobilin) recently have been developed (*77–79*).

## V. Outlook

HPLC has just recently made its debut in analysis of bilirubins and biliverdins, and undoubtedly will find broad application in bile pigment research. The role of this technique in the analysis of bilirubins in clinical laboratories, however, presently remains to be defined. Whereas an HPLC method for accurate and specific determination of unconjugated bilirubin and bilirubin mono- and diconjugates in serum is now available, and it appears that conventional diazo procedures do not permit accurate assessment of unconjugated and conjugated bilirubin, superiority of the HPLC assay in clinical diagnosis remains to be established before this new method is advocated for use in the routine clinical laboratory. Does specific determination of the unconjugated, monoconjugated and diconjugated bilirubin fractions in serum contribute to differential diagnosis of patients with hepatobiliary disease? Is measurement of conjugated bilirubin in serum of anicteric patients a sensitive indicator of subtle disturbances of the hepatic secretory apparatus? These, and other questions, need to be answered with the aid of the alkaline methanolysis HPLC assay before more effort is put into automation and/or adaptation of the present method to make it more compatible with the specific requirements of the clinical laboratory. Once pure, well-characterized, and stable reference bilirubin mono- and diglucuronide become available, it will be possible to also develop analytical methods for direct analysis of bilirubin and its conjugates by HPLC. In this approach, ion-pair reversed-phase chromatography appears to be particularly promising (*61*; Kabra, P. M., and Blanckaert, N., unpublished data).

Preparation of pure bilirubin mono- and diglucuronide standards remains a major challenge for bile pigment researchers, and these pigments are also badly needed in clinical laboratories for calibration purposes. Major obstacles are the present unavailability of a procedure for synthesis on a large scale and the pronounced lability of conjugated pigments, which easily undergo oxidative breakdown, hydrolysis, dipyrrole exchange, and conversion into positional isomers. Based on the possibility of achieving rapid isolation of the pigment fractions, with exclusion of light and oxygen, it is possible that preparative HPLC of bilirubin glucuronides may play an important role in future procedures for preparation of pure bilirubin mono and diglucuronide.

## Acknowledgments

I am indebted to Dr. F. Compernolle for a critical review of the manuscript. This work was supported by NIH grants AM-11275 and P50 AM-18520. The author is an Appointed Investigator of the Belgian National Research Council.

## References

*1.* Berk, P. D., and Javitt, N. B., *Am. J. Med.* **64,** 311 (1978).
*2.* Billing, B. H., *Gut* **19,** 481 (1978).
*3.* Fevery, J. Blanckaert, N., Degroote, J., and Heirwegh, K. P. M., Bilirubin conjugates: formation and detection, in *Progress in Liver Disease,* Popper, H., and Schaffner, F., eds., vol. 5, Grune and Stratton, New York, 1976, pp. 183–214.
*4.* Schmid, R., *Gastroenterology* **74,** 1307 (1978).
*5.* Schmid, R., and McDonagh, A. F., Hyperbilirubinemia, in *The Metabolic Basis of Inherited Disease,* Stanbury, J. B., Wyngaarden, J. B., and Frederickson, D. S., eds., McGraw-Hill, New York, 1978, pp. 1221–1257.
*6.* Bonnett, R., Nomenclature, in *The Porphyrins,* Dolphin, D., ed., vol. 1, Academic Press, New York, 1978, pp. 1–27.
*7.* Heirwegh, K. P. M., Fevery, J., Meuwissen, J. A. T. P., Degroote, J., Compernolle, F., Desmet, V., and Van Roy, F. P., *Methods Biochem. Anal.* **22,** 205 (1974).

*8.* Brodersen, R., and Theilgaard, J., *Scand. J. Clin. Lab. Med.* **24,** 395 (1969).
*9.* Fog, J., and Jellum, E., *Nature* **198,** 88 (1963).
*10.* Bonnett, R., Davies, J. E., Hursthouse, M. B., and Sheldrick, G. M., *Proc. Roy. Soc. Lond. B.* **202,** 249 (1978).
*11.* Mugnoli, A., Manitto, P., and Monti, D., *Nature* **273,** 568 (1978).
*12.* Lester, R., and Klein, P., *J. Clin. Invest.* **45,** 1839 (1966).
*13.* Blanckaert, N., Heirwegh, K. P. M., and Zaman, Z., *Biochem. J.* **164,** 229 (1977).
*14.* Jones, E. A., Shrager, R., Bloomer, J. R., Berk, P. D., Howe, R. B., and Berlin, N. I., *J. Clin. Invest.* **51,** 2450 (1972).
*15.* Heirweigh, K. P. M., Formation, metabolism and significance of bilirubin-IX glycosides, in *Conjugation Reactions in Drug Biotransformation,* Aitio, A., ed., Elsevier/North, Holland Biomedical Press, Amsterdam, 1978, pp. 67–76.
*16.* Blanckaert, N., Compernolle, F., Leroy, P., Van Houtte, R., Fevery, J., and Heirwegh, K. P. M., *Biochem. J.,* **171,** 203 (1978).
*17.* Compernolle, F., Van Hees, G. P., Blanckaert, N., and Heirwegh, K. P. M., *Biochem. J.* **171,** 185 (1978).
*18.* Kuenzle, C. C., Bilirubin conjugates in human bile, in *Metabolic Conjugation and Metabolic Hydrolysis,* Fishman, W. H., ed., vol. 3, Academic Press, New York, 1973, pp. 351–386.
*19.* Compernolle, F., *Biochem. J.* **175,** 1095 (1978).
*20.* Isselbacher, K. J., and McCarthy, E. A., *J. Clin. Invest.* **38,** 645 (1959).
*21.* Noir, B., and Nanet, H., *Biochim. Biophys. Acta.* **372,** 230 (1974).
*22.* Kondo, T., Kawai, T., Yamamoto, T., and Izawa, T., *Gastroenterol. Jap.* **6,** 217 (1971).
*23.* Ettner-Kjelsaas, H., and Kuenzle, C. C., *Biochim. Biophys. Acta.* **400,** 83 (1975).
*24.* Weber, A. P., and Schalm, L., *Acta. Med. Scand.* **177,** 519 (1965).
*25.* Jansen, P. L. M., Chowdhury, J. R., Fischberg, E. B., and Arias, I. M., *J. Biol. Chem.* **252,** 2710 (1977).
*26.* Blanckaert, N., Gollan, J., and Schmid, R., *Proc. Natl. Acad. Sci., USA* **76,** 2037 (1979).
*27.* Gordon, E. R., and Goresky, C. A., *Gastroenterol.* **75,** 966 (1978).
*28.* Blanckaert, N., Gollan, J., and Schmid, R., *J. Clin. Invest.* **65,** 1332 (1980).
*29.* Watson, C. J., *Ann. Intern. Med.* **70,** 839 (1969).
*30.* Bloomer, J. R., Berk, P. D., Howe, W. B., Waggoner, J. G., and Berlin, N. I., *Clin. Chim. Acta.* **29,** 463 (1970).
*31.* Billing, B. H., Cole, P. G., and Lathe, G. H., *Biochem. J.* **65,** 774 (1957).
*32.* Schmid, R., *J. Biol. Chem.* **22,** 881 (1957).
*33.* Talafant, E., *Nature* (London) **178,** 312 (1956).
*34.* Blanckaert, N., Kabra, P. M., Farina, F. A., Stafford, B. E., Marton, L. J., and Schmid, R., *J. Lab. Clin. Med.,* **96,** 198 (1980).
*35.* Ehrlich, P., *Centr. klin. Med.* **4,** 721 (1883).
*36.* Van den Bergh, A. A. H., and Muller, P., *Biochem. Z.,* **77,** 90 (1916).

37. Billing, B. H., Haslam, R., and Wald, N., *Ann. Clin. Biochem.,* **8,** 21 (1971).
38. Van Roy, F. P., and Heirwegh, K. P. M., *Biochem. J.* **107,** 507 (1968).
39. Hertz, H., Dybkaer, R., and Lauritzen, M., *Scand. J. Clin. Lab. Invest.* **34,** 265 (1974).
40. Jackson, S. H., and Hernandez, A. H., *Clin. Chem.* **16,** 462 (1970).
41. Liley, A., *Amer. J. Obstet. Gynecol.* **82,** 1359 (1961).
42. Ostrow, J. D., and Murphy, N. H., *Biochem. J.* **120,** 311 (1970).
43. Brodersen, R., and Jacobsen, J., Separation and determination of bile pigments, in *Methods of Biochemical Analysis,* Glick, D., ed., Vol. 17, Wiley-Interscience, New York, 1969, pp. 3–54.
44. Weber, A. P., and Schalm, L., *Clin. Chim. Acta.* **7,** 805 (1962).
45. Eberlein, W. R., *Pediatrics* **25,** 878 (1960).
46. Ostrow, J. D., and Boonyapisit, S. T., *Biochem. J.* **173,** 263 (1978).
47. Bratlid, D., and Winsnes, A., *Scand. J. Clin. Lab. Invest.* **28,** 41 (1971).
48. Schoenfield, L. J., Foulk, W. T., and Bollman, J. L., *Gastroenterology* **47,** 35 (1964).
49. Cole, P. G., Lathe, G. H., and Billing, B. H., *Biochem.* **57,** 514 (1954).
50. Kuenzle, C. C., Maier, C., and Ruttner, J. R., *J. Lab. Clin. Med.* **67,** 294 (1966).
51. Wolkoff, A. W., Scharschmidt, B. F., Plotz, P. H., and Berk, P. D., *Proc. Soc. Exptl. Biol. Med.* **152,** 20 (1976).
52. Billing, B. H., *J. Clin. Path.* **8,** 130 (1955).
53. Gordon, E. R., Chan, T. H., Samodai, K., and Goresky, C. A., *Biochem. J.* **167,** 1 (1977).
54. Thompson, R. P. H., and Hofmann, A. F.:*Clin. Chim. Acta.* **35,** 517 (1971).
55. Salmon, M., Fenselau, C., *Life Sci.* **15,** 2069 (1974).
56. Heirwegh, K. P. M., Fevery, J., Michiels, R., Van Hees, G. P., and Compernolle, F., *Biochem. J.* **145,** 185 (1975).
57. Noir, B. A., *Biochem. J.* **155,** 365 (1976).
58. Blumenthal, S. G., Taggart, D. B., Ikeda, R. M., Ruebner, B. H., and Bergstrom, D. E., *Biochem. J.* **167,** 535 (1977).
59. Jansen, F. H., and Billing, B. H., *Biochem. J.* **125,** 917 (1971).
60. Lim, C. K., *J. Liq. Chromatog.* **2,** 37 (1979).
61. Onishi, S., Itoh, S., Kawade, N., Isobe, K., and Sugiyama, S., *Biochem. J.* **185,** 281 (1980).
62. Yamaguchi, T., Yamaguchi, N., Nakajima, H., Komoda, Y., and Ishikawa, M., *Proc. Japan Acad.,* **55,** Ser. B., 89 (1979).
63. Blanckaert, N., *Biochem. J.* **185,** 115 (1980).
64. Scharschmidt, B. F., Blanckaert, N., Farina, F., Kabra, P., Weisiger, R., Marton, L., and Schmid, R., *Gastroenterology* **77,** A-39 (1979).
65. Rosenthal, P., Blanckaert, N., Kabra, P. M., and Thaler, M. M., *Gastroenterology* **77,** A-35 (1979).
66. Bratlid, D., and Winsnes, A., *Scand. J. Clin. Lab. Invest.* **31,** 231 (1973).

67. Blanckaert, N., Fevery, J., Heirwegh, K. P. M., and Compernolle, F., *Biochem. J.* **164,** 237 (1977).
68. Blumenthal, S. G., Stucker, T., Rasmusssen, R. D., Ikeda, R. M., Ruebner, B. H., Bergstrom, D. E., and Hanson, F. W., *Biochem. J.* **186,** 693 (1980).
69. Blanckaert, N., Heirwegh, K, P. M., and Compernolle, *Biochem. J.* **155,** 405 (1976).
70. Yamaguchi, T., Yamaguchi, N., Nakajima, H., Komoda, Y., and Ishikawa, M., *Proc. Japan Acad.* **55,** Ser. B., 84 (1979).
71. McDonagh, A. F., and Assisi, F.: *FEBS Lett.* **18,** 315 (1971).
72. Wooldridge, T. A., and Lightner, D. A., *J. Liq. Chromatog.* **1,** 653 (1978).
73. Lightner, D. A., Wooldridge, T. A., and McDonagh, A. F., *Proc. Natl. Acad. Sci. USA* **76,** 29 (1979).
74. Onishi, S., Itoh, S., Kawade, N., Isobe, K., and Sugiyama, S., *Biochem. Biophys. Res. Comm.* **90,** 890 (1979).
75. Brodersen, R., Detection of free bilirubin: a criterion for exchange transfusion, in *Liver Diseases in Infancy and Childhood,* Berenberg, S. R., ed., Martinus Nijhoff, The Hague, 1976, pp. 18–24.
76. Lu, K. C., Gooding, K. M., and Regnier, F. E., *Clin. Chem.* **25,** 1608 (1979).
77. Schoch, S., Lempert, U., Wieschoff, H., and Scheer, H., *J. Chromatog.* **157,** 357 (1978).
78. Fu, E., Friedman, L., and Siegelman, H. W., *Biochem. J.* **179,** 1 (1979).
79. Rasmussen, R. D., Yokoyama, W. H., Blumenthal, S. G., Bergstrom, D. E., and Reubner, B. H., *Anal. Biochem.* **101,** 66 (1980).

# Chapter 16

## Porphyrins

George R. Gotelli, Jeffrey H. Wall,
Pokar M. Kabra, and Laurence J. Marton

*Department of Laboratory Medicine*
*School of Medicine*
*University of California*
*San Francisco, California*

### I. Introduction

Historically the term porphyria has been used since it was coined in 1871 to describe a purple colored material extracted from pathological feces (*1*). The first case of porphyria was reported in 1874, (*2, 3*), but until the 1930 Nobel Prize winning work of Hans Fischer on the synthesis of protoporphyrin, there was little more than academic interest in porphyrin analysis. During the forty years between 1930 and 1970, the biosynthetic pathways leading to the formation of heme, and the details of porphyrin metabolism, were elucidated. During this time quantitative methods for porphyrins in biological fluids used complex and laborious solvent extraction techniques, requiring large sample volumes and hours to complete. We now know that these methods only partially separated the complex mixture of porphyrins found in biological fluids. These solvent extraction procedures fractionated the porphyrins into two broad groups, uroporphyrins (octacarboxylic) and coproporphyrins (tetracarboxylic). However, intermediate carboxylated porphyrin containing 2, 3, 5, 6, and 7 carboxyl groups are now known to exist in normal and pathlogical excreta, which were not differentiated, but which were included in the two broad uroporphyrin and copropophyrin groups.

When thin-layer chromatography (TLC) was applied to porphyrin analysis in the early 1960s (*4–6*), it was clearly demonstrated that urine and feces contained these intermediate carboxylated porphyrins, in addition to the previously known coproporphyrins and uroporphyrins. TLC also provided a convenient method for profiling these polycarboxylated porphyrins in pathological urine and feces. TLC methods, however, require the conversion of the isolated porphyrins to their corresponding methyl esters. Additionally, TLC methods are only semiquantitative. High-pressure (performance) liquid chromatography (HPLC) offers significant advantages over TLC and has substantially simplified the study and analysis of porphyrins.

## II. Urinary and Fecal Porphyrins by HPLC

The greatest effort to develop HPLC techniques has been directed towards the investigation of urinary and fecal porphyrins. Initial HPLC methods, borrowing from TLC, chromatographed porphyrin esters. Thus, Battersby et al. (*7*) in 1976 successfully separated synthetic isomers of type I, II, and III/IV (unresolved) coproporphyrin tetraethyl esters on a reversed phase column. However, they were unsuccessful in separating isomers of uroporphyrin, and did not apply their findings to clinical samples since they only investigated synthetic standards. In that same year, Adams and Vandemark (*8*) separated unesterified urinary porphyrins (protoporphyrin, uroporphyrin, and coproporphyrin) using a $C_{18}$ reversed-phase column. In addition, they obtained qualitative confirmation of the porphyrins using a stop-flow technique in which the eluted porphyrin was held in the fluorometer flow-cell while excitation and emission spectra were determined. Similarly, in 1976, Adams et al. (*9*) resolved the clinically important unesterfied isomers, types I and III, of both coproporphyrin and uroporphyrin. They also detected small quantities of protoporphyrin and deuteroporphyrin in urine. These investigators demonstrated the minimal sample preparation necessary for HPLC compared to that required by the laborious and complex solvent extraction techniques commonly used in other techniques. In their method, sample preparation consisted of evaporating the urine to dryness and reconstituting the residue in the mobile phase before injection into the HPLC.

Carlson and Dolphin (*10*) chromatographed both urinary and fecal polycarboxylated porphyrins as methyl esters using both reversed-phase partition and normal-phase adsorption chromatography. Their results with the reversed phase technique were disappointing, producing tailing peaks; however, separation of the polycarboxylated porphyrins with normal phase chromatography was satisfactory. Gray et al. (*11*) also used normal-phase adsorption chromatography of methyl esters in studying prophyrin excretion profiles in urine and feces. They successfully separated all of the polycarboxylated prophyrins, while achieving excellent resolution with isocratic elution (Figs. 1–3). Additionally, they partially resolved some porphyrins of the isocoproporphyrin series, and were able to separate the methyl esters of type I and type III isomers of coproporphyrin using a reversed-phase $C_{18}$ column. In an interesting report, Evans et al. (*12*) described a method of resolving porphyrin methyl esters using adsorption chromatography and gradient elution; however, they warned against the use of chlorinated hydrocarbons, which may contain traces of acid. The acid reportedly leeches copper from the stainless steel tubing, resulting in the formation of copper chelates of the porphyrins. Miller and Malina (*13*) took advantage of these copper chelates and developed a method that chromatographs the copper complexes of porphyrins using a short chain cyano column. Evans et al (*14*) reported an ion-exchange technique for free acid porphyrins, and separated the di-, tetra-, and penta-carboxylic acids. Finally in 1978, Bonnett et al. (*15*) chromatographed free acid porphyrins on a reversed-phase column using a mobile phase containing tetrabutylammonium phosphate as an ion-pairing reagent.

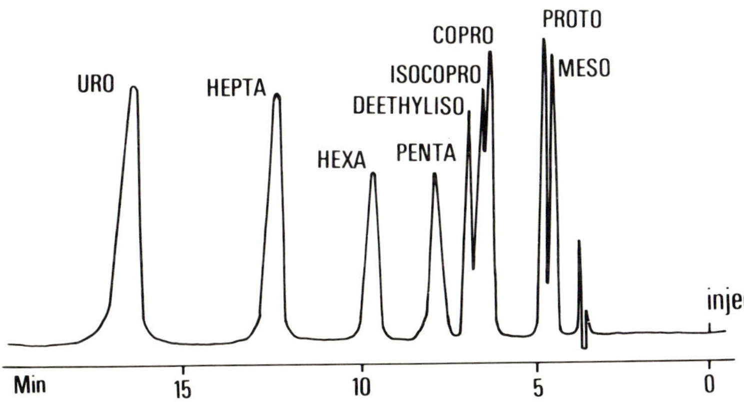


FIG. 1. Porphyrin reference standards.

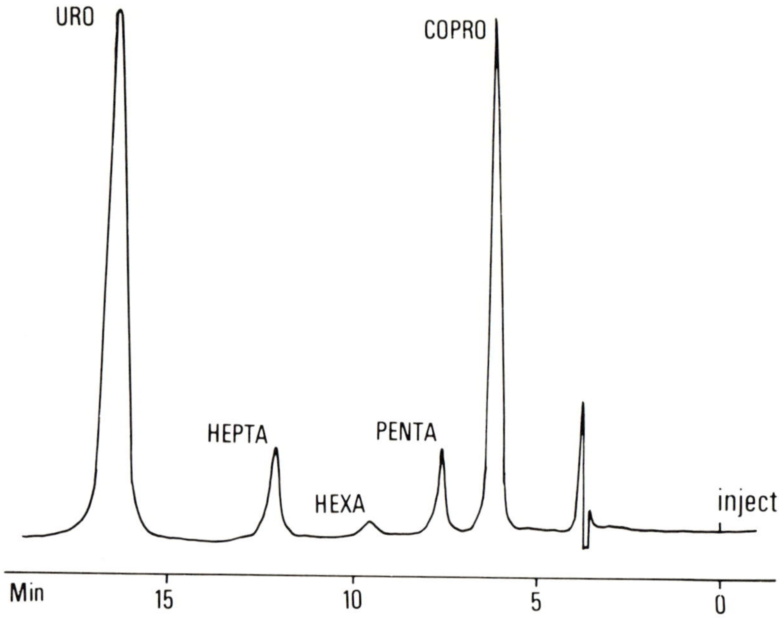


FIG. 2. Urinary porphyrins from a case of terminal, cirrhotic, erythrohepatic protoporphyria.

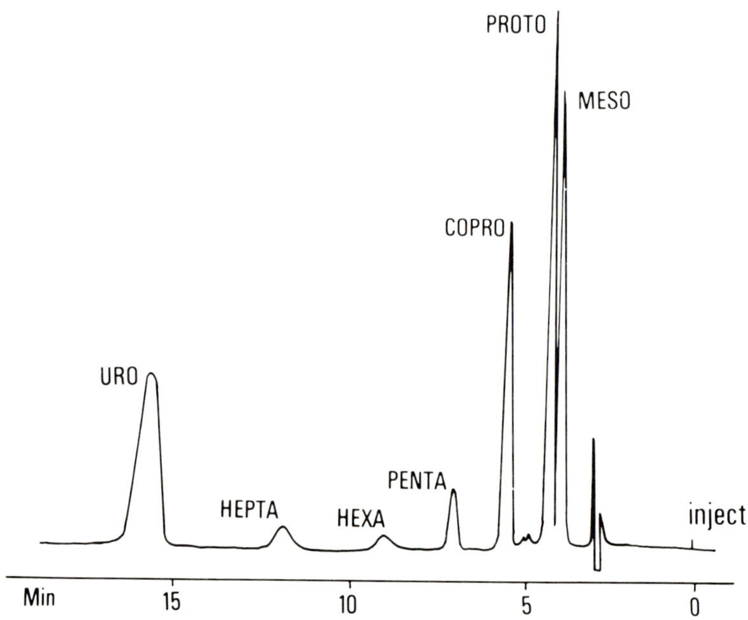


FIG. 3. Fecal porphyrins from the case in Fig. 2. (Figs. 1, 2, 3, reprinted with permission of authors).

The mobile phase was miscible with aqueous solutions and could be used to extract the porphyrins from a biological matrix.

From the successful applications of HPLC cited, it is apparent that urinary and fecal porphyrins can be separated by HPLC with relative ease. Normal-phase adsorption, reversed-phase partition, reversed-phase ion-pair, or ion-exchange chromatographic techniques can be used to achieve satisfactory separation of the clinically important porphyrins. Similarly, type I and type III isomers of coproporphyrin and uroporphyrin can be easily resolved. In addition, the di-, tri-, tetra-, penta-, hexa-, hepta-, and octacarboxylated porphyrins found in normal and pathological urine and feces can be separated, thus providing a characteristic porphyrin profile that can aid in the diagnosis of diseased states.

## III. Erythrocyte Porphyrins

Erythrocyte porphyrin analysis has received much less attention than urinary and fecal analysis over the past four decades. As a result, unlike the situation with urine and feces, there is little literature on the application of HPLC to the analysis of erythrocyte porphyrins.

### A. Extraction Methods

In 1932, Van den Bergh (*16*) described a method for analyzing red cell protoporphyrin that required 10–20 mL of whole blood, used an exhaustive 2-h solvent extraction, and required large volumes of ethyl acetate. Detection of the extracted porphyrins was by fluorometry. This original method has been modified many times, the modifications consisting of simplification and shortening of the extraction step, changes in the concentration of the hydrochloric acid used for back extraction, and a reduction in the volume of blood required. However, the principle of extraction of blood porphyrins into acidified ethyl acetate, followed by back extraction into dilute hydrochloric acid, has remained substantially unchanged for the past four decades.

It has been known for four decades that red cell protoporphyrin is increased in lead poisoning (*17*). However, this fact has been essentially ignored until the last decade. Since 1970 there has been renewed interest in erythrocyte porphyrin testing, stimulated largely by the major public health issue of childhood environmental lead poisoning. The data collected on environmental lead poisoning suggests that the effects of an increased lead burden on the body produces subtle but longlasting learning defects. However, if this increased lead burden could be detected sufficiently early, and intervention could occur, the subsequent CNS disturbances could be prevented. One of the first detectable effects of an increased lead burden on the body is the elevation of erythrocyte protoporphyrin. These accumulated facts have lead to a rapid improvement in assay methods for erythrocyte protoporphyrin during the last decade.

In 1973 Piomelli (*18*) described a two-step extraction procedure based upon the method of Van den Bergh (*16*). This simple, rapid, fluorometric method, requiring only 20 $\mu$L of whole blood, is probably the most common method for red cell protoporphyrin in use today. Briefly, 20 $\mu$L of whole blood is extracted with acidified ethyl acetate, the extracted porphyrins are back extracted into dilute hydrochloric acid, and these are subsequently quantitated by fluorometry. In attempts to simplify this two-step extraction procedure, Chisholm (*19*) reported a single-step acidified acetone extraction procedure; however, this method is reported to give poor extraction efficiency (*20*). In 1975 Chisholm and Brown (*21*) reported a fluorometric procedure, but later returned to the acidified ethyl acetate double extraction principle. Finally, Piomelli (*22*) reported a spot test using a blood-saturated filter paper as the sample. Clearly, erythrocyte porphyrin methods have been simplified, but the basic principle of the Van den Bergh methods has remained unchanged for 45 years.

When porphyrins are extracted from normal whole blood by acidified ethyl acetate, the extracted porphyrins are commonly called free erythrocyte protoporphrin (FEP). In fact, about 95% of the total porphyrin extracted is protoporphyrin IX (PPIX). The remaining 5% consists of coproporphyrin, traces of uroporphyrin, and small amounts of plasma porphyrins (mainly plasma PPIX). However, in lead poisoning and iron deficiency anemia, the predominant red cell porphyrin is zinc protoporphyrin IX (ZPPIX), as reported by Lamola and Yamane in 1974 (*23*). When ZPPIX is extracted from red blood cells by acidified ethyl acetate, the ZPPIX is dissociated to PPIX, and as a result the ZPPIX is measured as PPIX and reported as the FEP content. However, iron protoporphyrin (heme) does not dissociate, but in the presence of acid is converted to acid hematin, which interferes in the final spectrophotometric or fluorometric measurement. As a result, heme is removed by back-extracting the porphyrins into dilute hydrochloric acid before spectrophotometric or fluorometric measurement.

Following the report by Lamola and Yamane that ZPPIX is the predominant porphyrin in blood of patients with lead poisoning, methods for determining ZPPIX appeared in the literature. Lamola (*24*) introduced a non-extraction technique in which whole blood is diluted with a buffered detergent, and the ZPPIX in the red cells is measured without interference from PPIX. In 1977 Blumberg (*25*) specifically designed an instrument, the Hematofluorometer, to measure the ZPPIX content of unprocessed whole blood.

## B. HPLC Methods

Although the emphasis placed upon the early detection of childhood lead poisoning has resulted in simple, diagnostically valid screening procedures, few of these procedures are specific, and none can measure all of the red cell porphyrins simultaneously. The application of HPLC to the measurement of red cell porphyrins could presumably resolve these problems. In 1979, Cullbreth et al. (*26*), while studying the hydrolysis of PPIX esters, described a reversed-phase HPLC method that will separate PPIX and related porphyrins. In addition, they demonstrated that their method has sufficient sensitivity to measure PPIX in an acid extract of whole blood. However, they did not perform any other blood studied. This report appears to be the only HPLC method for erythrocyte porphyrins described in the recent literature.

Recently we have developed a HPLC method that will simultaneously determine red cell coproporphyrin, PPIX, and ZPPIX (*27*). This method combines reversed-phase ion-pair chromatography with fluorometric detection, and uses internal standardization. It can easily be adapted to pediatric samples, requiring only 10 $\mu$L of whole blood. Sample preparation requires minimal labor. Whole blood is added to a red cell lysing/porphyrin solubilizing reagent that contains an appropriate amount of the internal standard, uroporphyrin III octamethyl ester. The lysed blood can be injected onto the HPLC column immediately after a brief centrifugation to sediment the red cell ghosts. Chromatographic separation of the porphyrins is complete in 5 min (Fig. 4). Elution of the porphyrins from the column is achieved with an acetonitrile: tetrabutylammonium phosphate buffer (66/34 by vol) mobile phase, at a flow rate of 2 mL/min, and a temperature of 50°C. The eluted porphyrins are detected by fluorometry, using an excitation wavelength of 400 nm. Emitted fluorescent light of all wavelengths above 560 nm is allowed to reach the detector by using a 560 nm cutoff filter. This allows for the simultaneous detection of all porphyrins species in blood with a single sample injection. However, one can increase sensitivity and specificity by using the specific emission wavelength of the porphyrin of interest (Table 1).

This HPLC method was compared to the method of Piomelli (*18*). Because the acidified extraction reagent used in the Piomelli method dissociates ZPPIX to PPIX, the individual ZPPIX and PPIX values,

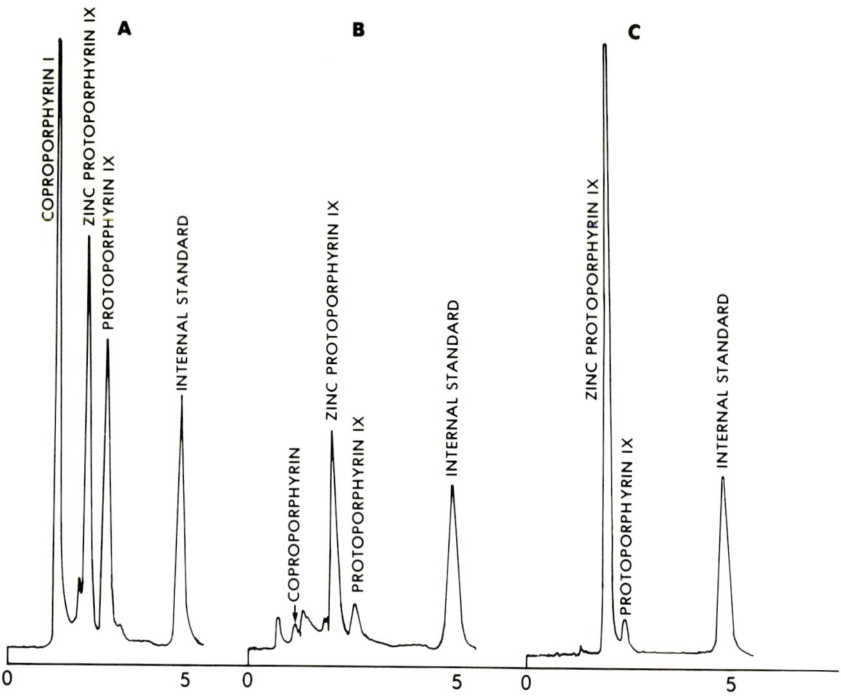


FIG. 4. (a) Porphyrin reference standards; (b) normal blood containing 390 μg/L of ZPPIX and 100 μg/L of PPIX; (c) Blood from lead poisoning containing 10,000 μg/L of ZPPIX and 690 μg/L of PPIX.

Table 1
Excitation and Emission Maxima of Porphyrins in Mobile Phase

| Porphyrin | Excitation, nm | Emission, nm |
|---|---|---|
| Coroporphyrin I | 392 | 615 |
| Zinc protoporphyrin IX | 414 | 586 |
| Protoporphyrin IX | 400 | 626 |
| Uroporphyrin III | 398 | 618 |
| Uroporphyrin III octamethylester | 398 | 578 |

and their combined value from the HPLC method was compared to the total FEP value obtained by the Piomelli method. Table 2 tabulates the results of a number of random blood samples. Although uroporphyrin can also be detected by this method, eluting just after the coproporphyrin peak, it was not noted, and only trace amounts of coproporphyrin were noted, in all blood samples tested.

Table 2
Blood Sample Comparison between Piomelli Method and HPLC Method

| Sample | Piomelli, μg/L | HPLC ZPPIX and PPIX, μg/L | ZPPIX, μg/L | PPIX, μg/L |
|---|---|---|---|---|
| 1 | 280 | 270 | 250 | 20 |
| 2 | 350 | 550 | 520 | 30 |
| 3 | 280 | 400 | 340 | 60 |
| 4 | 210 | 220 | 200 | 20 |
| 5 | 420 | 500 | 410 | 90 |
| 6 | 420 | 500 | 390 | 110 |
| 7 | 210 | 240 | 220 | 20 |
| 8 | 210 | 220 | 210 | 10 |
| 9 | 280 | 330 | 300 | 30 |
| 10 | 350 | 420 | 380 | 40 |
| 11 | 350 | 330 | 300 | 30 |
| 12 | 280 | 360 | 350 | 10 |
| 13 | 280 | 310 | 310 | 0 |
| 14 | 140 | 130 | 130 | 0 |
| 15 | 490 | 480 | 470 | 10 |
| 16 | 560 | 580 | 520 | 60 |
| 17 | 280 | 290 | 290 | 0 |
| 18 | 840 | 880 | 710 | 170 |
| 19 | 280 | 280 | 270 | 10 |
| 20 | 210 | 240 | 240 | 0 |
| 21 | 700 | 1,040 | 930 | 110 |
| 22 | 8,300 | 10,690 | 10,000 | 690 |

## C. Advantages of HPLC

Liquid chromatography offers significant advantages over traditional methods for measuring red cell porphyrins. Most notable is that individual erythrocyte prophyrins, ZPPIX, PPIX, coproporphyrin, and uroprophyrin can be quantitated on a single run. In addition, hematin does not interfere, thereby eliminating the need for back-extracting the porphyrins into dilute hydrochloric acid. This step is required by other techniques to eliminate the spectrophotometric interference and the fluorescence-quenching effect of hematin. The use of an ion-pair technique coupled with reverse-phase chromatography eliminates the need for esterification of the porphyrins; thus the free acids of porphyrins, as they occur in blood, can be measured. In addition, ion-pair chromatography uses reagents that are water soluble, facilitating solubilization of the porphyrins in biological materials, and uses mobile phases that are generally compatible with aqueous biological samples. Finally, the liquid chromatographic method does not require sample extraction, thus greatly simplifying the sample preparation needed before chromatography. These advantages may in some circumstances allow direct injection of unprocessed or minimally processed biological fluids, substantially reducing time and labor and allows the HPLC column to accomplish the separation that was previously performed by laborious manual extraction techniques.

The specificity, simplicity, and rapidity of HPLC techniques developed in the last decade makes HPLC the method of choice in porphyrin analytical techniques. These techniques may be expected to produce additional exciting developments in porphyrin analysis, and clearly demonstrate the potential usefulness of these assays in clinical situations.

## Acknowledgment

LJM is the recipient of NCI Research Career Development Award CA-00112.

## References

*1.* Hoppe-Seyler, F., *Medizin.-Chem. Untersuchusgen,* Tuebingen, 1871.
*2.* Schultz, J. H., Ein fall von Pemphigus leprosus, complicirt durch Lepra visceralis, Thesis, Greisswald, 1874.
*3.* Baumstark, F., *Arch. Ges. Physiol.* **9,** 568 (1874).
*4.* With, T. K., "A simplified system of clinical porphyrin analysis of urine and feces based on thin-layer chromatography in porphyrins in human diseases," paper presented at the 1st International Porphyrin Meeting, Freiburg/Br. 1975, Karger, Basel, 1976.
*5.* Holti, G., Rimington, C., Tate, B. C., and Thomas, G., *Quart. J. Med.* **27,** 1 (1958).
*6.* Elder, G. H., *J. Clin. Path.* **28,** 601 (1975).
*7.* Battersby, A. R., Buckley, D. G., Hodgson, G. L., Markwell, R. E., and McDonald, E., Separation of Porphyrin Isomers by HPLC—Biochemical and Biosynthetic Applications, in *High Pressure Liquid Chromatography in Clinical Chemistry,* Dixon, P. F., Gray, C. H., Lim, C. K., Stoll, M. S., eds., Academic Press, San Francisco, 1976.
*8.* Adams, R. F., and Van de Mark, F. L., *Clin. Chem.* **22,** 1180 (1976).
*9.* Adams, R. F., Slavin, W., and Rhys Williams, A., *Chromatog. Newsletter* **4,** 24 (Nov. 1976).
*10.* Carlson, R. E., and Dolphin, D., Application of HPLC to the Analysis of Clinically Important Porphyrins, in *High Pressure Liquid Chro-*

*matography in Clinical Chemistry,* Dixon, P. F., Gray, C. H., Lim, C. K., Stoll, M. S., eds., Academic Press, San Francisco, 1976.
11. Gray, C. H., Lim, C. K., and Nicholson, D. C., *Clin. Chimica Acta* **77,** 167 (1977).
12. Evans, N., Jackson, H., Matlin, S. A., and Towill, R., High-Performance Liquid Chromatographic Analysis of Porphyrins in Biological Materials, in *High-Pressure Liquid Chromatography in Clinical Practice,* Dixon, P. F., Gray, C. H., Lim, C. K., Stoll, M. S., eds., Academic Press, San Francisco, 1975.
13. Miller, V., and Malina, L., *J. Chromatogr.* **145,** 290 (1978).
14. Evans, N., Games, D. E., Jackson, A. H., and Matlin, S. A., *J. Chromatog.* **115,** 325 (1975).
15. Bonnett, R., Charalambides, A. A., Jones, K., Magnus, I. A., *Biochem. J.* **173,** 693 (1978).
16. Van den Bergh, A. A. H., Grotepass, W., and Revers, F. E., *Klin. Wschr.* **11,** 1534 (1932).
17. Vigliani, E., and Angeleri, C., *Klin. Wschr.* **15,** 700 (1936).
18. Piomelli, S., *J. Lab. Clin. Med.* **81,** 932 (1973).
19. Chisolm, J. J., Jr., Hastings, C. W., and Cheung, D. K. K., *Biochem. Med.* **9,** 113 (1974).
20. Hanna, T. L., Dietzler, D. N., Smith, C. H., Gupta, S., and Zarkowsky, H. S., *Clin. Chem.* **22,** 161 (1976).
21. Chisolm, J. J., Jr., and Brown, D. H., *Clin. Chem.* **21,** 1669 (1976).
22. Piomelli, S., *Clin. Chem.* **23,** 264 (1977).
23. Lamola, A. A., and Yamane, T., *Science* **186,** 936 (1974).
24. Lamola, A. A., Joselow, M., and Yamane, T., *Clin. Chem.* **21,** 93 (1975).
25. Blumberg, N. W., Eisenger, J., Lamola, A. A., and Zuckerman, D. M., *Clin. Chem.* **23,** 270 (1977).
26. Culbreth, P., Walter, G., Carter, R., and Burtis, C., *Clin. Chem.* **25,** 605 (1979).
27. Gotelli, G. R., Wall, J. H., Kabra, P. M., and Marton, L. J., *Clin. Chem.* **26,** 205 (1980).

# Chapter 17

## Organic Acids by Ion Chromatography

William E. Rich, Edward Johnson, and Louis Lois

*Dionex Corporation, Department of Research and Development, Sunnyvale, California*

and

Brian E. Stafford, Pokar M. Kabra, and Laurence J. Marton

*Department of Laboratory Medicine, University of California, School of Medicine, San Francisco, California*

### I. Introduction

The presence of increased levels of various organic acids in physiological fluids such as serum, plasma, and urine has been correlated with a variety of diseases (*1*). Although some are rare, others such as lactic acidosis and hyperoxaluria are more widespread (*2, 3*). The estimation of organic acids in biological fluids has long been an analytical problem owing to the nature of the samples and the hydrophilic behavior of the various acids.

The extremely complex composition of the sample matrix has made sample pretreatment almost a mandatory requirement of any chromatographic procedure regardless of the analytical technique (GC, GC–MS, HPLC) selected for final quantitation. The success or failure of the sample pretreatment is, in fact, the major factor in the overall success of any method to be developed. Sample pretreatments range from none (or direct injection) to solvent extraction, distillation, precipitation, absorption, ion-exchange, derivatization, etc. In general, owing to the range of functional groups that may be present in the various organic acids (phenolics, carboxylics, keto-acids, hydroxy acids, and hydroxy-keto acids), no technique has been found to achieve 100% recovery in all classes. However, in cases of severe forms of inborn errors of metabolism, the amounts of acids that accumulate are of such magnitude that less than complete recovery will in no way preclude diagnosing the disorder. If, however, one desires to apply the technique to "metabolic profiling" for rapid screening or highly accurate diagnostic analysis, a pretreatment that yields the highest possible recovery is mandatory.

Gas chromatographic methods require sample derivatization (*4*). HPLC methods are increasingly being applied, but most require extensive sample pretreatment and, in the case of nonchromophoric ions, lack sensitive detection (*5, 6*). Enzymatic methods are most frequently used, but there are many compounds that lack any enzymatic method.

Ion chromatography (IC) is a new technique that overcomes the problems of extensive sample pretreatment and detector sensitivity for ionic species that are nonchromophoric and have low p$K_a$ values. This method has proven itself extremely useful in the chromatographic analysis of inorganic ions such as $Cl^-$, $SO_4^{-2}$, $PO_4^{-3}$, etc., but has not been extensively used for determination of organic ions in complex biological matrices.

This article deals with the general aspect of ion chromatography (IC) and how coupled ion-exchange techniques, in this case, ion exclusion coupled to ion chromatography (ICE/IC), can be utilized to determine aliphatic organic acids in biological fluids. In particular, details of pyruvic acid and lactic acid determination in serum are presented. This includes interference, recovery, and linearity studies and a comparison of the results of serum analysis with clinical reference methods. Of special importance is the ability of IC to determine pyruvate and lactate simultaneously. Preliminary chromatographic data of other clinically important organic acids are also presented.

## II. Principles of ICE/IC

IC is a new technique, first described by Small, Stevens, and Bauman (*7*), that utilizes a unique low capacity, ion-exchange chromatographic column coupled with a high capacity ion-exchange suppressor column and conductivity detector. IC is a unique form of ion-exchange chromatography because it uses a high capacity ion-exchange "suppressor" column to reduce the background conductivity of the chromatographic eluent via ion-exchange action, thus allowing sensitive, conductimetric detection of ions having p$K_a$ values less than approximately 7. Small et al. (*7*) describe this technique and its applications in great detail.

Ion exclusion, here termed *I*on *C*hromatography *E*xclusion mode (ICE), is a well known ion-exchange chromatographic technique for separating strong acids as a class from weak acids (*8*). More recently, Turkelson (*9*) has reported the use of ICE in separating most of the Krebs cycle acids, demonstrating that ICE and also separate weak acids from each other. The theory of ICE has been discussed by Wheaton and Bauman (*8*). The adsorption of uncharged molecules by ion-exchange resins is, in part, responsible for the chromatographic separation of weak organic acids.

The concept of coupling ICE with IC to improve the chromatographic resolution of inorganic and/or organic acids in complex matrices was first presented by Rich et al. (*10*). This method is schematically represented in Fig. 1. The diluted serum or urine is injected into the ICE system. The ion exclusion separator column functions to class separate low p$K_a$ inorganic ions from higher p$K_a$ aliphatic organic acids, and further, from increasingly hydrophobic aromatic acids. This column also generally removes all cationic interfering ions by ion-exchange action. The suppressor column and electrical conductivity detector in combination, act as a very selective and sensitive detection system for anions, with p$K_a$'s below 7 eluting from the ICE system. In order to further improve the selectivity of the chromatographic system, an IC system is coupled to the ICE system via a low capacity anion-exchange concentrator column. Any ICE peak or peaks may be selectively trapped. Then, a high resolution IC chromatographic step further separates the ICE peak into its remaining components. By choosing different eluent strengths and/or gradient elution, a large variety of organic and inorganic ions can be accurately determined by this method. In effect, the ICE system acts as an on-line extraction and detection system that selectively removes ionic interferences and produces a limited separation according to increasing p$K_a$ and hydrophobicity. This minimizes the number of sample preparation steps and eliminates the need for internal standards since no extraction step is required. Table 1 lists chromatographic capacity factors ($k'$) and concentration ranges of a variety of organic acids found in serum.

Any ICE peak or region can be trapped on a low-capacity, anion-exchange concentrator column. This trapping process is only feasible

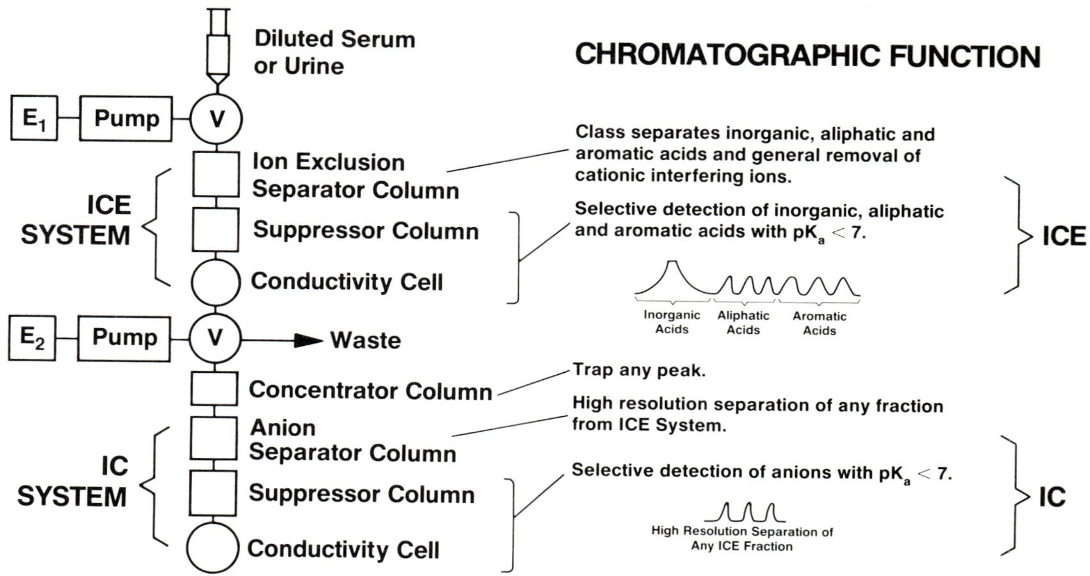


FIG. 1. Schematic representation and chromatographic function for *I*on *C*hromatography *E*xclusion mode coupled to *I*on *C*hromatography (ICE/IC).

Table 1
ICE Chromatographic $k'$ Data for Organic Acids Found in Serum

| Acid salt | $k'^a$ | Estimated Normal Serum Concentration, mg/dL[b] |
|---|---|---|
| Sulfate | 0 | 0.5–1.5 |
| Oxalate | 0.158 | 0.20–0.32 |
| Maleate | 0.607 | — |
| *Cis*-oxalacetate | 0.62 | 0.12 |
| Malate | 0.67 | 0.24–0.75 |
| Citrate | 0.746 | 1.30–1.67 |
| α-Ketoglutarate | 0.803 | 0.05–0.21 |
| α-Ketoisovalerate | 0.82 | 0.09–1.7 |
| Pyruvate | 0.86 | 0.2–1.0 |
| Lactate | 1.43 | 6.84–16.0 |
| Succinate | 1.52 | 1.0–6.0 |
| Formate | 1.59 | 0.425 |
| 3-Hydroxybutyrate | 1.708 | 0.14–0.99 |
| Acetoacetate | 1.85 | 0.055–0.26 |
| Acetate | 1.85 | 1.275 |
| DL-α-hydroxyisovalerate | 2.08 | — |
| Propionate | 2.227 | 0.085 |
| DL-α-hydroxy-*n*-valerate | 2.367 | — |
| Mavelonate | 2.47 | — |

[a]ICE eluent, 0.01 *N* HC1.
[b]Data taken primarily from Tietz and Norbert, *Fundamentals of Clinical Chemistry*, 2nd edition, Philadelphia, W. B. Saunders Co., 1976.

for low $pK_a$ acids since the ICE suppressor column is a cation exchange resin in the silver form. The silver acts to precipitate the hydrochloric acid in the ICE eluent without substantially effecting other inorganic (except halogens) and organic acids that elute from the system. The low $pK_a$ acids are then selectively trapped by ion-exchange action and further separated and detected in the IC system. Acids with $pK_a > 7$ are not appreciably dissociated at pH 7 and, consequently, cannot be detected or trapped.

Peaks trapped from the ICE system can be further separated by IC. Table 2 shows chromatographic $k'$ data for a variety of organic and inorganic ions under three different IC eluent strengths. Generally, as the IC eluent concentration increases, the k′ decreases. A general problem with IC, however, is that one or more ubiquitous inorganic

Table 2
IC Chromatographic $k'$ Data for Organic and Inorganic Acids Found in Serum

| Acid salt | Eluent concentrations, $k'$ | | |
|---|---|---|---|
| | 0.6 m$M$ $NaHCO_3$ | 3 m$M$ $NaHCO_3$/ /2.4 m$M$ $Na_2CO_3$ | 3 m$M$ $NaHCO_3$/ /6 m$M$ $Na_2CO_3$ |
| Propionate | 1.56 | 1.79 | 1.41 |
| Glutarate | × | 5.14 | 3.09 |
| 3-Hydroxybutyrate | 1.63 | × | × |
| DL-α-hydroxyiso-valerate | 2.0 | × | × |
| DL-α-hydroxy-*n*-valerate | 2.06 | × | × |
| Formate | 2.75 | 0.67 | × |
| α-Ketoisovalerate | 5.0 | × | × |
| Lactate | 1.75 | × | × |
| Pyruvate | 4.0 | 0.73 | × |
| Succinate | × | 5.64 | 2.49 |
| Malonate | × | 6.93 | 2.98 |
| Maleate | × | 8.33 | 3.59 |
| Tartarate | × | 9.84 | 4.01 |
| Oxalate | × | 14.69 | 5.99 |
| Chloride | × | 1.6 | 1.06 |
| Phosphate | × | 4.8 | 2.38 |
| Nitrate | × | 7.11 | 4.45 |
| Sulfate | × | 11.91 | 5.04 |

ions, such as phosphate, chloride, nitrate, and sulfate, have $k'$ values similar to many organic ions. This represents a major interference problem for IC when used alone in biological matrices. The problem is substantially resolved by ICE/IC. In general, ions that have similar $k'$ values in ICE have large differences in $k'$ values for IC and are easily resolved by ICE/IC. For example, α-ketoisovaleric acid and pyruvic acid have similar ICE $k'$ values and are not separated by ICE except under conditions of extremely high chromatographic efficiency. The IC $k'$ values differ by one $k'$ unit, resulting in baseline separation by IC. By careful choice of IC eluent conditions and selective ICE peak trapping, a highly accurate method for determination of any of the acids listed in Tables 1 and 2 can be quickly developed.

## III. Determination of Pyruvate and Lactate in Serum

Serum lactate analysis is becoming increasingly important in clinical investigation and diagnosis of lactic acidosis in diabetic patients (*11*) and in prognosis and diagnosis of acute myocardial infarction complicated by shock (*12*). The size of the infarct can be estimated from results of serum lactate concentration, either alone or in combination with other tests. Serum pyruvate is determined in the diagnosis of severe thiamine deficiency and heavy-metal poisoning (*13*). Lactate dehydrogenase (LDH, EC 1.1.1.27) enzymatic methods are most frequently used for lactate and pyruvate determination. Recently, Sunderson and Hinsch (*14*) reported a method for lactate and pyruvate determination that utilized immobilized enzymes in nylon tube reactors.

A chromatographic method offers several advantages over automated enzymatic methods, but the most important is accuracy. Unlike enzymatic methods, where near 100% selectivity is presumed, and interference studies are necessarily limited, chromatographic methods can be varied to optimize for either speed *or* resolution. In addition, chromatographic methods offer the analyst multiconstituent determination per sample. The determination of pyruvate and lactate in serum was chosen to test the feasibility of using ICE/IC as a clinical method for the determination of these organic acids as well as other low $pK_a$ acids in serum.

## IV. Materials and Methods

The standard LDH method was utilized for lactate determination. Pyruvate analyses were conducted by SmithKline Laboratories, Burlingame, California, using a Boeringer-Mannheim LDH kit. Serum samples sent for enzymatic assay were treated in the standard way by dilution with perchloric acid.

Reagents: Pyruvic acid, Type 11; sodium salt, crystalline, 99% (Sigma Chemical Co., Saint Louis, MO 63178); lactic acid, L-lactic acid; lithium salt, A grade (Calbiochem, La Jolla, CA); pyruvic acid stock solution, 20 mg/dL. Transfer 25 mg of the standard into a 100 mL volumetric flask, dilute to volume, and refrigerate. This solution should be prepared fresh each day.

Lactic acid stock solution, 100 mg/dL. Transfer 106 mg of the standard into a 100 mL volumetric flask, dilute to volume, and refrigerate. This solution should be prepared fresh each day.

ICE mobile phase, 0.01 *N* HCl; IC mobile phase, 0.66 m*M* $NaHCO_3$, except for succinate and sulfate analysis, which used 3 m*M* $NaHCO_3$ and 2.4 m*M* $Na_2CO_3$; IC regeneration solution, 0.5 *N* $H_2SO_4$; Acetonitrile (Burdick, Jackson Laboratories, Inc., Muskegon, MI 49442).

## A. Apparatus

Chromatographic System: Dionex Model 16 Ion Chromatograph (Dionex Corp., Sunnyvale, CA 94086). Injection valves were modified to accept 20 μL loops.

ICE columns: 9 × 250 mm separator #030508 followed by a suppressor #030581 and post suppressor #030582 to remove HCl from the eluent (all from Dionex Corp.).

IC columns: The separator and concentrator columns were specially prepared by Dionex Corp., R & D Dept. The anion suppressor column was a Dionex 3 × 250 mm #030366.

## B. Procedures

Approximately 30 min before analysis, the pumps were started and the analytical columns were equilibrated. The ICE flow rate was set at 0.86 mL/min and IC flow rate at 2.3 mL/min. The detection mode attenuation was set at 10 μmHO. A mixture of 20 mg/dL lactate and 1.5 mg/dL pyruvate was prepared, and then diluted 1/1 with acetonitrile. A 20-μL volume was injected into the ICE and IC systems to establish the pyruvate and lactate *K'* values. The standard was injected into the ICE system. The pyruvate region was trapped onto a concentrator column, and the trapped material injected into the IC system. The lactate region was also trapped onto the concentrator column and injected into the IC system after the last peak of interest from the ICE pyruvate fraction eluted from the IC system.

Venous blood was drawn and centrifuged for 5 min. The serum was diluted 1/1 with acetonitrile, vortexed, and centrifuged for 5 min. The supernatant was collected and 20 μL of the sample was injected into the ICE system. Coupled ICE/IC was performed. The concentrations of lactate and pyruvate were calculated from the peak heights.

The ICE suppressor column is disposable and is used up at a rate of 10–100 samples/column, depending on ICE eluent pH. The IC suppressor column is automatically regenerated every 8 h. Approximately 20 min are required for this process. The ICE separator column requires regeneration after several hundred samples, depending on the concentrations injected. This is easily accomplished by washing the column in the reverse direction with 1 *N* $H_2SO_4$ for a few minutes, then water. This will remove unwanted matrix ions that have been trapped on the top of the column. Small pre-columns containing ICE resin are also effective in prolonging column capacity.

## V. Results and Discussion

A typical ICE chromatogram obtained from serum is shown in Fig. 2. Note that several other peaks are also observed and given tentative identification based on ICE $k'$ data of known organic acids in serum.

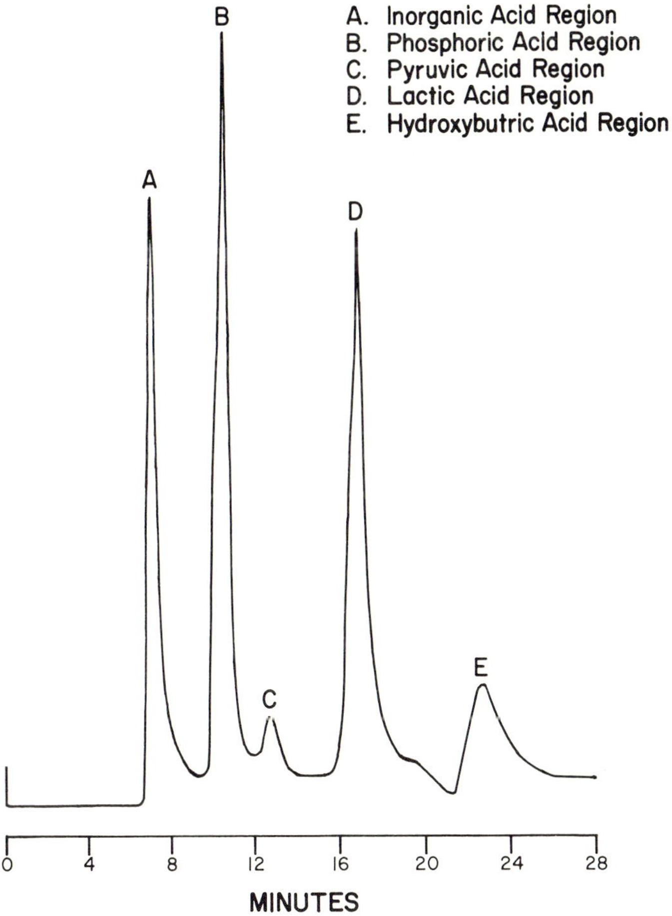


FIG. 2. ICE chromatogram of normal human serum. Each region contains acids of similar p$K_a$'s and hydrophobicity. p$K_a$ and hydrophobicity of the eluting peaks increase from A to E. The ICE eluent is 0.01 $N$ HCl.

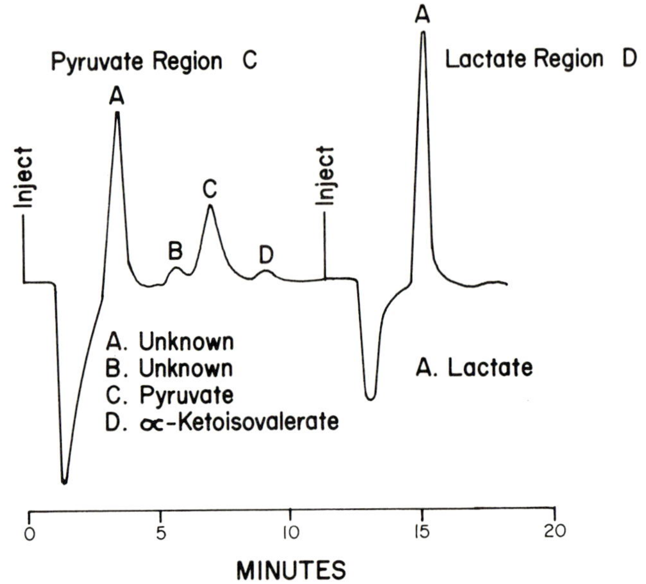


FIG. 3. IC chromatogram of pyruvate and lactate ICE regions following sequential trapping onto an IC concentrator column. Lactate is sequentially injected into the IC system approximately 12 min after pyruvate. The IC eluent is 0.66 m*M* $NaHCO_3$.

From Table 1, it is clear that succinate may partially co-elute with lactate and citrate, while α-ketoglutarate and α-ketoisovalerate co-elute with pyruvate in ICE. Figure 3 shows the IC chromatograms of the trapped pyruvate and lactate ICE regions. Regarding the pyruvate ICE region C, citrate and α-ketoglutarate are polyvalent acids and *do not* elute in IC using 0.66 m*M* $NaHCO_3$ eluent. Therefore, referring to Fig. 3, the first two IC peaks are unknown acids with $pK_a$ values between approximately 3.8 and 2.4. The third peak is pyruvate and the last peak is α-ketoisovalerate. In the lactate ICE region D, succinate and lactate are easily separated by IC using 0.66 m*M* $NaHCO_3$ eluent. Under these conditions, succinate is strongly retained. Figure 3 shows a typical IC chromatogram of the trapped lactate ICE region D from a serum sample. By increasing the IC eluent strength, polyvalent acids such as succinate can also be determined. Figures 4 and 5 illustrate this for succinate and sulfate, respectively. In the case of succinate, lactate elutes in the void volume, a second unknown acid elutes in 4 min, and succinate, generally found in concentrations of 1–6 mg/dL in serum, elutes in 16 min. In the case of sulfate, ICE peak A is trapped, and inorganic ions such as nitrate and sulfate can be determined.

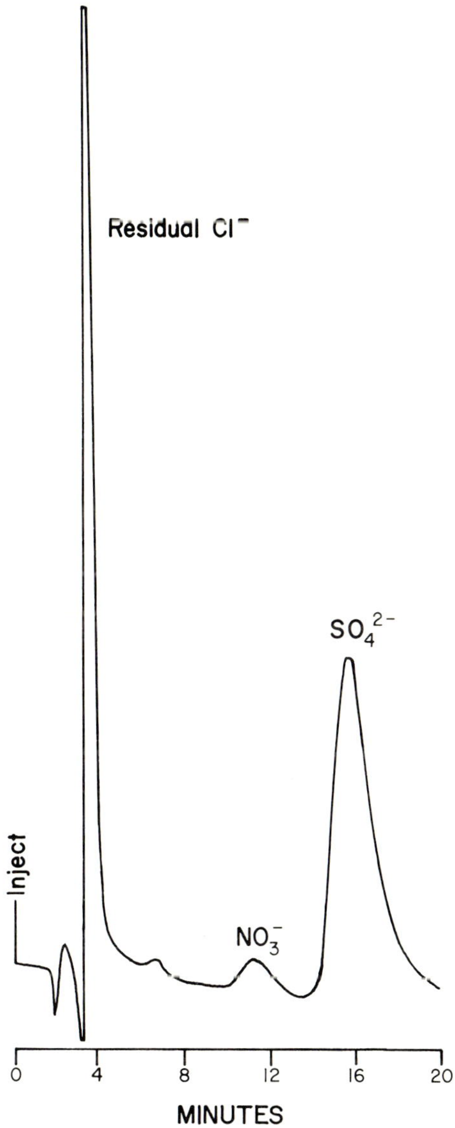


FIG. 4. IC chromatogram of ICE Region D using a 3 m*M* $NaHCO_3$/2.4 m*M* $Na_2CO_3$ IC eluent.

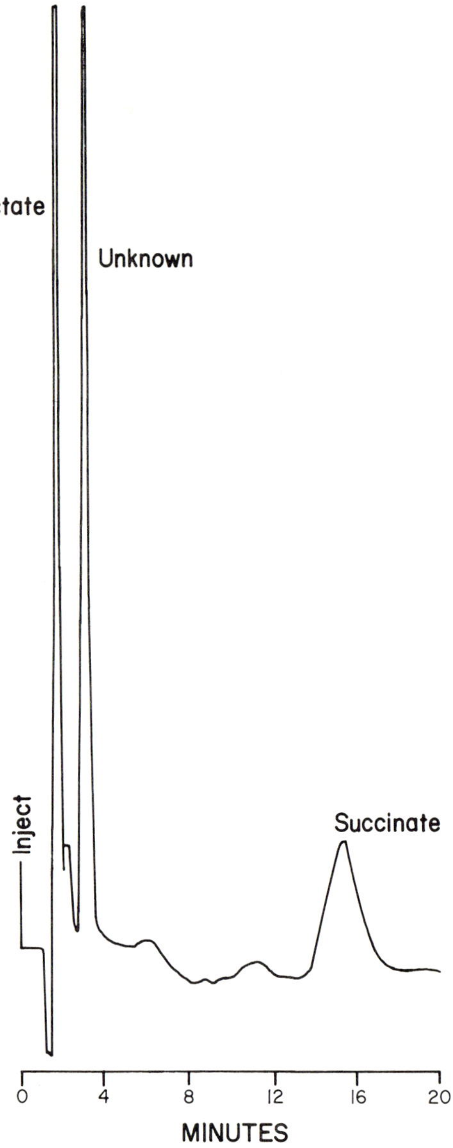


FIG. 5. IC chromatogram of inorganic ICE region A following trapping onto the IC concentrator column. The IC eluent is 3 m*M* $Na_2HCO_3$/2.4 m*M* $Na_2CO_3$.

Table 3
Pyruvate and Lactate Values Obtained by ICE, ICE/IC, and LDH Methods Using Serum from Normal Individuals

| | Pyruvate, mg/dL | | | Lactate, mg/dL | | |
|---|---|---|---|---|---|---|
| Normal serum | LDH | ICE | ICE/IC | LDH | ICE | ICE/IC |
| 1 | 0.17 | 2.0 | 0.46 | 7 | 7.2 | 8.4 |
| 2 | 0.16 | 1.3 | 0.73 | 12 | 12 | 13 |
| 3 | 0.15 | 1.0 | 0.60 | 11 | 10.4 | 11.4 |
| 4 | 0.19 | 1.33 | 0.86 | 14 | 11.8 | 12.6 |
| 5 | 0.23 | 1.5 | 0.83 | 15 | 15.4 | 17.2 |

Table 3 gives the pyruvate and lactate values found in the serum of five normal laboratory staff personnel. Both the ICE and ICE/IC values are listed along with comparative results obtained using standard clinical LDH methods. The results indicate that lactate ICE compares favorably with LDH and ICE/IC methods. Normal lactate levels range from 9 to 16 mg/dL. This indicates that ICE is reasonably effective in separating succinate and lactate, and that ICE/IC is not required for routine analysis. In the case of pyruvate, citrate has a normal concentration range of 1.3–1.7 mg/dL in serum, compared to 0.2–1 mg/dL for pyruvate. Consequently, ICE results are extremely high compared to those obtained by the LDH and ICE/IC methods. The normal range for serum pyruvate is consistent with the ICE/IC results. The LDH results for pyruvate were obtained from an outside laboratory after a delay of several days. During this time, the pyruvate may have polymerized, resulting in lower values compared to ICE/IC.

Table 4 shows analytical recovery data for pyruvate and lactate by ICE/IC. Samples of pooled serum spiked with lactate and pyruvate corresponding to values below and above the normal range found in human serum gave a linear regression equation for pyruvate: $y = 7.404x + 4.762$, correlation coefficient 0.994; and for lactate: $y = 1.868x + 48.816$, correlation coefficient 0.997.

A mixture of pyruvate and lactate was analyzed three times during a single day to determine precision. At 10 mg/dL lactate had a relative standard deviation (RSD) of 2.1%. At 1 mg/dL, pyruvate had a RSD of 3.3%.

The minimum detection limit for ICE/IC is limited by sample size and chromatographic resolution between a given set of ions. This is because the ICE/IC concentrator column allows concentration of a large number of ICE injections. As little as 100 $\mu$g/L of pyruvate can

Table 4
Analytical Recovery of Added Pyruvate and Lactate Pooled Serum by ICE/IC

| | Increase in control serum concentration, mg/dL | Measured concentration, mg/dL | Recovery, % |
|---|---|---|---|
| Pyruvate | Control | 0.4 | × |
| | 0.1 | 0.49 | 98 |
| | 0.5 | 0.90 | 100 |
| | 1.0 | 1.25 | 89 |
| Lactate | Control | 18.5 | × |
| | 2.0 | 21.06 | 103 |
| | 5.0 | 24.2 | 103 |
| | 10.0 | 28.2 | 99 |

easily be determined in a single 1 mL injection. Ten injections, followed by sequential concentration, would bring the minimal detectable level (MDL) to near 10 ng. In practice, the MDL is usually limited by chromatographic separation of trace ions from highly concentrated ions.

## VI. Conclusions

These preliminary results indicate that ICE/IC may prove to be an excellent method for establishing the accuracy of other clinical methods involving low p$K_a$ organic or inorganic ions. It is also clear that if low p$K_a$ multi-ion analysis per sample is desired, along with minimization of sample clean-up, ICE/IC should prove to be the preferred method over enzymatic techniques. The major advantages of ICE/IC over these methods are: (a) the ability to optimize for resolution or speed, (b) detection limits at the $\mu$g/L level, and (c) fast methods development and set-up time for low p$K_a$ ions infrequently analyzed.

## Acknowledgment

Laurence J. Marton is the recipient of NCI Research Career Development Award CA-00112.

## References


1. Snyder, L. R., Karger, B. L., and Giese, R. W., *Contemp. Top. Anal. Clin. Chem.* **2,** 230 (1978).
2. Chalmers, R. A., Lawson, A. M., and Borud, D., *Clin. Chem. Acta* **77,** 117 (1977).
3. Charransol, G., Barthelemy, C., and Peagrey, P., *J. Chromatogr.* **145,** 452 (1978).
4. Goodman, S. I., Helland, P., Stokke, O., and Jellum, E., *J. Chromatogr.* **142,** 497 (1977).
5. Molnar, I., and Horvath, C., *J. Chromatogr.* **143,** 391 (1977).
6. Miyagi, H., Miura, J., Takata, Y., and Canno, S., *Clin. Chem.* **25,** 1617 (1979).
7. Small, H., Stevens, T., and Bauman, W., *Anal. Chem.* **47,** 1801 (1975).
8. Wheaton, R. W., and Bauman, W. C., *Ind. Eng. Chem.* **45,** 228 (1953).
9. Turkelson, V. T., and Richards, M., *Anal. Chem.* **50,** 1420 (1978).
10. Rich, W., Smith, F., McNeill, L., and Sidebottom, T., Ion Exclusion Coupled to Ion Chromatography: Instrumentation and Application, in *Ion Chromatographic Analysis of Environmental Pollutants,* Vol. 2, Ann Arbor Science Press, Ann Arbor, Michigan, 1979, p. 17.
11. Wittmann, P., Haslbeck, M., Bachmann, W., and Mehnert, M., *Deutsch. Med. Wochenschr.* **102,** 5 (1977).
12. Afifi, A. A., Chang, P. C., Lin, V. Y., et al., *Am. J. Cardiol.* **33,** 826 (1974).
13. Laurence, D. R., *Clinical Pharmacology,* 4th ed., Churchill Livingstone, Edinburgh, 1973, p. 284
14. Sundaram, P. V., and Hinsch, W., *Clin. Chem.* **25,** 285 (1979).

# Chapter 18

# Major and Modified Nucleosides, RNA, and DNA

Charles W. Gehrke and Kenneth C. Kuo

*Experiment Station Chemical Laboratories, University of Missouri, Columbia, Missouri*

## I. Introduction

Most analytical chemists are well aware of the rapid rate of development of high-performance liquid chromatography (HPLC) over the past 5 years. A number of articles have been published in *Analytical Chemistry* on different topics in HPLC and many papers appear in the chromatographic journals. Some books also have been published covering this subject. HPLC has proved to be a very effective, broadly applicable chromatographic method for the separation and analysis of complex molecules in fields as diverse as biochemistry and environmental, pharmaceutical, medical, and polymer chemistry. HPLC is now having a major impact on the clinical and research aspects of medical biochemistry. Although the contributions of HPLC to other disciplines generally complements gas-liquid chromatography, this method is destined to play a much greater role in medical and biochemical research. This is because many of the biomolecules, owing to their molecular complexity and size, are thermally unstable or nonvolatile, preventing or complicating an analysis by GC. A major factor contributing to the powerful advances in biomedical liquid chromatography is the development of reversed-phase high-performance liquid chromatography (RP-HPLC) using *n*-alkyl and phenyl chemically bonded substrates.

Transfer ribonucleic acid (tRNA) has the most heterogeneous complex of nucleoside structures of all the nucleic acids. Up to 25% of its 76 or so nucleosides may be modified. The modifications, which number over 50, may be as simple as methyl group, or may be extremely complex. Borek has discovered that modifications are achieved after the synthesis of the primary sequence by enzymes that are specific for the species, base, site, and sequence of tRNA. Realization of the tremendous biological significance of tRNAs has stimulated research directed at the elucidation of the many functional aspects of these complex macromolecules.

The variety of functional roles implied by or ascribed to tRNA has led to widespread interest in this vital group of macromolecules. In addition to its critical role in protein synthesis, tRNA has been shown to have regulatory functions in transcription, reverse transcription, translation, inhibition of enzyme activity, and protein degradation (1–4). Such a variety of function implies considerable variability in structure, more than that provided by the anticodons. Borek (*5*) suggested that structural and conformational changes in the tRNAs arising from the addition of modifying moieties such as methyl groups could yield sufficient variability. These modifications are made after the synthesis of the macromolecule (*6–8*) by the addition of methyl groups from *S*-adenosyl methionine to specific base residues by specific methyltransferase enzymes (*3, 9, 10*).

Further, tRNAs participate in the regulation of amino acid biosynthesis and transfer amino acids to cell wall structure. Another possible function now being intensively investigated is the action of tRNA and tRNA-modifying enzymes in the control of cellular development and differentiation (*2, 35–37*). Both tumor and embryonic tissue contain some altered isoaccepting tRNAs (*38*). The control of the formation of these altered tRNAs may be the key to cancer prevention and treatment. It is likely that many more roles for tRNA will be found in the future as research progresses and better methods of analysis are developed. A sensitive, direct, rapid, and accurate method for the measurement of both major and modified nucleoside composition of often limited mammalian tissue samples of tRNA would advance our understanding of the biological significance of tRNA.

Modified nucleosides are found in the urine of both normal and cancerous animals and humans (*11–17*). Since there seems to be no mechanism for reincorporation of these post polymer-modified nucleosides into tRNA, their presence in urine is evidence of the extent of modification, as well as a measure of the turnover rate of tRNA (*18, 19*). Therefore, quantitation of modified nucleosides in urine could indicate changes in the tRNA profile during differentiation or tumor induction. Advantage has been taken of these excretion products to search for a biological marker(s) of cancer. Such a marker(s) would either be indicative of the presence of cancer or would parallel changes in tumor mass and thus be useful in following chemotherapy (*14, 15*).

Development of methods for the analysis of nucleic acid components has been a major thrust in our laboratory since 1967 with the early work utilizing gas-liquid chromatography (GLC) (*20–23*). The GLC methods we developed (*24–27*) have been used to monitor the levels of pseudouridine, $N^2$,$N^2$-dimethylguanosine and 1-methylinosine in urine. Further, reports by Waalkes et al. (*14, 15*) have indeed demonstrated that elevated levels of these markers do occur in the urine of cancer patients with Burkitt's lymphoma, lung, colon, breast, and other types of cancers.

Suits and Gehrke (*28*) demonstrated the potential of reversed-phase HPLC for the separation of nucleic acid bases and modified nucleosides; and recently Hartwick and Brown (*29*) reported on the evaluation of microparticle chemically bonded reversed-phase packings in the HPLC analysis of nucleosides and their bases. Gehrke et al. (*30*) have now completed a comprehensive study of the fundamental parameters relating the general effects of pH, ionic strength, polarity of solvents, flow-rate, and temperature of the mobile phase to the separation of nucleosides by reversed-phase HPLC.

Our investigations (*30, 32, 39, 41, 42*) and the work of Brown's group (*29, 43*) suggest that the versatility of the chromatographic method would be most useful in molecular biology and cancer research involving studies on the measurement of major and modified nucleosides in the urine and plasma of normals and cancer patients by HPLC. Hartwick and Brown in their major review article present high-pressure liquid chromatographic analysis of the nucleosides, bases, and other UV-absorbing compounds in biological materials with respect to the chromatography, sample preparation methods, peak identification by absorbance ratios, fluorescence, and enzymatic peak shifts, and the applications to urine, serum, and dialysates.

To provide the research tool for a better understanding of tRNA biochemistry and to follow the urinary excretion of modified nucleosides in medical research and the response of patients with cancer to chemotherapy, we describe a high-performance liquid chromatographic analytical method that can be readily placed in operation.

Our RP-HPLC method is rigorous, comprehensive, and has been

developed for the analysis of ribonucleosides in urine. An initial isolation of ribonucleosides with an affinity gel containing an immobilized phenylboronic acid was used to improve selectivity and sensitivity. Response for all nucleosides was linear from 0.1 to 50 nmol of injected material, and good quantitation was obtained for 25 μL or less of sample placed on the HPLC column. Excellent precision of analysis for urinary nucleosides was achieved on matrix dependent and independent samples, and the high resolution of the reversed-phase column allowed the complete separation of 18 nucleosides from other unidentified UV absorbing components at the 1-ng level. Supporting experimental data are presented on precision, recovery, chromatographic methods, minimum detection limit, retention time, relative molar response, sample cleanup, stability of nucleosides, boronate gel capacity, and application to analysis of urine from patients with leukemia and different cancers. This method is now being used routinely for the determination of the concentration and ratios of nucleosides in urine from patients with different types of cancer and in chemotherapy response studies.

Our HPLC method is characterized by the following features:

a. Sensitivity at the nanogram level.
b. High chromatographic resolution and selectivity.
c. Direct measurement of nucleosides with accuracy and precision.
d. Analysis is nondestructive and the high capacity of this chromatographic system allows easy isolation of pure nucleosides for further characterization.
e. Rapid separation and measurement in ca. 1 h after hydrolysis to nucleosides.
f. Quantitation without the use of radiolabeled compounds; however, labeled compounds are readily isolated and measured.

In other studies, the chromatographic conditions for the separation of urinary nucleosides was optimized using both isocratic and step-gradient conditions. The step-gradient system is more suitable for determining the nucleoside composition of tRNA hydrolysates, and the complete separation of the major ribo- and deoxyribonucleosides can be accomplished. Also, we have looked for nucleotides and oligonucleotides in normal and cancer patient urine and found none. In addition, we report a rapid isocratic system for the separation of $m_2^2G$ and $t^6A$, and for pseudouridine and other modified ribonucleosides in serum.

These sensitive and selective methods allow the rapid analysis of trace levels of nucleosides in complex matrices of biological material and in small samples of polynucleotide hydrolysates. They should serve as important tools in molecular biology and in clinical research.

## II. Experimental

### A. Apparatus

A modular HPLC system was used for the chromatographic studies. This consisted of a model 6000A Solvent Delivery System, a model U6K Universal Injector, and a model 440 two-channel Absorbance Detector (Waters Associates, Inc., Milford, MA USA.) The recorders used were a Honeywell Electronik 194 ABR and a Fisher Recordall Series 5000. The column consisted of two Waters μBondapak $C_{18}$ 300 × 4 mm columns connected in series.

The temperature of the column was maintained using a constant temperature circulating bath, Haake Model FJ (Saddle Brook, NJ USA), connected to an aluminum column jacket (*32, 39*).

Peak areas, retention times, and concentrations based on an external standard were calculated by a 3352B Laboratory Data System (Hewlett-Packard, Avondale, PA USA). The system consists of a 2100 computer with 24K memory, 18652A analog to digital converters (A/D), ASR33 teletype, and a 2748B high speed photo reader.

The columns used for the boronate gel were glass 5 × 150 mm (Fischer and Porter, Warminster, PA USA) modified by attachment of a 50 mL spherical reservoir to the top of the column.

The eluates from the boronate gel columns were lyophilized in Corex 25-mL screw cap, round-bottom centrifuge tubes (Corning Glass Works, Corning, NY USA) on a custom built lyophilizer capable of maintaining a pressure of 0.05–0.1 mm of Hg with cold trap at −60° C.

An Eppendorf Model 3200/30 microcentrifuge, Model 3300 rotary shaker, as well as various sizes of Eppendorf pipets (Brinkman Instruments, Inc., Westbury, NY USA) were used in the sample cleanup procedure.

A micro Gram-Atic Balance (Mettler Instrument Co., Hightstown, NY USA) was used to weigh milligram amounts of nucleosides for the calibration solutions.

### B. Chemicals

The nucleosides, nucleotides, and nucleic acid bases were obtained from the following sources: pseudouridine (ψ), cytidine (C), 3-methylcytidine ($m^3C$) inosine (I), 1-methylguanosine ($m^1G$), uridine 3′-

monophosphate (3′UMP), quanosine 2′- and 3′-monophosphate (2′- and 3′-GMP), adenosine 2′- and 3′-monophosphate (2′- and 3′-AMP) (Sigma Chemical Co., St. Louis, MO, USA); uracil (Ura), guanine (Gua), adenine (Ade),cytosine (Cyt), uridine (U), guanosine (G), adenosine (A), cytidine 5′-monophosphate (5′-CMP), uridine 5′-monophosphate (5′-UMP), adenosine 5′-monophosphate (5′-AMP), guanosine 5′-monophosphate (5′-GMP), (Mann Research Laboratories, New York, NY USA); 5-methyluridine ($m^5U$), 4-thiouridine ($s^4U$), 4-acetylcytidine ($ac^4C$), 2′-O-methylcytidine (Cm), 2′-O-methyluridine (Um), 2′-O-methyladenosine (Am) (P.L. Biochemicals, Inc., Milwaukee, WI USA); 1-methyladenosine ($m^1A$), $N^2$-methylguanosine ($m^2G$), $N^2$, $N^2$-dimethylguanosine ($m^2_2G$), and $N^6$-methyladenosine ($m^6A$), 5-methylcytidine ($m^5C$), 7-methylguanosine ($m^7G$), 1-methylinosine ($m^1I$) (Vega-Fox Biochemicals, Tucson, AZ USA).

Other chemicals were purchased from the following sources: ammonium acetate and formic acid ACS certified grade (Fisher Scientific Co., St. Louis, MO, USA) ammonium hydroxide, analytical reagent grade (Mallinckrodt, Inc., St. Louis, MO, USA), ammonium dihydrogen phosphate (J. T. Baker, Phillipsburg, NJ, USA). Hydrazide Biogel P-2, 200–400 mesh, lot no. 15569 (Bio-Rad Laboratories, Richmond, CA, USA), *m*-aminophenylboronic acid hemisulfate, succinic anhydride, and 1-ethyl-3(3-dimethylamino-propyl) carbodiimide hydrochloride (EDAC) (Aldrich Chem. Co., Milwaukee, WI, USA). All other chemicals were of the highest purity available. Methanol was obtained from Burdick and Jackson (Muskegon, MI, USA). All water used for the preparation of buffers and aqueous solutions was purified by a three-step process. The first step was reverse osmosis using an RO Pure apparatus (DO640 Barnstead Co., Boston, MA USA). A Nanopure D1794 four cartridge water purification system was then used. A charcoal cartridge for adsorption of organics, two mixed-bed ion-exchange cartridges for removal of cations and anions, and a filtration cartridge for removal of all particles larger than 0.22 $\mu$m were used. Finally, the Nanopure water was distilled in a Corning all glass still (Corning Glass Works, Corning, NY, USA).

## C. Enzymes

Pancreatic ribonuclease, ribonuclease CB, ribonuclease $T_2$, (Calbiochem, San Diego, CA, USA), and alkaline phosphatase, *E. coli* (Sigma Chemical Co., St. Louis, MO USA) were used for the hydrolysis of tRNA samples.

## D. HPLC Buffers

A stock buffer concentrate was prepared as 2 L of a 2.0 *M* solution of $NH_4H_2PO_4$. This concentrate was then sterilized by filtering through a Millipore GS-22 filter (0.22 μm) and stored in glass at 4°C. One liter of the working buffer was prepared daily by taking a 5.0 mL aliquot of the stock 2.0 *M* buffer solution and diluting it to 1.0 L with Nanopure distilled water in a volumetric flask. Then the pH was adjusted to 5.10 with a few drops of either a 1.0 *M* $H_3PO_4$ or 3.0 *M* $NH_4OH$ solution. If methanol was to be added to the buffer, the appropriate volume was added after ca. 200 mL of $H_2O$ had been added to the buffer concentrate, but before making to final volume with Nanopure distilled $H_2O$. All buffers were sterilized by filtering through a Millipore GS-22 filter (0.22 μm) before use. Stored buffers were maintained in a cold room at 4°C. If stored 24 h or longer, the buffer was refiltered through a 0.22 μm filter.

## E. Calibration Standard Solutions

Single compound stock solutions of nucleosides were exactly prepared to yield concentrations of about 1.00 μmol/mL in distilled Nanopure $H_2O$.

The exception to this concentration was $m^2G$, which was made up at 0.25 μmol/mL because of its low solubility. Standard solutions were stored at 4°C, except for $s^4U$ and $ac^4C$. These nucleosides were found to be relatively unstable and the solutions were frozen and stored at −20°C.

Calibration standards were made by dilution of aliquots of the single compound stock solutions to give a standard solution containing 200 μmol/L of C, U, G, and A; 20 μmol/L $m^1A$ and ψ; and 10 μmol/L for each of the other modified nucleosides. A 25 μL volume of this solution was used to calibrate the chromatography system.

## F. Enzymatic Hydrolysis of tRNA Sample to Ribonucleosides

A mixture of ribonucleases was made containing pancreatic ribonuclease 1 mg/mL, ribonuclease CB 500 units/mL, and ribonuclease $T_2$ 500 units/mL. tRNA samples were incubated with 5 μL of the ribonuclease mixture per 1 $A_{260}$ (approximately 50 μg) of tRNA for 8 h at 37°C. Following the ribonuclease digestion, 5 μL of a solution of alkaline phosphatase containing 12 mg/mL (144 units/mL) were added per 1 $A_{260}$ of tRNA with enough 0.5 *M* Tris buffer, pH 7.8, to make the solution 0.05 *M* in Tris. The mixture was then incubated for 4 h at 37°C. Following this treatment the solutions were diluted to accurately known concentrations and stored at −20°C until used for HPLC analysis. Aliquots of these solutions were used for direct HPLC analysis without further treatment, or an isolation of the ribonucleosides was made with a phenyl boronate substituted affinity gel (as described below) prior to HPLC analysis.

### G. Phenylboronate-Substituted Polyacrylamide Affinity Gel

An affinity gel with an immobilized phenylboronic acid functionality was used for isolation of ribonucleosides prior to HPLC separation and quantitation. The synthesis and use of this gel has been described in detail (*31, 32, 39, 40*). The tRNA enzymatic hydrolysate equivalent to 0.1–1.2 $A_{260}$ (approximately 5–60 $\mu$g) adjusted to pH 8.8 with 0.25 *M* $NH_4Ac$ buffer was placed on the 5 × 40 mm gel column. The column was washed sequentially with 1 × 1mL, then 2 × 3 mL 0.25 *M* $NH_4Ac$ buffer (pH 8.8) and the nucleosides then eluted with 5 mL 0.1 *M* HCOOH. The eluate was lyophilized to dryness, and redissolved in an accurately measured amount of distilled Nanopure water. Aliquots of this solution were then used for subsequent HPLC analysis.

### H. Samples, Collection, and Storage

The urine samples were collected at ice temperature. Aliquot samples were frozen and stored at −70°C. The normal control urines were from laboratory personnel. The cancer patients selected had advanced malignant disease, and at the time of the urine collection the patients were not receiving anti-neoplastic drugs or other antitumor therapy. The urine samples from the cancer patients were obtained through the courtesy of the following hospital services: Johns Hopkins University Medical School, Oncology Division, the National Cancer Institute Solid Tumor Service, the Cancer Research Center, Columbia, Missouri, Professor E. Borek of the University of Colorado Medical Center, Dr. Raymond Ruddon, Frederick Cancer Research Center, and Dr. John Speer, Penrose Medical Center, Colorado Springs.

### I. Cleanup of Urine Samples for Nucleoside Analysis by HPLC

The structure of the boronate derivatized polymer and the formation of the *cis*-diol boronate complex are presented in Fig. 1. An abbreviated urine sample cleanup schemat is given in Fig. 2, this is then followed by the detailed analytical method.

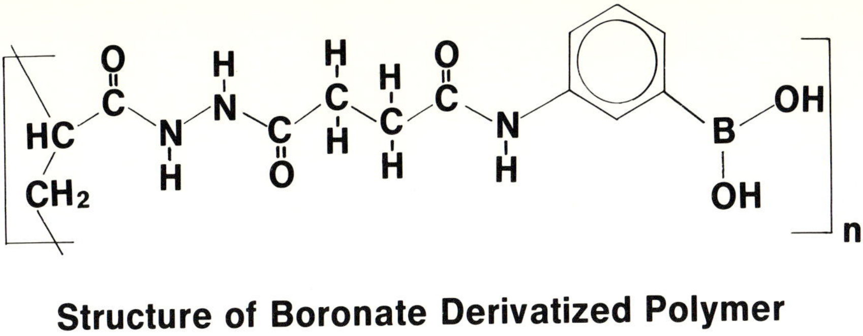


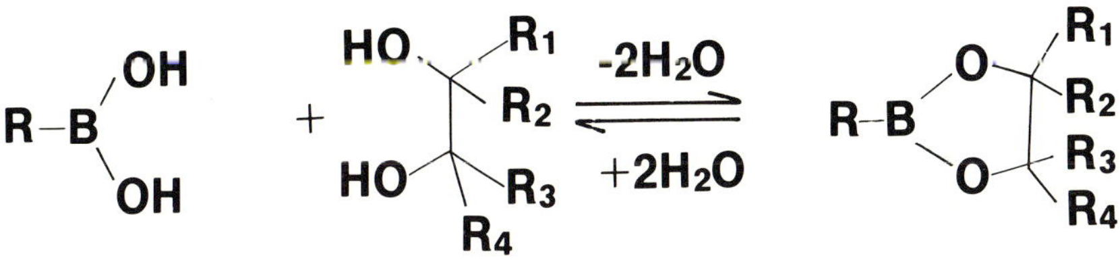

FIG. 1. Formation of *cis*-diol boronate complex (*31*).

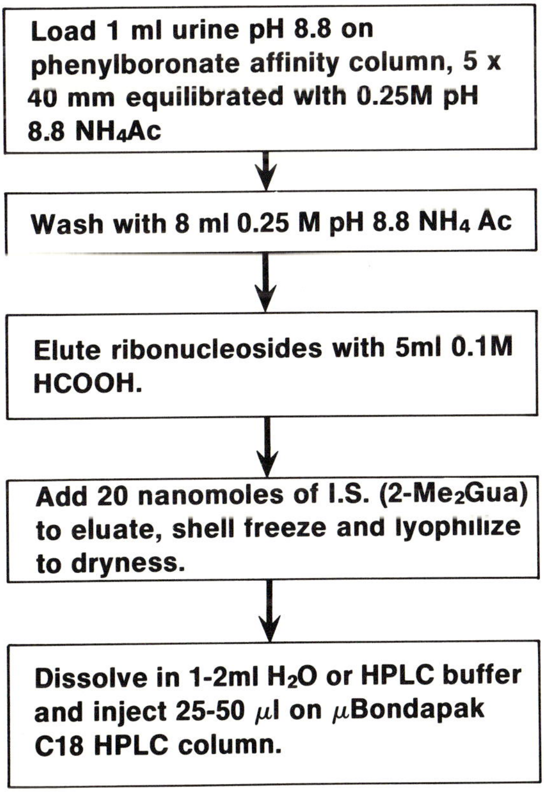


FIG. 2. Urine sample cleanup for HPLC ribonucleoside analysis.

## III. Analytical Procedure

### A. Column Preparation

1. Place ca. 1 mL 0.25 *M* ammonium acetate buffer, pH 8.8, in the column (Fischer and Porter No. 274-461, 150 mm × 5 mm, custom fitted with a 50 mL reservoir).

2. Slurry the resin in its 0.1 *M* sodium chloride storage solution and transfer to the column with a Pasteur pipet (Fisher Scientific Co. No. 13-678-5B).

3. Introduce the boronate (200–400 mesh) resin below the surface of the buffer in the column. Care must be taken to prevent the resin from contacting the sides of the reservoir as the resin adheres to glass.

4. Allow the column to begin draining and add resin to a height of 40 mm (bed volume 0.785 $cm^3$).

5. Rinse the resin with ca. 20 mL 0.25 *M* ammonium acetate, pH 8.8. No pressure is used on the column. All solutions are allowed to drain by gravity flow. The flow rate varies from column to column, averaging about 10 mL/h for the 0.25 *M* ammonium acetate buffer, pH 8.8, and about 20 mL/h for the 0.1 *M* formic acid solution.

6. Allow the buffer to drain to the top of the resin bed; then add 50 mL of 0.1 *M* formic acid rinse. The resin expands and contracts depending on the pH and ionic strength of the solution with which it is equilibrated. Formic acid causes the resin to contract visibly, but the bed volume is based on the initial volume of the resin in 0.25 *M* ammonium acetate buffer, pH 8.8.

7. Percolate ca. 10 mL of 0.25 M ammonium acetate buffer, pH 8.8, through the resin to equilibrate it with this buffer. The column is now ready for loading when the buffer has drained to the top of the resin bed.

### B. Sample Cleanup

8. The urine sample is thawed and shaken well to ensure sample homogeneity. Draw a 1.00 mL aliquot with a 1000 μL Eppendorf pipet (Brinkman Instruments, Inc., Westbury, NY) and place in a 1.5 mL Eppendorf microcentrifuge tube.

9. Add 300 μL of 2.5 *M* ammonium acetate buffer, pH 9.5, to the urine sample with a 100 μL Eppendorf pipet and mix the sample for 5 min on a vortex mixer (Eppendorf Model 3300 Rotary Shaker.).

10. Centrifuge the sample for 5 min at 12,000*g* (Eppendorf Model 3200/30 microcentrifuge).

11. Transfer the sample with a Pasteur pipet onto the column, being careful not to disturb the precipitate in the centrifuge tube.

12. Add 1 mL of the 0.25 *M* ammonium acetate buffer, pH 8.8, to the sample tube and mix for 5 min on the vortex shaker.

13. Centrifuge for 5 min at 12,000*g*.

14. Transfer the wash onto the column with the same Pasteur pipet.

15. Follow the sample and wash through the column with 4 mL of 0.25 *M* ammonium acetate buffer, pH 8.8.

16. Percolate an additional 3 mL of 0.25 *M* ammonium acetate buffer pH 8.8 through the column; after this wash has drained to the top of the resin bed, the column is ready for elution.

## C. Elution of Nucleosides

17. Use 5 mL of 0.1 *M* formic acid for the elution. Collect the eluate in a Corex 25-ml screw cap, round-bottom centrifuge tube (Corning Glass Works, Corning, NY) containing 0.50 mL (by 500 μL Eppendorf) of a 40 nmol/mL solution of $N^2$,$N^2$-dimethylguanine as internal standard.

18. Shell freeze the eluate and lyophilize. Redissolve the residue in 2 mL water. Complete solution is aided by mixing on a Vortex Genie mixer (Scientific Products, Evanston, IL).

19. After elution rinse the columns with ca. 20 mL of 0.1 *M* formic acid and store in the same solution.

20. Just prior to reuse, the columns are rinsed with ca. 10 mL of 0.1 *M* formic acid, and the process is repeated from Step 7.

## D. Reagents, Columns, and Supplies

1. Ammonium acetate buffer, 0.25 *M* (38.54 g/2 L) with pH adjusted to 8.8 with concentrated ammonium hydroxide. The ammonium acetate used was ACS certified grade from Fisher Scientific Co. and the ammonium hydroxide was analytical reagent grade from Mallinckrodt, Inc.

2. Ammonium acetate buffer, 2.5 *M* (385.4 g/2 L) with pH adjusted to 9.5 with concentrated ammonium hydroxide.

3. Formic acid, 0.1 *M* (10.33 g concentrated formic acid/2 L). The formic acid used was ACS certified grade from Fisher Scientific Co.

The solutions are made to nearly 2 L, pH adjusted, then diluted to volume.

## IV. Results: Reversed-Phase HPLC Analysis of Nucleosides

### A. Chromatography System

In our paper on the chromatography of the nucleosides *(30)*, we presented the fundamental relations among the general effects of pH, ionic strength, flow rate, polarity of solvents, and temperature of the mobile phase on the resolution of the major and minor nucleosides. Two quantitative chromatography systems were developed using a new internal standard, $Br^8G$.

Figure 3 shows the separation of 16 major and minor nucleosides achieved in less than 1 h by isocratic elution of the nucleosides from a bonded $C_{18}$ microparticulate reversed-phase partition column. An

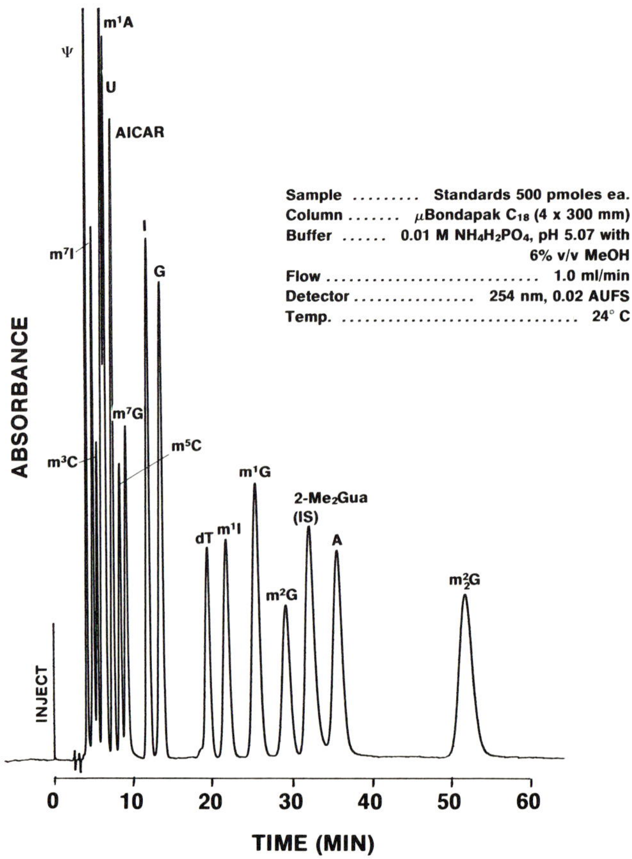


FIG. 3. Reversed-phase HPLC isocratic separation of nucleosides.

internal standard, $N^2,N^2$-dimethylguanine (2-$Me_2$Gua) was included for accurate quantitation of the nucleosides. The conditions were chosen to give optimum separation of the methylated purine nucleosides found in urine.

In our second chromatography system, thirteen nucleosides can be completely separated isocratically in less than 30 min with the chromatographic conditions given in Fig. 4. One percent (1%) methanol was employed mainly to achieve a separation of $m^1A$ and AICAR. An even better separation of most of the nucleosides in the early eluting group can be obtained without $CH_3OH$ in the 0.01 *M* $NH_4H_2PO_4$ buffer.

## B. Minimum Detection Limit

The high resolution of the reversed-phase HPLC column provides a narrow bandwidth and integrity of separation, thus giving high sensitivity, and allows the detection of 1–5 pmol amounts of the nucleosides. Figure 5 demonstrates an isocratic separation of nine nucleosides at the 1 ng level. This sensitivity is much more remarkable when one considers that absorbance detection is continuous, nondestructive, and does not require radiolabeling or derivatization.

## C. Retention Times and Relative Molar Response

A summary listing of the retention times and relative molar responses compared to $N^2,N^2$-dimethylguanine (2-$Me_2$Gua) for 20 nucleosides and related compounds are presented in Table 1. The eluent was 0.01 *M* $NH_4H_2PO_4$ buffer, pH 5.07, containing 6 v/v % methanol. A 4 × 300 mm $\mu$Bondapak $C_{18}$ column was used with a flow rate of 1.0 mL/min. The relative molar response $RMR_{N/IS}$ values are given for comparative purposes and must be determined in each laboratory. Usually three independent analyses are made with further confirmation of $RMR_{N/IS}$ daily in routine analytical work. In our laboratory the calculated $RMR_{N/IS}$ values for eight nucleosides were obtained from three HPLC instruments with less than 2% difference. All the RMR values remained essentially constant over a 3-month period.

## D. Precision of HPLC Analysis, Standards

The reversed-phase HPLC internal standard method gives excellent precision for standards at concentrations normally found in urine and for small samples of biological materials (Table 2). Repeated injections of 50 $\mu$L of each of these four solutions (0.1–1nmol each) of six ribonucleosides gave an average relative standard deviation of

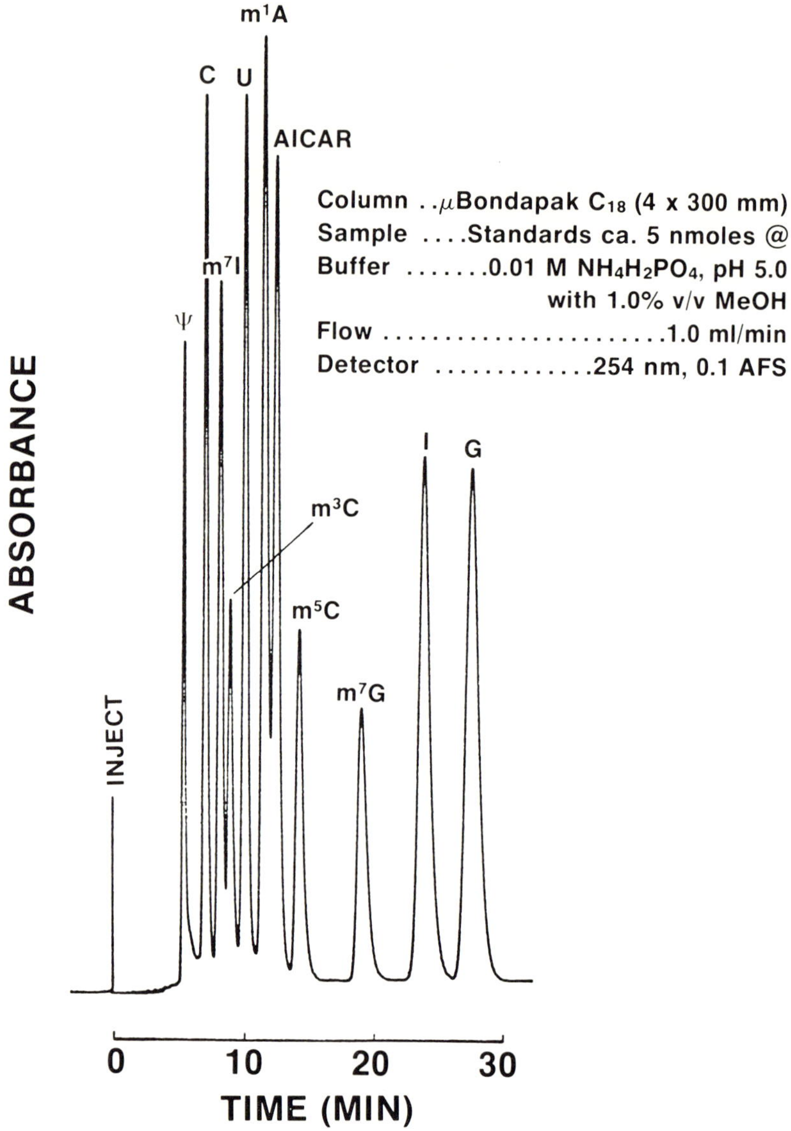


FIG. 4. Reversed-phase HPLC isocratic separation of nucleosides.

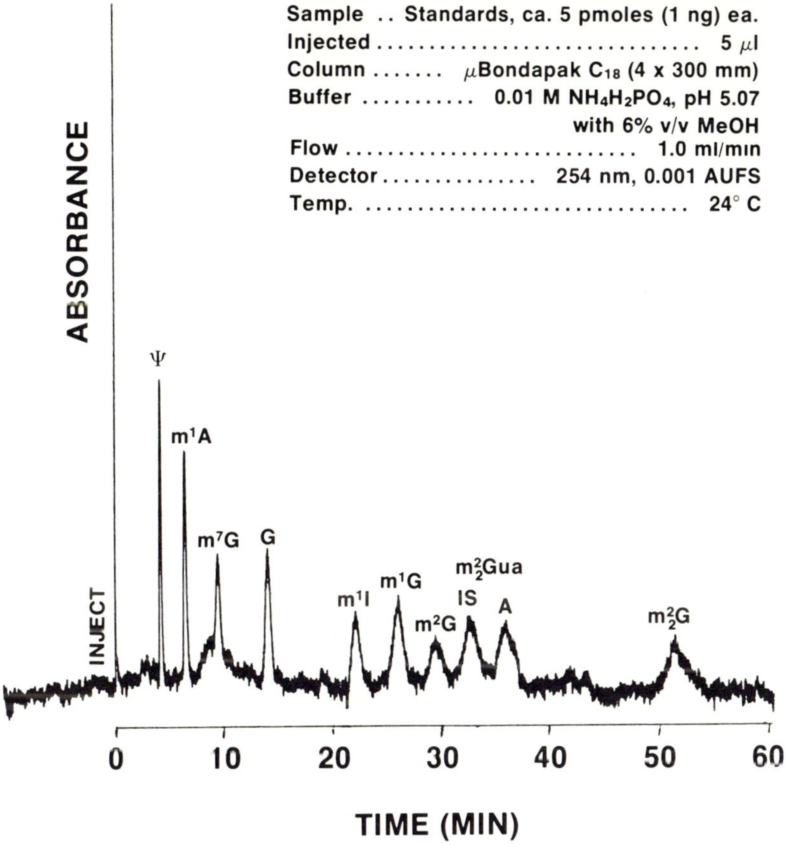


FIG. 5. Reversed-phase HPLC isocratic separation of nucleosides.

0.8–3.0%. These data show excellent precision of HPLC analysis for standards over a range of 2.50–20.00 nmol of nucleoside/mL. The relative standard deviations were 3.0% or less.

## E. Linearity

The HPLC analytical method gave linear response curves for all nucleosides determined, as shown in Fig. 6 for ψ and $m^1A$. Response for all nucleosides was found to be linear from 0.1 to 50 nmol injected. Also, good quantitation was obtained with 25 μL or less of urine placed on the HPLC column. The wide linear dynamic range for the different nucleosides is more than adequate for the analysis of nucleosides in biological samples.

Table 1
Relative Molar Response of Nucleosides in HPLC Analysis[a]

| Nucleoside | Retention time (min)[b] | RMR[c] |
|---|---|---|
| $\psi$ | 4.20 | 0.485 |
| $m^7I$ | 5.22 | 0.425 |
| $m^3C$ | 5.45 | 0.213 |
| $m^1A$ | 6.18 | 0.759 |
| U | 6.48 | 0.644 |
| AICAR[d] | 7.43 | 0.684 |
| $m^5C$ | 8.78 | 0.357 |
| $m^7G$ | 9.11 | 0.460 |
| I | 11.88 | 0.769 |
| G | 13.75 | 0.921 |
| $m^1I$ | 21.32 | 0.683 |
| $m^1G$ | 24.05 | 0.971 |
| $m^2G$ | 28.51 | 0.978 |
| A | 34.81 | 0.963 |
| $m_2^2G$ | 49.16 | 1.14 |
| dC | 6.52 | 0.422 |
| dU | 8.86 | 0.608 |
| dG | 17.94 | 0.863 |
| dT | 20.42 | 0.540 |
| dA | 44.31 | 0.909 |
| 2-$Me_2$Gua($IS_1$) | 31.18 | 1.000 |

[a]Eluent is 0.01 *M* $NH_4H_2PO_4$, pH 5.07, containing 6% MeOH.

[b]Uncorrected for void volume, pumping rate, 1.0 mL/min. Void volume = 3.08 mL.

[c]Relative molar response compared to IS, 2-$Me_2$Gua at 254 nm = 1.000.

[d]5-Aminoimidazole-4-carboxamide riboside.

## F. Urine Sample Cleanup for HPLC Ribonucleoside Analysis

Owing to the complex nature of most biological fluids, a rapid preliminary class separation of ribonucleosides was made prior to HPLC analysis. This was accomplished by use of an affinity gel column. The column was packed with a modified polyacrylamide gel having an immobilized phenylboronic acid functionality covalently

Table 2
Precision of HPLC Analysis of Nucleosides Using Internal Standard Method[a]

| Nucleoside | Nucleoside standards, nmol/mL 1 | 2 | 3 | 4 | RSD, % Av. |
|---|---|---|---|---|---|
| $M^1A$ | 19.50 | 10.06 | 4.99 | 2.66 | |
| RSD, % | 1.69 | 2.39 | 0.80 | 7.14 | 3.01 |
| G | 18.68 | 9.42 | 4.76 | 2.40 | |
| RSD, % | 1.50 | 1.17 | 0.84 | 1.25 | 1.19 |
| $m^1I$ | 19.42 | 9.51 | 5.15 | 2.68 | |
| RSD, % | 1.18 | 0.84 | 0.39 | 0.75 | 0.79 |
| $m^1G$ | 19.31 | 9.81 | 5.03 | 2.54 | |
| RSD, % | 1.04 | 0.92 | 0.80 | 0.39 | 0.79 |
| A | 20.23 | 10.28 | 5.26 | 2.66 | |
| RSD, % | 1.19 | 1.36 | 0.57 | 1.50 | 1.16 |
| $m_2^2G$ | 19.52 | 10.06 | 5.25 | 2.76 | |
| RSD, % | 0.77 | 1.69 | 0.95 | 4.34 | 1.94 |

[a]Each value is the mean of five or more analyses.

linked by a spacer arm of succinic acid to the polymer backbone (Fig. 1). This type of boronate resin introduced by Uziel (*31*) can selectively bind *cis*-diols as boronate complexes under mild alkaline conditions, pH 8.8, and the complex can be easily broken by simply reducing the pH. The stability of the complex varies with conformation of the sugar residue and is maximal with diols having the same conformation as ribose (*33*).

The urine sample cleanup procedure is outlined in Fig. 2 and described in detail in Section II. It differs from the procedure originally described by Uziel (*31*) in several ways. At Uziel's suggestion, the boronate gel was synthesized from 200 to 400 mesh Hydrazide Bio-Gel P-2 instead of the coarser 100–200 mesh material. The amount of hydrazide substitution of this starting material was also decreased to 1.2 mEq/g of dry gel instead of ca. 6 mEq/g. The smaller mesh size Bio-Gel minimizes the amount of shrinkage of the gel and gives a higher column packing density. Thus the channelling effect is reduced and the column efficiency is improved. Also, in the synthesis, a 10% excess of *m*-aminophenyl boronic acid was used to achieve complete coupling with the succinylated gel. With these changes in gel synthesis

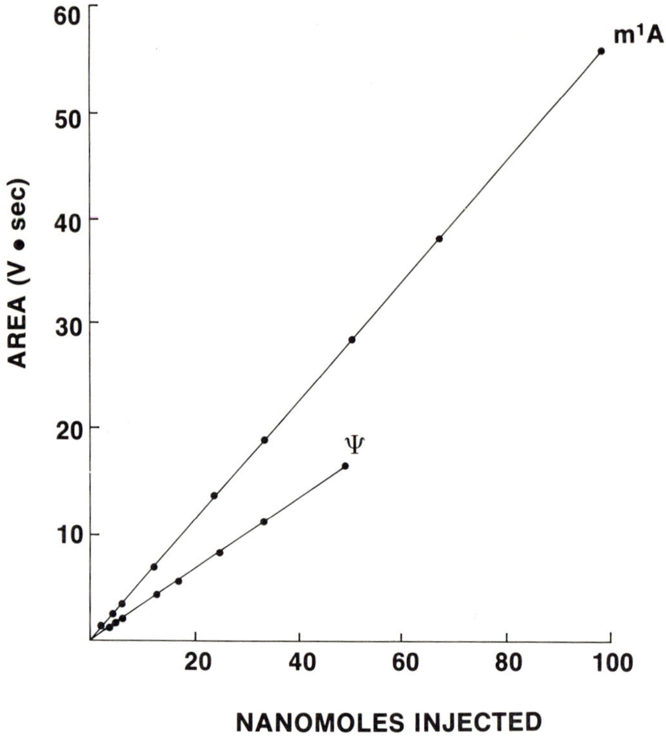


FIG. 6. Linearity of HPLC analysis for ψ and m¹A.

the columns are now washed with a total of 8 mL of pH 8.8 0.25-*M* ammonium acetate to ensure complete recovery of pseudouridine (ψ). All of the nucleosides were also eluted with 5 mL of 0.1 *M* formic acid. The use of both the lower pH and ionic strength of 0.1 molar formic acid gave a much more efficient elution, and essentially quantitative recovery for the ribonucleosides examined (Table 3).

The elution of ψ and eight other nucleosides from the affinity phenyl-boronate gel is shown in Fig. 7. These nucleosides were added to a pooled control urine at a level of 100 nmol/mL of urine for ψ and 40 nmol each/mL of urine for the other eight nucleosides. With the synthesis method that we used, a highly efficient gel was obtained and a larger volume of ammonium acetate wash could be used before breakthrough of the nucleosides into the eluate. This allows an excellent cleanup from other extraneous interfering urine components without any loss of ψ. Uziel also stated that with his boronate gel ψ was eluted earlier than the other nucleosides. We used 0.1 *M* HCOOH to plug elute all of the nucleosides in the same eluate volume of 5 mL

Table 3
Recovery of Nucleosides Added to Pooled Control Urine

| Nucleoside | nmol/mL[a] | | | Average recovery, % |
|---|---|---|---|---|
| | Urine + spike | Urine | Spike recovered | |
| $m^1A$ | 23.30 | 17.38 | 5.92 | 92 |
| $m^7G$ | 10.05 | 5.69[b] | 4.36 | 88 |
| G | 13.47 | 8.91[b] | 4.56 | 92 |
| $m^1 1$ | 15.46 | 10.48 | 4.98 | 99 |
| $m^1G$ | 10.66 | 5.64 | 5.02 | 101 |
| $m^2G$ | 10.69 | 5.46 | 5.23 | 100 |
| A | 7.38 | 2.55 | 4.83 | 98 |
| $m_2^2G$ | 15.72 | 11.28 | 4.44 | 100 |

[a]Each value an average of four runs.
[b]An unknown peak eluted with G and $m^7G$ and were integrated together.

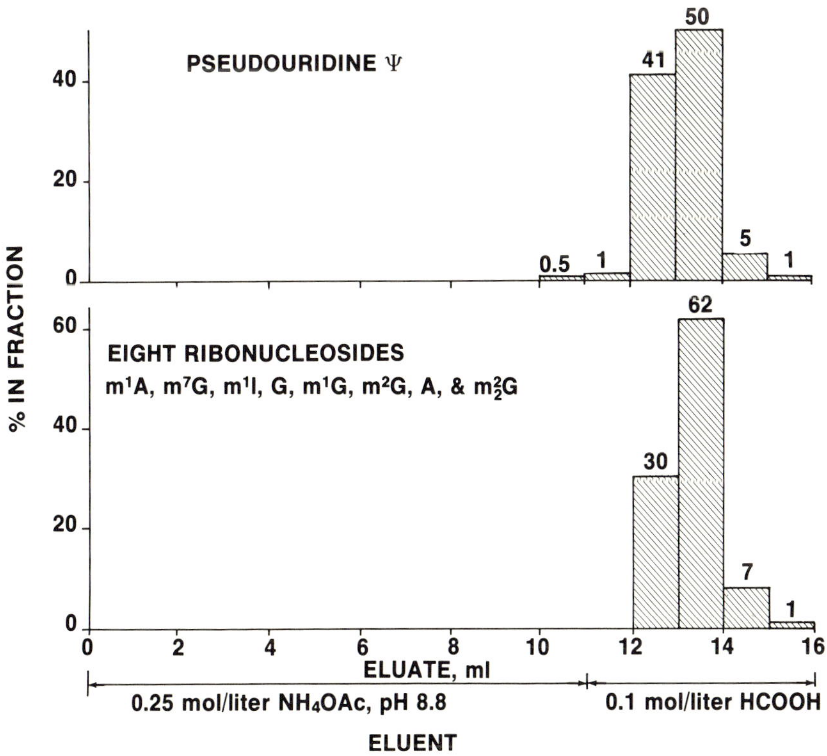


FIG. 7. Elution of nucleosides from phenylboronate gel.

(Fig. 7). One molar acetic acid can be used as the eluent, but 30 mL were required for quantitative elution of all nucleosides.

## G. Stability of Nucleosides

The stability of the nucleosides at low pH was of concern since elution from the boronate gel was made with 0.1*M* HCOOH. A study was made in which nine nucleosides were dissolved in 0.1*M* HCl and 0.1*M* HCOOH and stored at room temperature for 5 days. The solutions contained 10.0 nmol/mL for each of the ribonucleosides. HPLC analyses were made each day and the recovery determined by comparing the experimentally determined concentration on each of the 5 days with the known concentration of each standard nucleoside solution. No loss of any nucleoside was observed over the 5 day period (Table 4).

## H. Capacity, Recovery, and Stability of Gel

The recovery for our phenylboronate gel was determined to about a level of 1000 nmol of total nucleosides added to 1 mL of urine then placed on the affinity column. A pooled normal urine was spiked with 5, 25, 50, and 100 nmol each of nine nucleosides/1 mL of urine, and these spiked samples were then passed through four identical gel-affinity columns. The eluates were chromatographed by HPLC and the recoveries ascertained. The average recoveries for the 9 nucleosides

Table 4
Stability of Nucleosides in Acidic Solutions as a Function of Time

| | Recovery[a] and standard deviation | | | |
|---|---|---|---|---|
| Nucleoside | 0.1*N* HCl, % | $\sigma$ | 0.1*N* HCOOH, % | $\sigma$ |
| $\psi$ | 98.7 | 0.5 | 99.0 | 1.0 |
| $m^1A$ | 99.7 | 0.6 | 102.0 | 2.6 |
| $m^7G$ | 98.7 | 2.3 | 97.7 | 2.1 |
| G | 99.0 | 1.7 | 98.7 | 2.3 |
| $m^1I$ | 98.5 | 2.1 | 96.5 | 3.5 |
| $m^1G$ | 98.7 | 2.5 | 98.7 | 2.3 |
| $m^2G$ | 99.3 | 1.5 | 100.7 | 0.6 |
| A | 99.3 | 1.5 | 99.0 | 1.0 |
| $m^2_2G$ | 101.0 | 2.6 | 99.7 | 2.1 |

[a]Each recovery value is an average of the results for the 5 days.

were: 25 mmol/mL (101%), 50 nmol/mL (104%), and 100 nmol/mL (104%). Recovery of nucleosides added to a pooled control normal urine at a level of 5 nmol/mL each, as given in Table 3, was found to be excellent.

It is important that the performance of this chromatographic method should be routinely monitored by determining the percentage recovery. This was done by adding nucleosides to 1.0 mL of urine to give approximately twice the original urinary nucleoside concentrations, then the nucleosides were isolated on the gel column and determined by reversed-phase HPLC.

The percent recovery was calculated as follows:

$$\text{Recovery (\%)} = \frac{\begin{bmatrix}(\text{nmol}_N \text{ found in spiked sample} - \\ \text{nmol}_N \text{ found in sample}\end{bmatrix}}{[\text{nmol}_N \text{ added}]} \times 100$$

As an example:

$$\text{Recovery (\%) for } m^2_{2G} = \frac{(20.0 \text{ nmol} - 10.0 \text{ nmol})}{(10.0 \text{ nmol})} \times 100$$

## I. Calculation of Nucleoside Concentration

The experimental areas of the peaks were integrated by a HP-3352B Laboratory Data System (Hewlett-Packard, Avondale, PA) and the amount of each nucleoside was calculated by the computer as follows:

$$\text{Nucleoside nmol/mL of sample} = (\text{area}_N/\text{area}_{IS})_{\text{sample}}\ (1/\text{RMR}_{N/IS})\ (\text{nmol}_{IS}/\text{mL}_{\text{sample}})$$

where

$$\text{RMR}_{N/IS} = [\text{area/nmol/mL}]_N\ [\text{nmol/mL/area}]_{\text{standard}}$$

The RMR values for each of the nucleosides were determined by at least three independent analyses of calibration standards of the nucleoside, with subsequent determinations of the RMR daily. In the above expression of $\text{RMR}_{N/IS}$, the concentration terms must be given as nmol/mL.

In separate studies we have verified the maximum capacity of these small gel columns (0.8 mL) and found with our experimental conditions that the affinity gel would retain 40–50 μmol of nucleosides without breakthrough. However, pH is a critical factor in the proper functioning of the gel, and the pH of the sample solution must be adjusted to between 8.5 and 9.4 for good results. High capacity of the

Table 5
Isolation of Ribonucleosides from Phenylboronate Gel Affinity Column[a]

| | Nucleoside and recovery, % | | | | | | | | |
|---|---|---|---|---|---|---|---|---|---|
| $N = 10$ | $\psi$ | $m^1A$ | $m^7G$ | G | $m^1I$ | $m^1G$ | $m^2G$ | A | $m^2_2G$ |
| $\bar{x}$ | 101.2 | 100.7 | 88.1 | 97.6 | 99.4 | 97.8 | 96.5 | 94.0 | 96.2 |
| RSD, % | 2.9 | 4.7 | 6.6 | 2.8 | 2.7 | 2.2 | 1.9 | 2.4 | 4.4 |
| | Three Months Later | | | | | | | | |
| $\bar{x}$ | 103.9 | 91.0 | 88.4 | 98.5 | 97.7 | 97.9 | 96.2 | 97.4 | 98.9 |
| RSD, % | 1.5 | 1.4 | 2.3 | 2.4 | 2.1 | 2.7 | 1.7 | 1.1 | 2.8 |

[a]Each nucleoside added at a level of about 10 nmol.

gel is also needed in preparative isolation of nucleosides on a fairly large scale and to ensure the accuracy of routine analysis. The concentration of nucleosides varies widely in different biological samples, and thus it is most important to know the gel capacity limits.

Ten different affinity columns have been repeatedly used in urine analysis over a period of 3 months (30 analyses/column). These columns showed no deterioration and quantitative recovery was still obtained. Table 5 gives the recoveries and RSD (%) for the isolation of nine nucleosides on ten gel columns at zero time and after these same ten columns had been used daily for a period of three months. Quantitative recovery and good precision were achieved for both the initial study and after 3 months of column use.

### J. Precision of Urinary Nucleoside Analysis—Matrix Dependent and Independent

Urinary nucleosides were determined using 1 mL samples of the matrix dependent urine and a pooled control urine. The nucleosides were isolated using different 5 × 40 mm boronate gel affinity columns as described above, then separated and quantitated by the HPLC system using 2-$Me_2$Gua as internal standard. Samples equivalent to 25 $\mu$L of urine were used for each HPLC analysis, and a chromatogram for such an analysis is shown in Fig. 8.

The matrix dependent and independent precision for the HPLC analysis of the six urinary nucleosides is given in Table 6. The values for the matrix-dependent samples were obtained from four different urine samples, each analyzed independently twice, whereas the data for the matrix-independent samples were obtained by analyzing independently four times the sample of pooled control urine. Excellent precision of analysis was achieved.

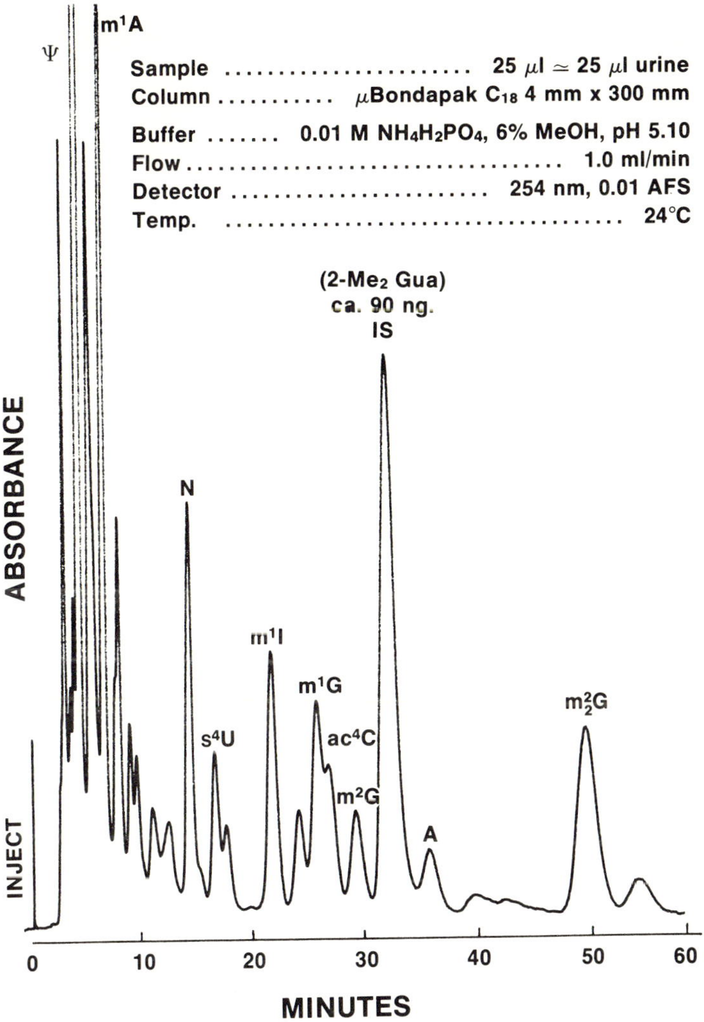


FIG. 8. Reversed-phase HPLC isocratic separation of nucleosides in control urine.

## K. Precision of Retention Times

The retention time was found to be independent of the sample matrix. Excellent precision of retention time for nine nucleosides was obtained in routine analysis over a 2-day period for ten different urine samples having different matrices (Table 7). The RSD values ranged from 0.16 to 1.35%. After the elution of $m_2^2G$ the mobile phase was changed from 5% $CH_3OH$ in the $NH_4H_2PO_4$ buffer to 50% $CH_3OH$ in water to elute

Table 6
Precision of HPLC Analysis of Urinary Nucleosides

| | $\bar{x}^a$ and RSD, % | | | | | |
|---|---|---|---|---|---|---|
| | $\psi$ | $m^1A$ | $m^1I$ | $m^2G$ | A | $m_2^2G$ |
| Matrix dependent[b] $N = 4$ | | | | | | |
| $\bar{x}$ | 139.4 | 16.5 | 6.43 | 5.73 | 1.59 | 11.40 |
| $\sigma$ | 2.85 | 0.628 | 0.179 | 0.151 | 0.05 | 0.290 |
| RSD, % | 2.04 | 3.80 | 2.79 | 2.63 | 3.14 | 2.54 |
| Matrix independent[c] $N = 4$ | | | | | | |
| $\bar{x}$ | 26.78 | 6.97 | 4.68 | 1.14 | 1.32 | 4.66 |
| $\sigma$ | 0.024 | 0.066 | 0.029 | 0.012 | 0.045 | 0.068 |
| RSD, % | 0.09 | 0.95 | 0.61 | 1.10 | 3.42 | 1.47 |

[a] $\bar{x}$ values are nmol/mL.
[b] $\sigma = [\Sigma(x_1 - x_2)^2/2P]^{1/2}$, where $P$ = number of pairs of analyses on *different* samples.
[c] $\sigma = [\Sigma(\bar{x} - x)^2/(N - 1)]^{1/2}$ for a pooled control urine.

Table 7
Precision of Urinary Nucleoside Analysis

| | Retention time, min[a] | | | | | | | | |
|---|---|---|---|---|---|---|---|---|---|
| $N = 10$ | $\psi$ | $m^1A$ | G-N | $m^1I$ | $m^1G$ | $m^2G$ | 2-$Me_2$Gua | A | $m_2^2G$ |
| $\bar{x}$ | 4.21 | 6.72 | 14.61 | 22.87 | 26.68 | 30.54 | 33.39 | 36.63 | 53.60 |
| $\sigma$ | 0.007 | 0.016 | 0.179 | 0.076 | 0.360 | 0.166 | 0.114 | 0.186 | 0.248 |
| RSD, % | 0.16 | 0.24 | 1.22 | 0.33 | 1.35 | 0.38 | 0.34 | 0.51 | 0.46 |

[a] Eluent was 5% MeOH in 0.01 *M* $NH_4H_2PO_4$, pH 5.10 buffer.

the strongly retained components. The 50% MeOH/water solution was pumped for 8 min at 1.0 mL/min then, the column was equilibrated for 20 min with the 0.01*M* $NH_4H_2PO_4$ analysis buffer before injection of the next sample. Total analysis time was 90 min for a complete run. Somewhat higher RSD values were noticed for G-N and $m^1G$; this may occur because of different ratios of two unresolved peaks in different samples, thus resulting in a slight change in the retention time.

## L. Analysis of Leukemia and Breast Cancer Urine

Chromatograms of the analysis of urinary ribonucleosides in patients with advanced leukemia and breast cancer are shown in Figs. 9 and 10.

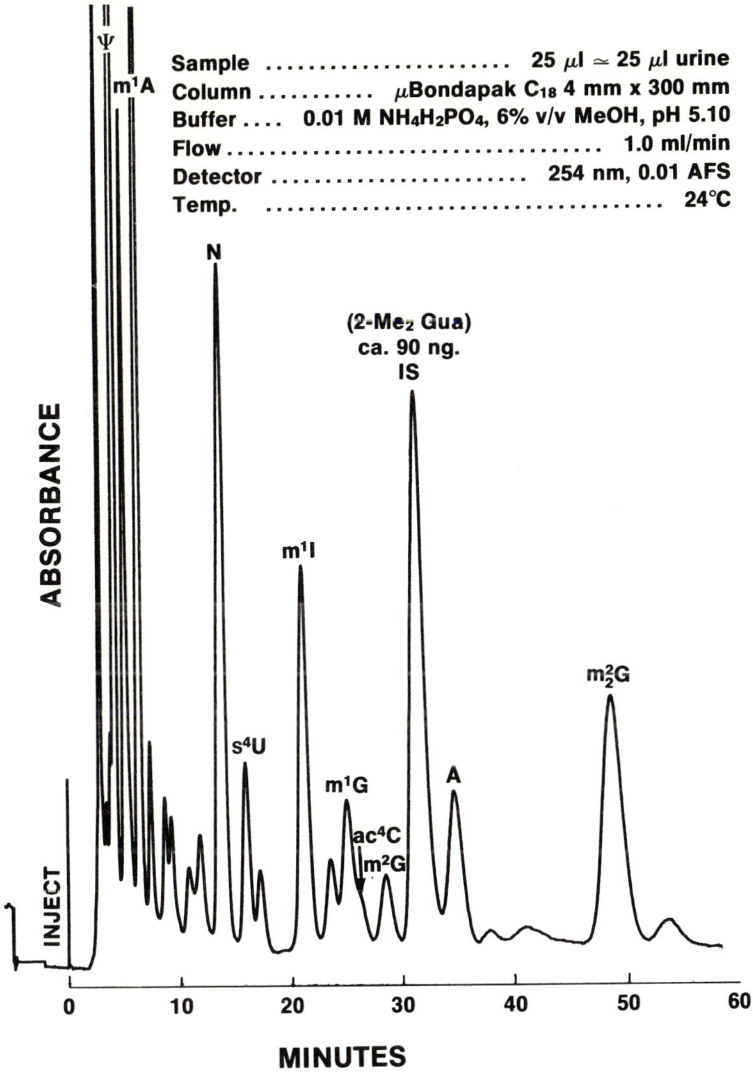


FIG. 9. Reversed-phase HPLC isocratic separation of nucleosides in leukemia urine.

A spike leukemia urine chromatogram is shown in Fig. 11; Fig. 9 is the same urine without nucleosides added. Elevated levels of N (not identified, $m^1I$, A, $ac^4C$, and $m_2^2G$ can be readily observed in the leukemia and breast cancer urine chromatograms; $ac^4C$ is elevated in the breast, and not in the leukemia, sample. A major component, probably a nucleoside, N, has been observed to be elevated in many

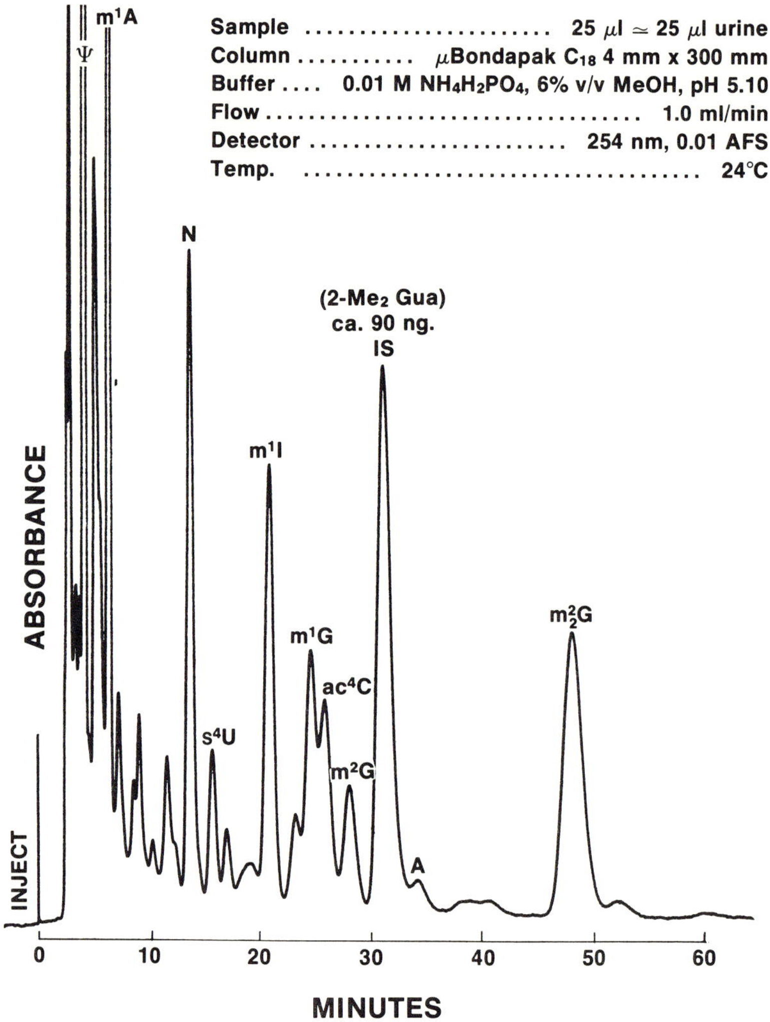


FIG. 10. Reversed-phase HPLC isocratic separation of nucleosides in breast cancer urine.

types of cancer urine samples. Its identity has not been determined at this time.

HPLC analyses have been completed on 10 normal, 10 colon, 15 breast, and 9 leukemia cancer subjects. An in-depth collection study has also been completed showing that random collected urine samples can be substituted for the 24-h total collections when the levels of nucleosides are expressed independently of the volume as a ratio of

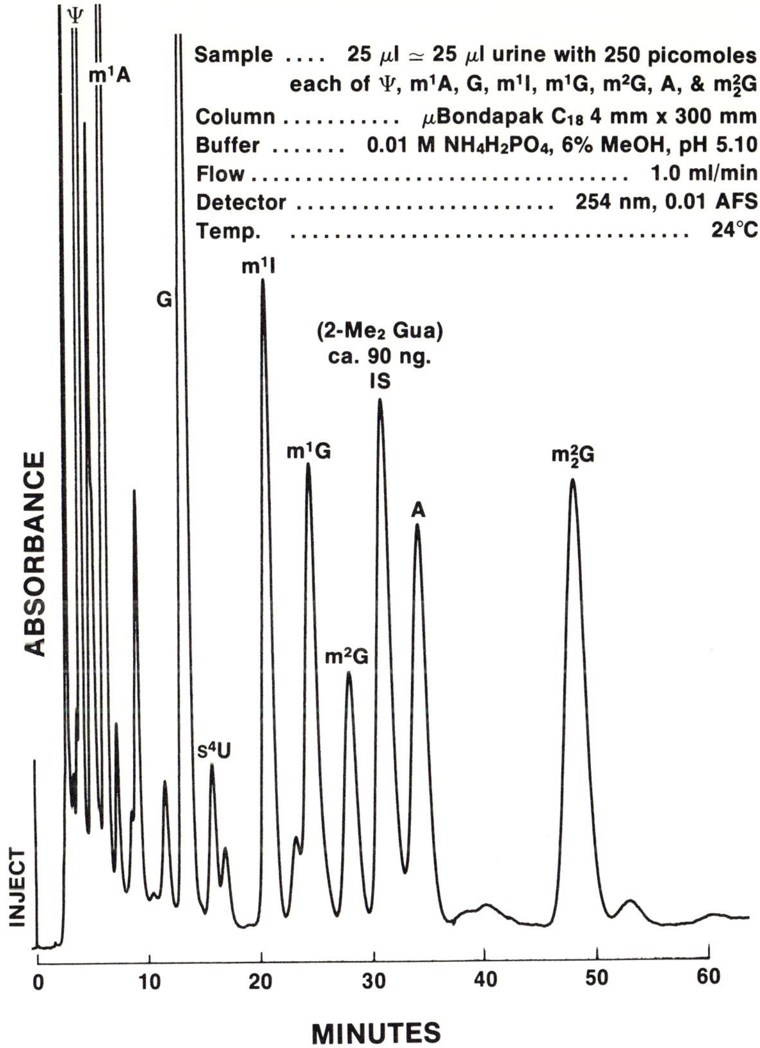


FIG. 11. Reversed-phase HPLC isocratic separation of nucleosides in leukemia urine spiked with nucleosides.

nanomoles of nucleoside per micromole of creatinine (*34*). Using the urinary nucleoside-to-creatinine ratio as a measure of elevation, we have found that greater than 90% of ten patients each with advanced colon, breast, and leukemia cancer have $m^1I$ and $m^2_2G$ values greater than two sigma above the normal average. The nucleoside A was elevated only in colon cancer and not in the leukemia and breast cancer. Elevation of $m^2G$ was noted only in breast cancer. These observations suggest that urinary nucleosides may be indicative and specific for different types of cancer.

## V. Discussion

The reversed-phase partition mode of high performance liquid chromatography with ultraviolet absorption detection combined with a highly selective affinity gel isolation technique gives a rapid and sensitive quantitative method for the simultaneous analysis of many nucleosides in biological samples. In our first paper (*30*) on the chromatography of nucleosides we discussed the fundamental relations among the general effects of pH, ionic strength, flow rate, polarity of solvents, and temperature of the mobile phase on the resolution of the major and minor nucleosides. This fundamental information can be used as a guide to select the appropriate chromatographic conditions to achieve the separation of nucleosides in various types of samples.

To achieve the required precision and accuracy, an isocratic elution of the nucleosides was employed to obviate the baseline drift, retention time changes, and variability of flow rate that are common with gradient elution methods. Thus improved reproducibility was achieved at low levels. Two isocratic chromatography systems were developed giving the separation of 17 nucleosides.

In application investigations we first made a class separation of the ribonucleosides on the boronate gel column, which greatly improves the selectivity of the method and extends its reliability to handle the complex matrices of most biological fluids. We have confirmed that nucleic acid bases, deoxyribonucleosides, deoxyribonucleotides, urinary pigments, and most other interfering ultraviolet light absorbing compounds are not retained by the boronate gel. A few ribonucleotide monophosphates are retained by the boronate column, and some are eluted with the ribonucleosides. However, they are separated on the HPLC column and do not interfere with the quantitation of the nucleosides. In this way, an excellent cleanup of the sample was achieved.

In the past year, we have applied this method in the analysis of a large number of urine samples. Similar nucleoside elution patterns were observed in all urine samples; however, the relative amounts present varied widely. More than 20 chromatographic peaks are present. At this time we have identified only 10 of the nucleosides, mainly because of the nonavailability of pure synthetic compounds.

These, including $m^6A$, $m^6_2A$, and isopentenyl adenosine ($i^6A$), are strongly retained by the HPLC column and require a 45% methanol buffer to elute them in reasonable time and sensitivity. A few select urines showed a trace level of $m^6A$ and no $m^6_2A$.

The values obtained for $\psi$, $m^1I$, and $m^2_2G$ are comparable to those obtained by GLC (*24*). However, with the HPLC method a number of other nucleosides and related compounds can be quantitated in one chromatographic run from the same sample, and their total or relative concentrations compared. This gives us a powerful tool for identifying potential nucleoside biochemical marker(s) of cancerous growth. Further, this rapid and selective method can be used to study inborn errors in purine and pyrimidine metabolism as well as determine concentrations of major and modified nucleosides in hydrolysates of tRNA, in cell extracts, biological fluids, or in following metabolism in cell cultures.

In further studies, we have also developed a rapid, highly precise, and accurate method for pseudouridine in urine. An HPLC analysis can be completed in 8 min after isolation of $\psi$ from the gel affinity column.

## Optimization of Nucleoside Separations

Based on our earlier described studies, we have developed four sets of chromatographic parameters for the RP-HPLC separation of nucleosides. These include (a) an improved isocratic method, (b) a rapid method for $\psi$, (c) a two buffer step-gradient method, and (d) a rapid method for $m^2_2G$. These four methods are now described.

Earlier we published experimental conditions for a single column isocratic separation of seventeen nucleosides in less than 1 h (*32, 39*). Our improved method for the isocratic separation (a) and analysis of urinary nucleosides is presented for standards in Fig. 12 and urine in Fig. 13. By elevating the temperature to 35°C and doubling the column length to 600 mm, we obtained a more efficient separation than presented earlier (*32, 39*) of the known urinary nucleosides from unidentified components present in urine.

In addition, our routine analysis of nucleosides in urine has been improved by the use of 8-bromoguanosine ($Br^8G$) as the internal standard. This internal standard elutes at a clear portion of the chromatogram, thus eliminating any separation problems that might occur as a result of small changes in the separation characteristics of the column. A further improvement in the reliability of the method is achieved with absorption measurements at 254 and 280 nm. Thus, false elevations of nucleosides by coelution of other components are

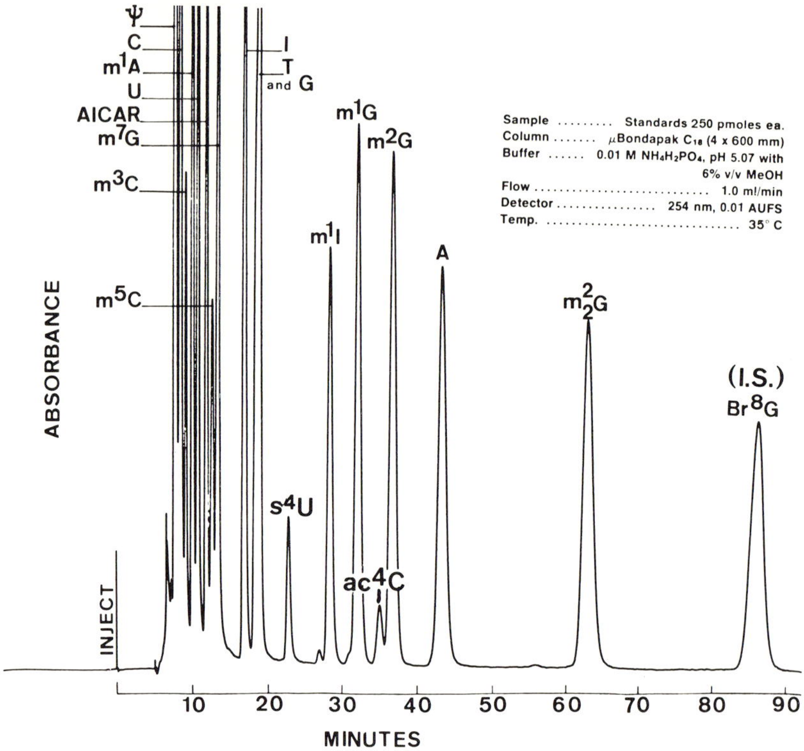


FIG. 12. Reversed-phase isocratic separation of nucleosides.

detectable. The molar absorbance ratios of nucleosides at 254 and 280 nm under these conditions has been measured. However, the separation of ψ in some urine samples is not optimal, since unknown components coelute with this molecule. Therefore, a rapid method (b) for the analysis of only ψ was developed to overcome this problem (*40*). Thus, to obtain a complete profile of urinary nucleosides under isocratic conditions, two chromatographic analyses are required.

A still more efficient separation of urinary nucleosides was achieved using a two buffer step-gradient elution (c). Figure 14 shows the separation of seventeen nucleosides with $Br^8G$ as the internal standard. The use of this chromatographic system gives a complete analysis of nucleosides in urine and is demonstrated in Fig. 15.

The high selectivity of this chromatography system is again demonstrated in the separation of the corresponding major ribo- and deoxyribonucleosides (Fig. 16). These large molecules are multifunctional and have a difference of only one hydroxyl group for a hydrogen; however, complete separation was easily achieved. In

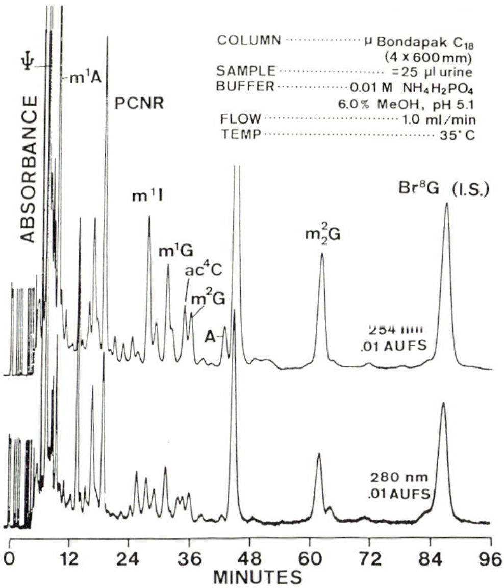


FIG. 13. Reversed-phase isocratic separation of nucleosides in urine.

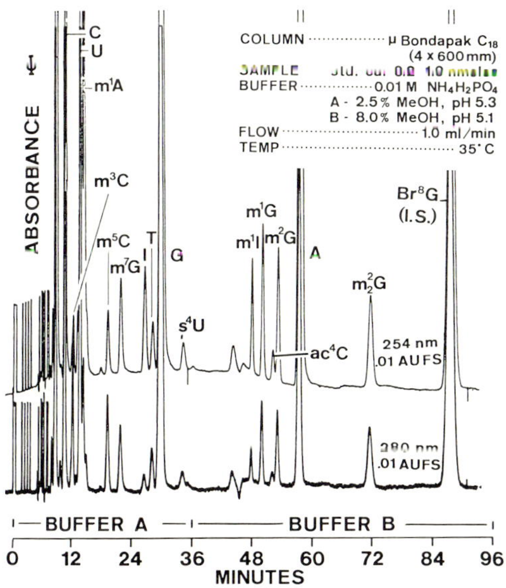


FIG. 14. Reversed-phase step gradient separation of nucleosides.

research, this chromatography will be most useful in verifying the cross contamination of RNA and DNA isolates, and can be used for the composition analysis of DNA (*41*).

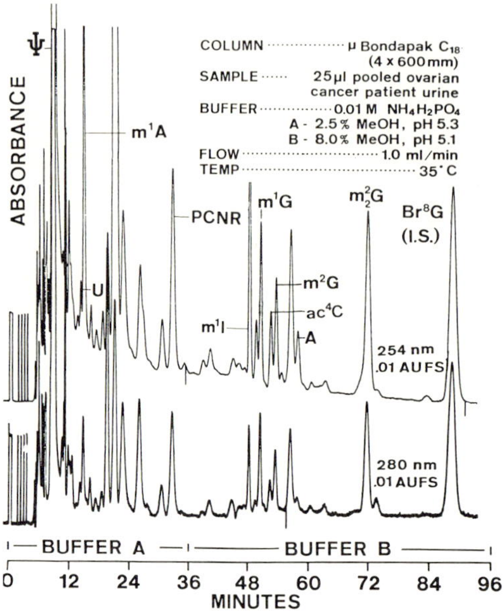


FIG. 15. Reversed-phase step-gradient separation of nucleosides in urine.

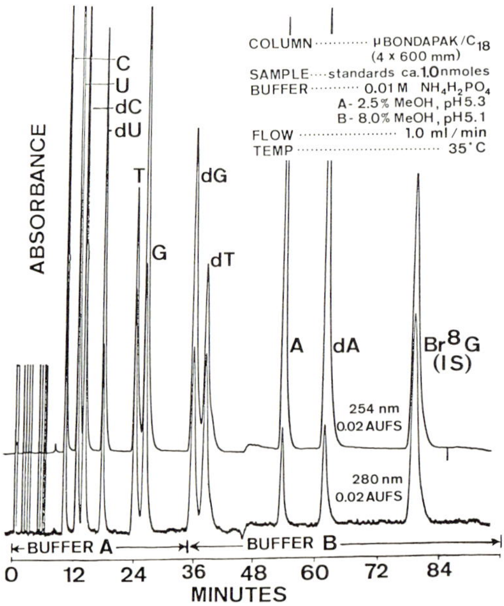


FIG. 16. Reversed-phase HPLC chromatography for major deoxyribo- and ribonucleosides.

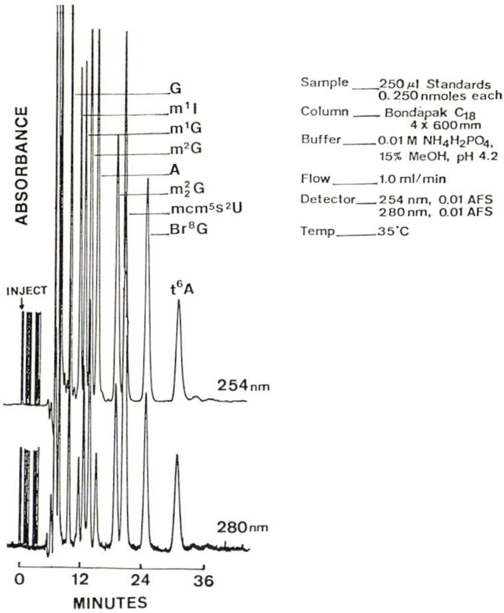


FIG. 17. Reversed-phase HPLC chromatography for $m_2^2G$, $mcm^5s^2U$, and $t^6A$.

One approach to the study of potential biologic markers of cancer has been to study the turnover rate of tRNA. Since $m_2^2G$ is unique to tRNA, a rapid method for the analysis of $m_2^2G$ in urine would be useful in studying tRNA turnover rates. A rapid isocratic separation of $m_2^2G$ (d) from a number of other nucleosides is shown in Fig. 17. The chromatographic conditions presented in Fig. 17 were the only conditions we have found that permit separation of $m_2^2G$ from $mcm^5s^2U$. The elution position of $t^6A$ is also presented. The analysis of a urine sample for $m_2^2G$ with this rapid chromatographic system is shown in Fig. 17. For this analysis to be performed correctly, the pH of the elution buffer must be precisely adjusted to 4.20. A pH of 4.2 is the only pH for which $m_2^2G$ was completely separated from the other components in urine.

## References


1. Borek, E., Transfer RNAs as regulatory molecules: an assessment after a decade, in *Control Proesses in Neoplasia,* Mehlman, M., and Hanson, W. R., eds., Academic Press, New York, 1974, p. 147–161.
2. Rich, A., and RajBhandary, U. L., *Ann. Rev. Biochem.* **45,** 805 (1976).
3. Sharma, O. K., Kerr, S. J., Lipshitz-Wiesner, R., and Borek, E., *Fed. Proc.* **30,** 167 (1971).

4. Turkington, R. W., *J. Biol. Chem.* **244,** 5140 (1969).
5. Borek, E., *Cancer Res.* **31,** 596 (1971).
6. Starr, J. L., *Biochim. Biophys. Acta* **61,** 676 (1962).
7. Biswas, B. A., Edmonds, M., and Abrams, R., *Biochem. Biophys. Res. Commun.* **6,** 146 (1961).
8. Fleissner, E., and Borek, E., *Biochem.* **2,** 560 (1963).
9. Srinivasan, P. R., and Borek, E., *Proc. Natl. Acad. Sci. USA* **49,** 529 (1963).
10. Mandel, L. R., and Borek, E., *Biochem.* **2,** 560 (1963).
11. Weissman, D., Bromberg, P. A., and Guttman, A. B., *J. Biol. Chem.* **224,** 407 (1957).
12. Adams, W. S., Davis, F., and Nakatani, M., *Am. J. Med.* **28,** 726 (1960).
13. Park, R. W., Holland, J. F., and Jenkins, A., *Cancer Res.* **22,** 469 (1962).
14. Waalkes, T. P., Gehrke, C. W., Bleyer, W. A., Zumwalt, R. W., Olweny, C. L. M., Kuo, K. C., Lakings, D. B., and Jacobs, S. A., *Cancer Chemotherapy Reports* **59,** 721 (1975).
15. Waalkes, T. P., Gehrke, C. W., Zumwalt, R. W., Chang, S. Y., Lakings, D. B., Tormey, D. C., Ahman, D. L., and Moertel, C. G., *Cancer* **36,** 390 (1975).
16. Mandel, L. R., Srinivasan, P. R., and Borek, E., *Nature (London)* **209,** 586 (1966).
17. McFarlane, E. S., and Shaw, G. J., *Can. J. Microbiol.* **14,** 185 (1968).
18. Dlugajcyk, A., and Eiler, J. J., *Proc. Soc. Exp. Biol. Med.* **123,** 453 (1966).
19. Borek, E., Baliga, B. S., Gehrke, C. W., Kuo, K. C., Belman, S., Troll, W., and Waalkes, T. P., High Turnover Rate of tRNA in Tumor Tissue, *Cancer Res.* **37,** 398 (1977).
20. Gehrke, C. W., Stalling, D. L., and Ruyle, C. D., *Biochem. Biophys. Res. Commun.* **28,** 869 (1967).
21. Gehrke, C. W., and Ruyle, C. D., *J. Chromatogr.* **61,** 45 (1968).
22. Gehrke, C. W., and Lakings, D. B., *J. Chromatogr.* **61,** 45 (1971).
23. Lakings, D. B., and Gehrke, C. W., *Clin. Chem.* **18,** 810 (1972).
24. Chang, S. Y., Lakings, D. B., Zumwalt, R. W., Gehrke, C. W., and Waalkes, T. P., *J. Lab. Clin. Med.* **83,** 816 (1974).
25. Gehrke, C. W., and Patel, A. B., *J. Chromatogr.* **123,** 335 (1976).
26. Gehrke, C. W., and Patel, A. B., *J. Chromatogr.* **130,** 103 (1977).
27. Patel, A. B., and Gehrke, C. W., *J. Chromatogr.* **130,** 115 (1977).
28. Suits, R. D., and Gehrke, C. W., Reversed-phase liquid chromatographic separation of nucleic acid bases from DNA and RNA hydrolysates, 18th West Central States Biochemistry Conference (1975).
29. Hartwick, R. A., and Brown, P. R., *J. Chromatogr.* **126,** 679 (1976).
30. Gehrke, C. W., Kuo, K. C., and Zumwalt, R. W., *J. Chromatogr.* **188,** 129 (1980).
31. Uziel, M., Smith, L. H., and Taylor, S. A., *Clin. Chem.* **22,** 1451 (1976).
32. Davis, G. E., Suits, R. D., Kuo, K. C., Gehrke, C. W., Waalkes, T. P., and Borek, E., *Clin. Chem.* **23,** 1427 (1977).
33. Boeseken, J., *Advan. Carbohydrate Chem.* **4,** 189 (1949).

34. Borek, E., Waalkes, T. P., Gehrke, C. W., and Kuo, K. C., The dynamics of excretion of modified nucleosides by normal subjects and cancer patients, in preparation.
35. Agris, P. F., and Soll, D., The modified nucleosides in transfer RNA, in *Nucleic Acid-Protein Recognition,* H. Vogel, ed., Academic Press, New York, 1977, pp. 321–344.
36. Clark, B. F. C., *Prog. Nuc. Acid Res. Mol. Biol.* **20,** 1 (1977).
37. Rich, A., and Schimmel, P. R., *Accts. Chem. Res.* **10,** 385 (1977).
38. Kuchino, Y., and Borek, E., Similar patterns of tRNA structure and tRNA methyltransferases in embryonic and tumor tissue, in *Onco-Developmental Gene Expression,* Academic Press, New York, 1976.
39. Gehrke, C. W., Kuo, K. C., Davis, G. E., Suits, R. D., Waalkes, T. P., and Borek, E., *J. Chromatogr.* **150,** 455 (1978).
40. Kuo, K. C., Gehrke, C. W., McCune, R. A., Waalkes, T. P., and Borek, E., *J. Chromatogr. Biomed. Appl.* **145,** 383 (1978).
41. Gehrke, C. W., Kuo, K. C., Ehrlich, M., and McCune, R. A., Quantitative determination of major and modified deoxynucleosides in DNA, *Nucleic Acids Res.* **8,** 4763 (1980).
42. Davis, G. E., Gehrke, C. W., and Kuo, K. C., *J. Chromatogr.* **173,** 281 (1979).
43. Brown, P. R., and Hartwick, R., *Serum Analysis for Nucleosides,* CRC Press, L. Hammer, Ed., in press.

# Chapter 19

# Polyamines

Laurence J. Marton

*Department of Laboratory Medicine*
*and Brain Tumor Research Center*
*University of California School of Medicine*
*San Francisco, California*

## I. Introduction

The polyamines spermidine [$NH_2(CH_2)_3NH(CH_2)_4NH_2$] and spermine [$NH_2(CH_2)_3NH(CH_2)_4NH(CH_2)_3NH_2$] and their diamine precursor, putrescine [$NH_2(CH_2)_4NH_2$], have been the subject of intense study relative to their potential as tumor markers during the past decade (*1*). These compounds have been implicated in numerous biochemical reactions and have been clearly associated with cellular growth processes. A number of publications have reviewed our present knowledge regarding these compounds, including their relationship to human disease (*2–6*).

Earlier hopes that assaying these compounds in physiological fluids would be useful in screening for cancer have been abandoned by most because the inadequate sensitivity and specificity of the test does not allow us to define a useful predictive value (*1*). Nevertheless, this assay has been proposed as useful for the long-term monitoring of tumor growth in a number of cancers. In at least one type of tumor, medulloblastoma, this assay is the most sensitive diagnostic test for recurrence presently available (*7, 8*). The potential for using this test diagnostically in patients with suspected tumor, along with other diagnostic techniques, and in monitoring the short-term efficacy of therapy, has also been studied (*1*); and although definitive data is not yet available, further investigation seems warranted.

Recent data has implicated the polyamines in a variety of diseases other than cancer, such as cystic fibrosis and renal disease (*5, 6*). Though the usefulness of monitoring the fluid levels of these compounds in these diseases is even less well-defined than in cancer, the "potential" usefulness of these assays seems clear.

The need for definitive clinical studies is obvious; however, such studies are frequently shunned for methodological reasons. Furthermore, even when clinical utility is shown, few clinical laboratories find themselves able to offer the assay. A review of available techniques reveals that significant effort has gone into developing adequate laboratory approaches to measuring these compounds in physiological fluids. A variety of techniques is presently available, including thin layer chromatography, gas-liquid chromatography, gas-liquid chromatography–mass spectrometry, enzymatic analysis, ion exchange chromatography utilizing an amino acid analyzer, high performance liquid chromatography (HPLC), and radioimmunoassay (for a review of methods, see refs. *9, 10*). Each technique has inherent advantages and disadvantages that will be discussed briefly below.

The majority of in-depth clinical studies have utilized some variation of the amino acid analyzer technique. Briefly, it is possible to measure either total (acid hydrolyzed) or free polyamines in all physiologic fluids with a minimum of sample preparation. Most current instruments are automated—and expensive—and the chromatographic procedure is relatively lengthy (in the range of 1 h per analysis). Less expensive instruments are becoming available, but the methods used for physiologic fluids with these instruments have not had extensive field testing. The standard amino acid analyzer techniques, in which the polyamines are quantitated after reaction with either ninhydrin or *o*-phthalaldehyde, have had extensive use and are quite reliable and reproducible both intra- and interlaboratory. Because of the growing interest in the acetylated polyamines and their relationship to cancer, current amino acid analyzer techniques may be modified to include measurement of these compounds. Such a method has already been described (*11*), but its length will limit its use for anything other than preliminary clinical evaluations.

HPLC techniques have thus far not gained wide acceptance for clinical purposes. There have been difficulties in sample preparation for physiologic fluids. Because it is a most promising approach to the simultaneous analysis of both the free and acetylated polyamines, and instrumentation is relatively inexpensive and available, this technology deserves continued interest and development.

Thin layer chromatography with fluorometric detection has been used extensively for many years and is a well-defined, sensitive technique (*9, 10*). It has not gained wide acceptance for clinical studies primarily because of its tediousness relative to other available techniques. Nevertheless, it is inexpensive, reliable, and capable of separating the free and acetylated polyamines.

Gas-liquid chromatography has not been of particular value for clinical purposes because of the caution and time necessary for sample preparation. Gas-liquid chromatography–mass spectrometry has been used successfully for the analysis of free and acetylated polyamines (*9*), but the availability of instruments and the expertise necessary to operate them reliably is limiting.

Enzymatic analysis has found limited utility for clinical purposes. One available assay measures total spermidine and spermine (*12*) and another measures only putrescine (*13, 14*). Because putrescine appears to be the compound of interest in monitoring brain tumors (*15*), this assay probably deserves further study. Preliminary data comparing results from this assay, for cerebrospinal fluid putrescine, to those from the amino acid analyzer technique (*14*), appear promising.

Radioimmunoassay offered real hope as a clinically useful tool. Fairly specific assays have been developed for spermidine and spermine (*16, 17*), but to date assays are not available for putrescine or the acetylated compounds. In addition, there may be some cross reactivity between the "specific antibodies" and a variety of unknown, perhaps polyamine-containing, compounds (*18*). Nevertheless, the techniques are available, and appropriate clinical studies may prove their utility.

## II. High Performance Liquid Chromatographic Methods

Though no fully developed HPLC assay for polyamines in biofluids has been published, a number of interesting studies indicate potential routes for future work. Clearly, several criteria must be met before an HPLC assay will be able to replace presently available clinical assays. Completion of interference, linearity, sensitivity, reproducibility, and recovery studies, and comparative studies with presently accepted techniques, are obviously necessary. The assay should also be flexible in accommodating a variety of physiological fluids, both nonhydrolyzed and hydrolyzed. With increasing interest in the acetylated polyamines, these compounds should also be assayable. At the very least, these compounds should be included in interference studies. None of the HPLC studies that will be reviewed in this chapter meet all of these criteria.

A complete review of all HPLC literature will not be presented. With several exceptions, only studies that utilized physiological fluids will be mentioned. With all published HPLC methods the polyamines must be derivatized for detection. Sensitivity is problematic for certain fluids (e.g., CSF and serum), and fluorescent derivatizing reagents are frequently utilized. Fluorescence also imparts somewhat greater selectivity. Derivatization can either be pre- or post-column. An excellent review of many reagents used in derivatizing amines has been published by Seiler (*19*). The following methods summary will be grouped according to the derivatizing reagent used.

## A. Fluorescamine

Fluorescamine, 4-phenylspiro-[furan-2(3H),1′-phthalan]-3,3′-dione, forms fluorescent products when reacted with primary amino groups (*19*). It has been used extensively for the assay of amines, amino acids, and peptides. Veening et al. (*20*), in 1974, described an HPLC method for hydrolyzed urine utilizing this derivatizing reagent. Derivatization was accomplished post-column, and the ion exchange chromatographic separation of the polyamines utilized a NaCl gradient. Though reasonable sensitivity was obtained for putrescine, poor sensitivity was obtained for spermidine and spermine. The study was preliminary in nature and no interference or comparative data were described.

In 1976, Samejima et al. (*21*), described a pre-column derivatization technique that is followed by separation of the derivatized compounds by reversed-phase HPLC. Both hydrolyzed and nonhydrolyzed urines can be assayed. Sample preparation includes a pre-HPLC cleanup utilizing a CM-Cellulose column. Diamines and polyamines are eluted separately from the CM-Cellulose column, and each fraction, following derivatization, is subjected to independent HPLC analysis utilizing gradient elution. The need for two HPLC separations for each sample is an obvious drawback. In addition, the method appears to be less than satisfactory for spermine, no comparative study was described, and reproducibility data were not mentioned. On the positive side, linearity and interference studies were done. Acetyl compounds were included in the interference study, but it appeared as if they were poorly resolved from other peaks.

Kai et al. (*22*), in 1979, published a method for serum, both hydrolzed and nonhydrolyzed, utilizing reversed-phase gradient chromatography and pre-column derivatization. Their method also includes a column cleanup procedure utilizing Cellex-P; however, both di- and polyamines were eventually separated by a single HPLC run. They incorporate an internal standard into their method, and interestingly, utilize nickel ion to reduce reactivity of the fluorescamine with biogenic amines other than the polyamines. They described recovery, sensitivity, and linearity studies, but failed to include the acetylated polyamines in their interference studies. An additional drawback of their method is the fact that the reagent blank, when carried through the procedure, revealed small but significant interfering peaks, possibly secondary to the Cellex-P chromatography.

## B. Tosyl Chloride

Tosyl chloride, *p*-toluenesulfonyl chloride, reacts with both primary and secondary amino groups, forming ultraviolet-absorbing products. This derivatizing agent has been used by Hayashi et al. (*23*) to assay for the polyamines in hydrolyzed urine. Their method utilized two cleanup columns, Amberlite IRA-410 and Dowex 50, prior to tosylation. Reversed-phase chromatography is utilized for analytical separation. Though their study appears promising, interference, linearity, sensitivity, reproducibility, and comparative studies were not reported.

## C. Dansyl Chloride

Dansyl chloride (Dns-Cl), 5-dimethylaminonaphthalene-1-sulfonyl chloride, reacts with primary and secondary amino groups, forming fluorescent products. This reagent has been extensively studied by Seiler (*9, 10*) as a derivatizing agent for the TLC separation of the polyamines. As previously mentioned, this method can be used for clinical studies. A number of investigators, including Seiler, have utilized this reagent in HPLC studies.

Abdel-Monem and Ohno (*24*), in 1975, described a normal-phase chromatographic separation of the polyamines, including the acetylated compounds, with detection at 280 nm, rather than with fluorescence. Though this paper only described the separation of standard solutions, in a subsequent article the same authors described its application to unhydrolyzed urine (*25*). Following derivatization, the sample is subjected to two dimensional TLC prior to HPLC. Obviously this method is tedious, and in addition, no reproducibility or comparative studies were reported.

Seiler et al. (*26*) reported the separation and quantitation of Dns derivatives of the polyamines from tissue samples using reversed-phase gradient chromatography and fluorescence detection. A silica column cleanup procedure was utilized. Though physiological fluids were not included in this study, the careful chromatography described may be one reasonable starting place for development of a clinically useful method.

Saeki et al. (*27*) described a method utilizing Dns-Cl for red blood cell polyamines. They also report on plasma levels, but they do not describe the sample preparatory procedure utilized for obtaining these levels. Chromatography is by gradient elution from a reversed-phase column. Detection is by fluorometry. It is somewhat worrisome that they were not able to detect putrescine in normal red blood cells, since Cooper et al. (*28*) were able to measure these levels utilizing an amino acid analyzer.

Vandemark et al. (*29*) reported on the separation of hydrolyzed urinary polyamines, following derivatization with Dns-Cl, by reversed-phase chromatography with gradient elution. Detection was by fluorometry. Their study was preliminary in nature, the authors stating the need for further sample cleanup. However, the chromatography looks promising, and this method may be another reasonable starting point for further development.

### D. Benzoyl Chloride

Benzoyl chloride has been proposed as an ultraviolet absorbing derivatizing reagent for the polyamines by Redmond and Tseng (*30*). They describe a reversed-phase, isocratic chromatographic separation of standard solutions of the polyamines following derivatization. Detection was at 254 nm. Intrigued by the simplicity of this approach, we attempted to modify their procedure for use with hydrolyzed and unhydrolyzed urine (Kabra, Lo, and Marton, unpublished results). Suffice it to say that considerable developmental effort will be necessary.

## III. Amino Acid Analyzer Methods

Amino acid analyzer techniques clearly deserve review in this chapter. As mentioned previously, the majority of in-depth clinical studies thus far conducted have utilized some variation on this approach. One of the earliest uses of this technique for physiological fluids was published by Bremer and Kohne in 1971 (*31*). Their method was rather lengthy, but it served to stimulate a number of groups to utilize their approach in shorter procedures. A number of methods were soon published utilizing Beckman amino acid analyzers, cation exchange resins, and ninhydrin as a post-column colorimetric derivatizing reagent (*11, 32, 33*). The most extensive, but also lengthiest, procedure separated the free and acetylated polyamines from many other biogenic amines (*11*). A highly refined method, utilizing ninhydrin, was published by Gehrke et al. in 1977 (*34*). This method has had extensive field testing, and is useful for conducting clinical studies. In 1975 our group published an amino acid analyzer technique utilizing a Durrum D-500 instrument, cation exchange chromatography, and *o*-phthalaldehyde as a fluorescent post-column derivatizing reagent (*35*). The Durrum instrument is a high pressure (performance) instrument, and with the addition of *o*-phthalaldehyde sensitivity was increased significantly. A modification of this procedure is now used in a number of laboratories for a variety of physiological fluids. We have replaced the original sodium elution buffers with potassium buffers (*7*) for improved reliability. Four independent laboratories, one using Gehrke's method, three using ours or modifications of ours, were compared on a series of unknown urine specimens; the results were remarkably uniform. For those who can afford the equipment, and who can live with the slow throughput, these methods are quite reliable and useful.

A new twist in the amino acid analyzer approach is that of Shipe and Savory (*36*). They use an Aminco "Aminalyzer" with cation exchange chromatography and *o*-phthalaldehyde post column to assay for polyamines in plasma and erythrocytes. It seems feasible that this approach might be applied to a number of available HPLCs.

## IV. Conclusions

We have recently outlined what we consider a rational approach to studying the utility of monitoring physiological fluid polyamines in cancer (*1*). Similarly, studies are presently being conducted evaluating the relationship of polyamine fluid levels to a variety of other diseases. In the midst of these studies, one must note the absence of clinically useful HPLC methods, if amino acid analyzer techniques are discounted. It is intriguing that such chemically simple compounds pose such a difficult analytical problem. Though significant strides have been made in showing the potential of HPLC polyamine assays, this chapter will have served its purpose if it acts as an indicator of the difference between an analytical technique and a clinically useful method.

## Acknowledgments

A portion of the work described in this chapter was supported by NCI grants CA-13525 and CA-15515. The author is the recipient of NCI Research Career Development Award CA-00112.

## References

1. Marton, L. J., and Seidenfeld, J., Approaches to the study of polyamines as cancer markers, *Polyamines in Biology and Medicine,* Dekker, New York, in press.
2. Cohen, S. S., *Introduction to the Polyamines,* Prentice-Hall, Englewood Cliffs, NJ, 1971.
3. Bachrach, U., *Function of Naturally Occurring Polyamines,* Academic Press, New York, 1973.
4. Russell, D. H., ed., *Polyamines in Normal and Neoplastic Growth,* Raven Press, New York, 1973.
5. Campbell, R. A., Morris, D. R., Bartos, D., Daves, G. D., and Bartos, F., eds., *Advances in Polyamine Research,* Vols. 1 and 2 Raven Press, New York, 1978.
6. Morris, D. R., and Marton, L. J., eds., *Polyamines in Biology and Medicine,* Dekker, New York, in press.
7. Marton, L. J., Edward, M. S., Levin, V. A., Lubich, W. P., and Wilson, C. B., *Cancer Res.* **39,** 993 (1979).
8. Marton, L. J., Edward, M. S., Levin, V. A., Lubich, W. P., and Wilson, C. B., CSF Polyamines: A new and important means of monitoring medulloblastoma. *Cancer,* in press.
9. Seiler, N., *Clin. Chem.* **23,** 1519 (1977).
10. Bachrach, U., Analytical methods for polyamines, in *Advances in Polyamine Research,* vol. 2, Campbell, R. A., et al., eds., Raven Press, New York, 1978, p. 5.
11. Tabor, H., Tabor, C. W., and Irreverre, R., *Anal. Biochem.* **55,** 457 (1973).
12. Bachrach, U., and Reches, B., *Anal. Biochem.* **17,** 38 (1966).
13. Harik, S. I., Pasternak, G. W., and Snyder, S. H., *Biochim. Biophys. Acta* **304,** 753 (1973).
14. Harik, S. I., and Marton, L. J., *Arch. Neurol.* in press.
15. Marton, L. J., Heby, O., Levin, V. A., Lubich, W. P., Crafts, D. C., and Wilson, C. B., *Cancer Res.* **36,** 973 (1976).
16. Bartos, D., Campbell, R. A., Bartos, F., and Grettie, D. P., *Cancer Res.* **35,** 2056 (1975).
17. Bartos, F., Bartos, D., Dolney, A. M., Grettie, D. P., and Campbell, R. A., *Res. Commun. Chem. Path. Pharm.* **19,** 295 (1978).
18. Campbell, R. A., Grettie, D. P., Bartos, F., Bartos, D., and Marton, L. J., *Proc. Dialysis Transplant Forum.* Printed by Charbray Printers,

Lovettsville, VA for the Amer. Soc. for Artificial Internal Organs, Schreiner, G. E., ed., **8,** 194 (1978).

*19.* Seiler, N., *J. Chromatogr.* **143,** 221 (1977).

*20.* Veening, H., Pitt, W. W., and Jones, G., *J. Chromatog.* **90,** 129, (1974).

*21.* Samejima, K., Kawase, M., Sakamoto, S., Okada, M., and Endo, Y., *Anal. Biochem.* **76,** 392 (1976).

*22.* Kai, M., Ogata, T., Haraguchi, K., and Ohkura, Y., *J. Chromatog.* **163,** 151 (1979).

*23.* Hayashi, T., Sugiura, T., Kawai, S., and Ohno, T., *J. Chromatog.* **145,** 141 (1978).

*24.* Abdel-Monem, M. M., and Ohno, K., *J. Chromatog.* **107,** 416 (1975).

*25.* Abdel-Monem, M. M., and Ohno, K., *J. Pharm. Sci.* **66,** 1089 (1977).

*26.* Seiler, N., Knodgen, B., and Eisenbeiss, F., *J. Chromatog.* **145,** 29 (1978).

*27.* Saeki, Y., Vehara, N., and Shirakawa, D., *J. Chromatog.* **145,** 221 (1978).

*28.* Cooper, K. D., Shukla, J. B., and Rennert, O. M., *Clin. Chim. Acta* **82,** 1 (1978).

*29.* Vandemark, F. L., Schmidt, G. J., and Slavin, W., *J. Chromatog. Sci.* **16,** 465 (1978).

*30.* Redmond, J. W., and Tseng, A., *J. Chromatog.* **170,** 479 (1979).

*31.* Bremer, H. J., and Kohne, E., *Clin. Chim. Acta* **32,** 407 (1971).

*32.* Marton, L. J., Russell, D. H., and Levy, C. C., *Clin. Chem.* **19,** 923 (1973).

*33.* Gehrke, C. W., Kuo, K. C., and Zumwalt, R. W., *J. Chromatog.* **89,** 231 (1974).

*34.* Gehrke, C. W., Kuo, K. C., and Ellis, R. L., *J. Chromatog.* **143,** 345 (1977).

*35.* Marton, L. J., and Lee, P. L. Y., *Clin. Chem.* **21,** 1721 (1975).

*36.* Shipe Jr., J. R., and Savory, J., *Ann. Clin. Lab. Sci.* **10,** 128 (1980).

# Index

D



E

## H



## I

## T